ENVIRONMENTAL HAZARDS

The Fluid Dynamics and Geophysics of Extreme Events

LECTURE NOTES SERIES
Institute for Mathematical Sciences, National University of Singapore

Series Editors: Louis H. Y. Chen and Ser Peow Tan
Institute for Mathematical Sciences
National University of Singapore

ISSN: 1793-0758

Published

Vol. 12 Harmonic Analysis, Group Representations, Automorphic Forms and Invariant Theory — In Honor of Roger E. Howe
edited by Jian-Shu Li, Eng-Chye Tan, Nolan Wallach & Chen-Bo Zhu

Vol. 13 Econometric Forecasting and High-Frequency Data Analysis
edited by Roberto S Mariano & Yiu-Kuen Tse

Vol. 14 Computational Prospects of Infinity — Part I: Tutorials
edited by Chitat Chong, Qi Feng, Theodore A Slaman, W Hugh Woodin & Yue Yang

Vol. 15 Computational Prospects of Infinity — Part II: Presented Talks
edited by Chitat Chong, Qi Feng, Theodore A Slaman, W Hugh Woodin & Yue Yang

Vol. 16 Mathematical Understanding of Infectious Disease Dynamics
edited by Stefan Ma & Yingcun Xia

Vol. 17 Interface Problems and Methods in Biological and Physical Flows
edited by Boo Cheong Khoo, Zhilin Li & Ping Lin

Vol. 18 Random Matrix Theory and Its Applications: Multivariate Statistics and Wireless Communications
edited by Zhidong Bai, Yang Chen & Ying-Chang Liang

Vol. 19 Braids: Introductory Lectures on Braids, Configurations and Their Applications
edited by A Jon Berrick, Frederick R Cohen, Elizabeth Hanbury, Yan-Loi Wong & Jie Wu

Vol. 20 Mathematical Horizons for Quantum Physics
edited by H Araki, B-G Englert, L-C Kwek & J Suzuki

Vol. 21 Environmental Hazards: The Fluid Dynamics and Geophysics of Extreme Events
edited by H K Moffatt & E Shuckburgh

Vol. 22 Multiscale Modeling and Analysis for Materials Simulation
edited by Weizhu Bao & Qiang Du

*For the complete list of titles in this series, please go to
http://www.worldscibooks.com/series/LNIMSNUS

Lecture Notes Series, Institute for Mathematical Sciences, National University of Singapore **Vol. 21**

ENVIRONMENTAL HAZARDS

The Fluid Dynamics and Geophysics of Extreme Events

Editors

H K Moffatt
University of Cambridge, UK

Emily Shuckburgh
British Antarctic Survey, UK

World Scientific

NEW JERSEY • LONDON • SINGAPORE • BEIJING • SHANGHAI • HONG KONG • TAIPEI • CHENNAI

Published by

World Scientific Publishing Co. Pte. Ltd.
5 Toh Tuck Link, Singapore 596224
USA office: 27 Warren Street, Suite 401-402, Hackensack, NJ 07601
UK office: 57 Shelton Street, Covent Garden, London WC2H 9HE

British Library Cataloguing-in-Publication Data
A catalogue record for this book is available from the British Library.

Cover photograph:
Typhoon image by NASA (from the Visible Earth catalog of images)

Lecture Notes Series, Institute for Mathematical Sciences, National University of Singapore — Vol. 21
ENVIRONMENTAL HAZARDS
The Fluid Dynamics and Geophysics of Extreme Events

ISBN-13 978-981-4313-28-5
ISBN-10 981-4313-28-9
ISBN-13 978-981-4366-99-1 (pbk)
ISBN-10 981-4366-99-4 (pbk)

Typeset by Stallion Press
Email: enquiries@stallionpress.com

Printed in Singapore by Mainland Press Pte Ltd.

Contents

Foreword

The Institute for Mathematical Sciences at the National University of Singapore was established on 1 July 2000. Its mission is to foster mathematical research, both fundamental and multidisciplinary, particularly research that links mathematics to other disciplines, to nurture the growth of mathematical expertise among research scientists, to train talent for research in the mathematical sciences, and to serve as a platform for research interaction between the scientific community in Singapore and the wider international community.

The Institute organizes thematic programs which last from one month to six months. The theme or themes of a program will generally be of a multidisciplinary nature, chosen from areas at the forefront of current research in the mathematical sciences and their applications.

Generally, for each program there will be tutorial lectures followed by workshops at research level. Notes on these lectures are usually made available to the participants for their immediate benefit during the program. The main objective of the Institute's Lecture Notes Series is to bring these lectures to a wider audience. Occasionally, the Series may also include the proceedings of workshops and expository lectures organized by the Institute.

The World Scientific Publishing Company has kindly agreed to publish the Lecture Notes Series. This Volume, "Environmental Hazards: The Fluid Dynamics and Geophysics of Extreme Events," is the twenty-first of this Series. We hope that through the regular publication of these lecture notes the Institute will achieve, in part, its objective of promoting research in the mathematical sciences and their applications.

January 2011

Louis H. Y. Chen
Ser Peow Tan
Series Editors

Preface

Natural environmental hazards, and their potentially disastrous consequences, have been increasingly prominent over the last decade. Chief among these are perhaps the great Sumatra-Andaman tsunami, triggered by the earthquake of 26 December 2004, which devastated large parts of the coastline of the Indian Ocean; Hurricane Katrina in the Gulf of Mexico in August 2005 with its deadly consequences for the city of New Orleans; and the catastrophic flooding in Pakistan following the exceptional monsoon rains of July–August 2010. Such geophysical phenomena have their origin in the dynamics of ocean and atmosphere on the large scales on which Coriolis effects associated with the Earth's rotation can be of dominant importance. In seeking to mitigate the disastrous consequences of such natural hazards, it is necessary to understand the fundamental fluid dynamical principles that underlie these awe-inspiring phenomena of nature. The extent to which climate change may influence the frequency and intensity of such phenomena is of course a matter of great concern, with major political implications at a global level.

It will be no surprise therefore that one of the current priority areas of the International Council for Science (ICSU) is "Natural and Human-Induced Environmental Hazards and Disasters;" and it was under this heading that a grant was awarded to two of ICSU's International Scientific Unions (IUTAM, the International Union of Theoretical and Applied Mechanics, and IUGG, the International Union of Geodesy and Geophysics) to hold a two-week Spring School (19 April–2 May 2009) on the subject "Fluid Dynamics and Geophysics of Environmental Hazards." The School, supported by ICSU's Regional Office for Asia and the Pacific Region (ROAP) in Kuala Lumpur, was aimed at graduate students and young post-docs in mathematics, physics or engineering, from Asia and the Pacific Region, with the aim of encouraging them to undertake research in this field. It was held at the Institute for Mathematical Sciences (IMS) of the National University of Singapore, attracting some 50 students from

Australia, Indonesia, Philippines, Vietnam, Malaysia, China, Japan, Korea, Bangladesh, Pakistan, India, Sri Lanka, Georgia, and Iran, as well as a number from Singapore itself (see Fig. 1).

Nine short courses of lectures were presented during morning sessions of the School; Chapters 1–9 of this volume contain the written versions of these lectures. Seminars on relevant topics were also held; one of these, on "Rogue Waves" is included in Chapter 10.

By way of supplementary activity related to the lecture courses, the students undertook research activity on nine different projects proposed by the lecturers. For this purpose, the students were divided into groups, four or five students in each group. The students worked on these projects, with guidance from the lecturers, in afternoon sessions during the first week of the School, and made presentations of their results during the afternoon sessions of the second week. Their reports are available on the School website http://www.ims.nus.edu.sg/Programs/09fluidss/index.htm. The students were uniformly enthusiastic about this style of project work, which promoted an unusual degree of international and interdisciplinary collaborative activity, and opened up research projects for the students to pursue in more depth in the future.

Three posters were prepared in advance of the School in both English and Chinese versions, for wide circulation to schools and universities. We are grateful to Andrew Burbanks (University of Portsmouth, UK) for help in the design of these posters, to Weizhu Bao (NUS) who provided the Chinese translations, and to World Scientific who printed the posters and donated them free of charge for the benefit of the School. Versions of these posters are reproduced in the following pages (see Figs. 2 to 4).

We wish to express our thanks also to Louis Chen, Director of IMS, for his constant support and encouragement and for the financial support provided by IMS for the School; and to the local organizing committee, particularly its co-chairs Boo Cheong Khoo (NUS) and Pavel Tkalich (NUS). Finally, we thank Sue Liu and Emma Boland (DAMTP, Cambridge) who have provided invaluable assistance in text preparation; and Sarah Haynes of World Scientific for her patience and understanding throughout the publication process.

September 2010

Keith Moffatt
University of Cambridge, UK

Emily Shuckburgh
British Antarctic Survey, Cambridge, UK
Editors

Fig. 1. Participants of the Spring School. *Front row, from left to right*: A. W. Jayawardena, Gerd Tetzlaff, Pavel Tkalich, Kerry Emmanuel, Ser Peow Tan, Louis Chen, Keith Moffatt, Tieh Yong Koh. Photo by Jolyn Wong.

Fig. 2. Tsunami poster: Chinese version.

Fig. 3. Typhoon poster.

Fig. 4. Monsoon poster.

Chapter 1

A Brief Introduction to Vortex Dynamics and Turbulence

H. Keith Moffatt

Department of Applied Mathematics and Theoretical Physics
University of Cambridge
Wilberforce Road, Cambridge, UK
h.k.moffatt@cam.ac.uk

The emphasis in this short introductory chapter is on those fluid dynamical phenomena that are best understood in terms of convection and diffusion of vorticity, the curl of the velocity field. Vorticity is generated at fluid boundaries and diffuses into the fluid where it is subject to convection, stretching, and associated intensification. Far from boundaries, viscous effects may be negligible, and then vortex lines are transported with the fluid. Vortex rings, which propagate under their own self-induced velocity, are a widely observed phenomenon, and a fundamental ingredient of fluid flow. Stretching and intensification is best illustrated by the "Burgers vortex" (the simplest model for a hurricane) in which these processes are in equilibrium with viscous diffusion. Instabilities of Kelvin–Helmholtz type are all-pervasive in highly sheared flow, and inexorably lead to transition to turbulence. In turbulent flow, the vorticity is random, but these fundamental processes still dictate many features of the flow. Fully three-dimensional turbulence is characterized by a cascade of energy through a broad spectrum from large scales to very small scales at which kinetic energy is dissipated by viscosity, a scenario that leads to the famous $(-5/3)$ Kolmogorov spectrum. These topics are reviewed and discussed with a view to geophysical applications. The phenomena of intermittency and concentrated vortices as revealed by direct numerical simulation are also briefly discussed.

1. Introduction

Vortex (or vorticity) dynamics is concerned with the manner in which swirling flows evolve in fluids when viscous (i.e., internal friction) effects are relatively weak, and can be neglected in a first approximation. Such flows are controlled largely by inertial effects. An understanding of vortex dynamics is an essential preliminary to a consideration of turbulent flows in which the vorticity distribution is a highly complex function of position. Its time evolution is most easily understood through the statement that "vortex lines are frozen in the fluid," i.e., they are transported with the flow like material curves of fluid particles. This is not quite the whole story however, because, insofar as the flow may be treated as incompressible, the vorticity is intensified as the vortex lines are transported, in proportion to the stretching of vortex line elements. This stretching is very persistent in a turbulent flow, leading to very strong intensification of vorticity coupled with progressive decrease of the scale of variation of the flow, an effect usually described in terms of an "energy cascade." This cascade to small scales is ultimately controlled by viscosity, no matter how weak this physical property of the fluid may be; and one of the remarkable properties of turbulent flow is that the rate of dissipation of energy by viscosity is independent of the value of viscosity even in the limit as this tends to zero, and this is because the smallest scales of the flow adjust in just such a way as to dissipate the kinetic energy at the very rate at which it cascades down from the larger scales.

The central role of vorticity in describing fluid motion was recognized by Hermann von Helmholtz (1858), who first recognized the above crucial "frozen-in" property. The 150th anniversary of the publication of this seminal paper was marked by the IUTAM symposium *150 Years of Vortex Dynamics*, recently held at the Technical University of Denmark (Aref, 2010; the 50 papers contained in this volume provide an indication of the huge scope and applications of the subject). The theory of vorticity was taken up and enthusiastically developed by William Thomson (later Lord Kelvin) (1867; 1869 and many subsequent papers), who proposed that the atomic structure of the various elements might be explained in terms of knotted vortex tubes, whose "knottedness" would be conserved under frozen field evolution. Such structures turned out to be dynamically unstable, and Kelvin was ultimately obliged to abandon his theory of "vortex-atoms." Nevertheless, his pioneering investigations opened up the new field of hydrodynamic instability, providing important clues concerning the ubiquity of turbulent, as opposed to laminar, flows in all large-scale natural systems. Figure 1 shows Helmholtz and Kelvin around 1870, when both were at the height of their powers and creativity.

Fig. 1. Hermann von Helmholtz (left) and William Thomson (Lord Kelvin): The early pioneers of vortex dynamics.

2. Vorticity and the Biot–Savart Law

Let $\mathbf{u}(\mathbf{x}, t)$ be the velocity field in a fluid which fills all space. This is of course an idealization, relevant when we consider fluid behavior that is uninfluenced by remote fluid boundaries. We shall suppose further, for simplicity, that the fluid has uniform density ρ, and that it (or rather the flow) is incompressible, i.e., $\nabla \cdot \mathbf{u} = 0$. Under this approximation, sound waves are filtered out of the governing Navier–Stokes equations. The vorticity field $\boldsymbol{\omega}(\mathbf{x}, t)$ is defined by

$$\boldsymbol{\omega} = \nabla \times \mathbf{u}(\mathbf{x}, t)\,, \tag{2.1}$$

so that immediately $\nabla \cdot \boldsymbol{\omega} = 0$. We can conveniently think of "vortex tubes" in the flow, i.e., the set of vortex lines passing through any small material surface element δA. The "circulation" round such a tube is

$$\Gamma = \oint_C \mathbf{u} \cdot d\mathbf{x} = \iint_{\delta A} \boldsymbol{\omega} \cdot \mathbf{n}\, dA\,, \tag{2.2}$$

where C is a closed curve circling the tube once, and this is evidently constant, independent of the particular cross-section of the tube that is chosen (Fig. 2(a)). It is frequently stated that vortex lines must either be

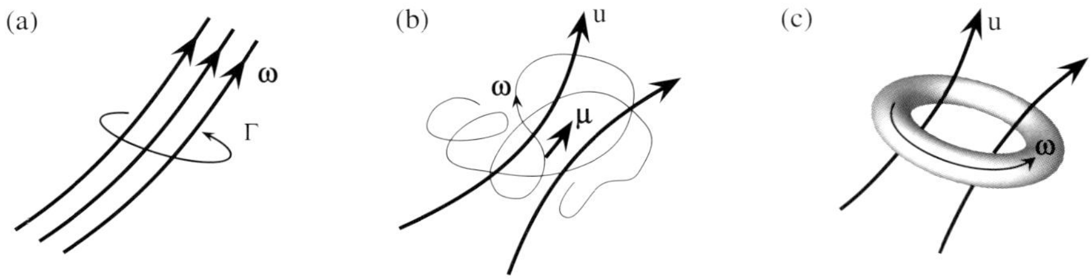

Fig. 2. Vorticity configurations and induced velocity fields: (a) Vortex tube with circulation Γ; (b) localized vorticity field, and induced velocity, dipolar at a large distance; (c) vortex ring and its induced velocity.

closed curves or end on a fluid boundary, but this is incorrect: It is now known that in a general three-dimensional flow, the vortex lines are chaotic, and any two neighboring vortex lines will in general diverge exponentially (a good example may be found in the "ABC"-flow studied by Dombre *et al.*, 1986). For this reason, the concept of a vortex tube must be treated with caution, particularly in a turbulent flow in which the cross-section of any instantaneous vortex tube will become seriously deformed if followed far enough along its length.

By virtue of the incompressibility condition $\nabla \cdot \mathbf{u} = 0$, we may introduce a vector potential $\mathbf{A}(\mathbf{x}, t)$ for $\mathbf{u}$, such that $\mathbf{u} = \nabla \times \mathbf{A}$, $\nabla \cdot \mathbf{A} = 0$. Then we have immediately $\boldsymbol{\omega} = \nabla \times (\nabla \times \mathbf{A}) = -\nabla^2 \mathbf{A}$. If the vorticity distribution is localized (and by this, we usually mean that $|\boldsymbol{\omega}|$ decreases exponentially rapidly outside some bounded region), then the appropriate solution of this Poisson equation is

$$\mathbf{A}(\mathbf{x}, t) = \frac{1}{4\pi} \int \frac{\boldsymbol{\omega}(\mathbf{x}', t)}{|\mathbf{x} - \mathbf{x}'|} dV' . \tag{2.3}$$

The corresponding velocity field is then

$$\mathbf{u}(\mathbf{x}, t) = \nabla \times \mathbf{A} = -\frac{1}{4\pi} \int \frac{(\mathbf{x} - \mathbf{x}') \times \boldsymbol{\omega}(\mathbf{x}', t)}{|\mathbf{x} - \mathbf{x}'|^3} dV' . \tag{2.4}$$

This is the "Biot–Savart law," giving the velocity field $\mathbf{u}(\mathbf{x}, t)$ "induced" by the vorticity field $\boldsymbol{\omega}(\mathbf{x}, t)$. It is this velocity field that transports the vorticity field, a nonlinear feedback that encapsulates the central difficulty of the dynamics of fluids.

If, as supposed, the vorticity field is localized, then for $|\mathbf{x}| \gg |\mathbf{x}'|$, (where $\mathbf{x}'$ is any point within the vortical region), Eq. (2.3) may be manipulated to give

$$\mathbf{A}(\mathbf{x}) \sim -(\boldsymbol{\mu} \times \nabla)\frac{1}{r}\,, \tag{2.5}$$

where

$$\boldsymbol{\mu} = \frac{1}{8\pi}\int \mathbf{x} \times \boldsymbol{\omega}\, dV\,, \tag{2.6}$$

and $r = |\mathbf{x}|$. The corresponding asymptotic behavior of $\mathbf{u}$ is

$$\mathbf{u} \sim \nabla(\boldsymbol{\mu} \cdot \nabla)\frac{1}{r}\,, \tag{2.7}$$

an irrotational velocity field associated with an (apparent) dipole $\boldsymbol{\mu}$ located at $r = 0$. (The result is independent of the origin chosen for $\mathbf{x}$; proof: An exercise for the reader!) The situation is as sketched in Fig. 2(b). Equation (2.7) shows that the velocity field associated with an arbitrary localized vorticity distribution is dipolar at a large distance, of order r^{-3} as $r \to \infty$.

The most familiar example of a localized vorticity distribution is provided by the "vortex ring" for which the vorticity field is axisymmetric and confined to a torus, the vortex lines being circles around the axis of the torus (Fig. 2(c)). Such vortex rings may be produced and visualized by tapping a smoke-filled box so that air is ejected impulsively through a suitably shaped orifice; both the vortex ring and the smoke are then transported together by the self-induced velocity field. This was the basis of Tait's (1867) demonstration which so impressed Kelvin, who proceeded to calculate the speed of propagation V of a vortex ring of radius R, starting from the Biot–Savart law (2.4), and on the assumption that the vorticity is uniformly distributed across the "core" of the vortex of small core radius a; his result, recorded in an appendix to Tait (1867), was

$$V = \frac{\omega a^2}{2R}\left(\log\frac{8R}{a} - \frac{1}{4}\right). \tag{2.8}$$

Vortex rings generated by the method of Tait (exploiting the retarding effect of viscosity in the boundary layer inside the orifice) can travel a considerable distance before being dispersed as a result of instability or through the direct action of viscosity. Vortex rings appear to be ubiquitous in nature, the most striking example being the vortex/steam rings emitted in volcanic eruptions. See, for example, the beautiful photographs by Marco Fulle of this phenomenon at http://www.swisseduc.ch/stromboli/etna/etna00. A fine example of the persistence of vortex rings (visualized with bubbles at their core), and the playful manner in which dolphins can interact with them can be found at http://www.metacafe.com/watch/1041454/dolphinplaybubblerings.

3. The Euler Equation and its Invariants

We take as a starting point the Navier–Stokes equations for a viscous incompressible fluid in their familiar form

$$\frac{\partial \mathbf{u}}{\partial t} + \mathbf{u} \cdot \nabla \mathbf{u} = -\frac{1}{\rho}\nabla p + \nu \nabla^2 \mathbf{u}\,, \tag{3.1}$$

$$\nabla \cdot \mathbf{u} = 0\,, \tag{3.2}$$

where ρ is the fluid density (here assumed constant), and ν is the kinematic viscosity of the fluid. If, for the moment, we neglect viscous effects entirely, we simply set $\nu = 0$, giving the equations obtained by Euler (1755).

$$\frac{\partial \mathbf{u}}{\partial t} + \mathbf{u} \cdot \nabla \mathbf{u} = -\frac{1}{\rho}\nabla p\,, \tag{3.3}$$

$$\nabla \cdot \mathbf{u} = 0\,. \tag{3.4}$$

It is remarkable that, despite the fact that these Euler equations were discovered more than 250 years ago (Eyink *et al.*, 2008), we still do not know whether the solutions that evolve from smooth initial conditions of finite energy remain smooth for all time; or conversely, whether there exist any smooth finite-energy initial conditions for which the solution of the Euler equations becomes singular at finite-time. This "finite-time singularity problem" may seem a rather esoteric issue, of more interest to mathematicians than to geophysicists or engineers; but in fact it lies at the heart of the problem of turbulence, having an obvious bearing on the mechanism of dissipation of energy at the smallest scales of motion, and it is therefore a problem that merits serious study. It is known that if a singularity occurs at some finite-time t_c, say, then the time-integral of the maximum value of the vorticity must diverge as $t \to t_c$ (Beale *et al.*, 1984). This result places the focus of investigation firmly on the behavior of the vorticity field in general three-dimensional situations. We shall suppose in what follows that the velocity and vorticity fields do in fact remain smooth for all time, unless otherwise stated.

The Euler equation (3.3) may be written in the equivalent form

$$\frac{\partial \mathbf{u}}{\partial t} = \mathbf{u} \times \boldsymbol{\omega} - \nabla\left(\frac{p}{\rho} + \frac{1}{2}\mathbf{u}^2\right), \tag{3.5}$$

from which, taking the curl, we immediately obtain the "vorticity equation"

$$\frac{\partial \boldsymbol{\omega}}{\partial t} = \nabla \times (\mathbf{u} \times \boldsymbol{\omega})\,. \tag{3.6}$$

This is the equation that implies that the vortex lines behave like material lines, and are therefore transported with the fluid. Kelvin proved, on the basis of this equation, that the circulation, defined as in Eq. (2.2),

$$K = \oint_C \mathbf{u} \cdot d\mathbf{x}\,, \tag{3.7}$$

but now for any material (i.e., "Lagrangian") circuit C that moves with the, fluid, is constant. By virtue of (2.2), K is also the flux of vorticity through C; hence any flow that stretches a vortex tube and (by incompressibility) decreases its cross-section must proportionately intensify the vorticity in the tube. In fact, if $\delta\mathbf{x}$ is an element of a vortex line which moves with the fluid, then $|\boldsymbol{\omega}| \propto |\delta\mathbf{x}|$. (The corresponding result for compressible flow is that $|\boldsymbol{\omega}| \propto \rho|\delta\mathbf{x}|$.)

There are four known invariants of the Euler equations, namely momentum $\mathbf{P}$, angular momentum $\mathbf{M}$, (kinetic) energy E, and helicity $\mathcal{H}$. One might naively suppose that the momentum should be given by $\mathbf{P} = \int \rho\mathbf{u}dV$, the integral being over the whole fluid domain. This integral is, however, at best only conditionally convergent, due to the slow $O(r^{-3})$ decrease of $\mathbf{u}$ at infinity. One may calculate the momentum of any given flow by supposing that the corresponding vorticity distribution is established from a state of rest by an impulsive force distribution at the moment under consideration (Saffman, 1995); the result is that

$$\mathbf{P} = \frac{1}{2}\int \rho\mathbf{x} \times \boldsymbol{\omega}\; dV\,, \tag{3.8}$$

an integral that is certainly convergent for any localized vorticity distribution. It may also be verified directly from (3.6) that $\mathbf{P}$ is indeed constant. Note that $\mathbf{P} = 4\pi\boldsymbol{\mu}$, so that the dipole moment of a localized vorticity distribution is constant in time. This result is true also for viscous evolution under the Navier–Stokes equations, the reason being that under the influence of viscosity, momentum is neither created nor destroyed, but merely redistributed by the process of diffusion.

Similarly, the correct expression for angular momentum may be obtained in the form

$$\mathbf{M} = \frac{1}{3}\int \rho\mathbf{x} \times (\mathbf{x} \times \boldsymbol{\omega})\; dV\,, \tag{3.9}$$

and this integral is also constant under either Euler or Navier–Stokes evolution.

The kinetic energy (divided by density ρ) is given by the convergent integral

$$E = \frac{1}{2}\int \mathbf{u}^2 \, dV\,, \tag{3.10}$$

and this is constant under Euler evolution. However, under Navier–Stokes evolution, we have

$$\frac{dE}{dt} = -\nu \int \boldsymbol{\omega}^2 \, dV\,, \tag{3.11}$$

the right-hand side representing the rate of dissipation of energy by viscosity. The integral on the right is called the "enstrophy" of the flow, and is usually denoted by the symbol Ω:

$$\Omega = \int \boldsymbol{\omega}^2 \, dV\,, \quad \frac{dE}{dt} = -\nu\Omega\,. \tag{3.12}$$

Like vorticity itself, the enstrophy has a persistent tendency to increase in turbulent flow, a process ultimately controlled by viscosity.

Finally, the helicity $\mathcal{H}$ is given by

$$\mathcal{H} = \int \mathbf{u}\cdot\boldsymbol{\omega} \, dV, \tag{3.13}$$

and this also is an invariant of the Euler equations (Moreau, 1961; Moffatt, 1969). Like energy, it is a quadratic functional of the velocity field, but, unlike energy, it is not sign-definite; actually it is a "pseudo-scalar," changing sign under change from a right- to left-handed frame of reference; this is why we use the non-mirror-symmetric symbol $\mathcal{H}$ to denote it. By the Schwartz inequality, it is bounded in magnitude:

$$|\mathcal{H}| \leq E\Omega\,, \tag{3.14}$$

with equality only if $\boldsymbol{\omega}$ is everywhere parallel to $\mathbf{u}$. Such "Beltrami" flows are evidently flows of maximal helicity. The helicity is conserved even in compressible flows provided these satisfy the barotropic condition that pressure is a function only of density (and not for example of temperature), i.e., $p = p(\rho)$. In fact, helicity is conserved under precisely the same conditions under which Kelvin's circulation theorem is satisfied and vortex lines are frozen in the fluid. The physical interpretation of helicity is topological in character: This integral represents the "degree of linkage" of the vortex lines of the flow, a quantity that should certainly be preserved under frozen-field evolution. The interpretation is most transparent for the case of two *simply linked* vortex tubes of circulations Γ_1 and Γ_2; for this configuration, it emerges that

$$\mathcal{H} = \pm 2n\Gamma_1\Gamma_2\,, \tag{3.15}$$

where n is the (Gauss) linking number of the two tubes, and the plus or minus sign is chosen according as the linkage is right- or left-handed (assuming of course, as is conventional, that we use a right-handed frame of reference). This topological interpretation has been extended to flows for which the vortex lines are chaotic (the generic situation) by Arnol'd (1974).

4. The Stretched Vortex of Burgers (1948)

In a turbulent flow, each constituent vortex tube (or portion of a vortex tube) is subject to the stretching associated with all other vortices in the flow. It is natural therefore to consider an idealized situation in which this stretching is as simple as possible, i.e., axisymmetric, uniform, and steady. We consider a vorticity distribution with just one component

$$\boldsymbol{\omega} = (0, 0, \omega(r))\,, \tag{4.1}$$

where we use cylindrical polar coordinates (r, ϕ, z) with $r^2 = x^2 + y^2$, and we suppose this subjected to the action of "uniform axisymmetric straining flow" with constant rate of strain $\gamma\,(> 0)$:

$$\mathbf{U} = (-2\gamma r, 0, \gamma z)\,. \tag{4.2}$$

In the absence of this strain, the vortex would diffuse under the action of viscosity; the strain and associated vortex stretching counteracts this effect and a steady state is possible. Note that the additional velocity induced by the vortex is given, from Eq. (2.1), by

$$\mathbf{u} = (0, v(r), 0)\,, \tag{4.3}$$

where

$$v(r) = \frac{1}{r}\int_0^r \omega(r')r'\,dr'\,, \tag{4.4}$$

and that this additional velocity has no effect on the vorticity distribution (because $\nabla \times (\mathbf{u} \times \boldsymbol{\omega}) = 0$).

The vortex therefore evolves according to the equation

$$\frac{\partial \boldsymbol{\omega}}{\partial t} = \nabla \times (\mathbf{U} \times \boldsymbol{\omega}) + \nu \nabla^2 \boldsymbol{\omega}\,; \tag{4.5}$$

this equation has only a ϕ-component, which reduces to

$$\frac{\partial \omega}{\partial t} = \frac{\gamma}{2r}\frac{\partial (r^2\omega)}{\partial r} + \frac{\nu}{r}\frac{\partial}{\partial r} r \frac{\partial \omega}{\partial r}\,. \tag{4.6}$$

The steady solution, with boundary conditions $\omega(0) = \omega_0$, $\omega \to 0$ as $r \to \infty$, is

$$\omega(r) = \omega_0 \exp -(\gamma r^2/4\nu)\,, \tag{4.7}$$

a Gaussian vorticity distribution, with total flux of vorticity

$$\Gamma = 2\pi \int_0^\infty \omega(r) r\, dr = 4\pi\omega_0\nu/\gamma\,. \tag{4.8}$$

The associated velocity component $v(r)$ is given, from (4.4), by

$$v(r) = \frac{\Gamma}{2\pi r}\left(1 - \exp\left(-\frac{\gamma r^2}{4\nu}\right)\right)\,. \tag{4.9}$$

The circulation round a circle of radius r is $2\pi r v(r)$, and this tends to the constant Γ for $r > \delta$ where $\delta = \nu/\gamma$ is a measure of the radius of the tube. The structure of this vortex is sketched in Fig. 3.

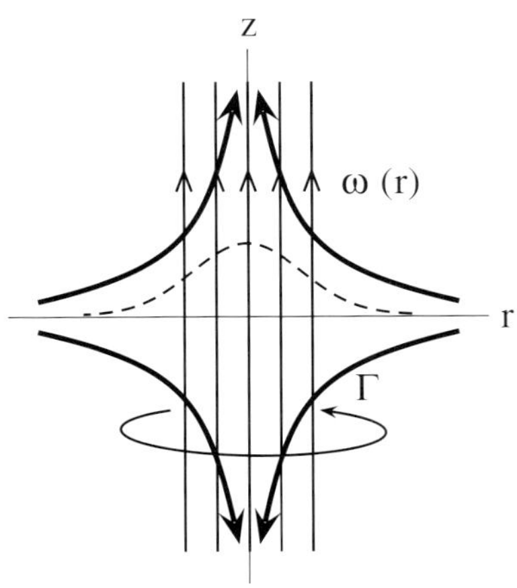

Fig. 3. The stretched Burgers vortex with circulation Γ and Gaussian vorticity profile.

A remarkable feature of this vortex, as noted by Burgers (1948), is that the corresponding rate of dissipation of energy per unit length of vortex, namely

$$\Phi = 2\pi\nu \int_0^\infty \omega^2 r\, dr = \Gamma^2\gamma/8\pi\,, \tag{4.10}$$

is independent of ν (for fixed circulation Γ) even in the limit as $\nu \to 0$. In this limit, $\delta \to 0$, $\omega_0 = O(\delta^{-2})$, and the Gaussian distribution of vorticity tends to a delta-function. Thus, the vorticity is indeed singular in the limit, yet the rate of dissipation of energy per unit length of vortex remains finite.

If the strain field is non-axisymmetric, of the form

$$\mathbf{U}(x,y,z) = (\alpha x, \beta y, \gamma z), \quad \text{with } \alpha < \beta \leq 0 < \gamma\,,\ \alpha+\beta+\gamma = 0\,, \tag{4.11}$$

the problem becomes much more complicated, and the behavior is strongly influenced by the value of the appropriate Reynolds number, here $Re_\Gamma = \Gamma/\nu$. When $Re_\Gamma \gg 1$, as relevant in the context of turbulence, and when $\beta < 0$, the rapid spin within the vortex is sufficient to minimize departures from axisymmetry, and the solution (4.7) is still valid at leading order, the small departures from axisymmetry in the contours of constant ω having an interesting topological structure (Moffatt *et al.*, 1994).

The particular situation when $\beta = 0$ provides a stretched vortex sheet localized near the plane $x = 0$, also with Gaussian structure. This two-dimensional solution has been generalized by conformal mapping techniques to provide a wide class of exact solutions of the Navier–Stokes equations exhibiting a fascinating range of "floral" vortical patterns (Bazant and Moffatt, 2005). For such two-dimensional solutions however, the maximum vorticity in each sheet increases in proportion to $\nu^{-1/2}$ as $\nu \to 0$, and the rate of dissipation of energy per unit area of the vortex sheets is $O(\nu^{1/2})$, thus vanishing in the limit $\nu = 0$, in striking contrast to the axisymmetric case. This is one reason why vortex tubes, rather than vortex sheets, are the more promising candidates for the role of typical structures within a turbulent flow.

5. Kelvin–Helmholtz Instability

In consideration of the instabilities to which fluid flows are subject, we should distinguish between "fast" instabilities, i.e., those that are of purely inertial origin and have growth rates that do not depend on viscosity, and "slow instabilities," which are essentially of viscous origin, and whose growth rates therefore tend to zero as the viscosity ν tends to zero, or equivalently as the Reynolds number $Re = UL/\nu$ tends to infinity. Examples of fast instabilities are the "Rayleigh–Taylor instability" that occurs when a heavy layer of fluid lies over a lighter layer; the "centrifugal instability" (leading to "Taylor vortices") that occurs in a fluid undergoing differential rotation when the circulation about the axis of rotation decreases with radius; and the "Kelvin–Helmholtz instability" that occurs in any region of rapid shearing of the fluid. The best-known example of a slow instability is the instability of pressure-driven "Poiseuille flow" between parallel planes, which is associated with subtle effects of viscosity in "critical layers" near the boundaries; the "dynamo instability" of magnetic fields in electrically conducting fluids is also diffusive in origin (through magnetic diffusivity rather than viscosity), and may therefore also be classed as a slow instability.

Here, we shall focus on the Kelvin–Helmholtz instability, idealized as the instability of a tangential discontinuity of velocity, which we may take to be

$$\mathbf{U} = (\mp U/2, 0, 0) \quad \text{for } y > \text{or} < 0\,. \tag{5.1}$$

The vorticity is then concentrated on the sheet $y = 0$, and given by the delta-function

$$\boldsymbol{\omega} = (0, 0, U\delta(y))\,. \tag{5.2}$$

We suppose that this sheet is subjected to the sinusoidal perturbation

$$y = \eta(x,t) = \eta(t)\exp ikx\,, \tag{5.3}$$

with $k > 0$, the real part of Eq. (5.3) being understood. All perturbations may similarly be supposed proportional to $\exp ikx$. The flow is assumed to be irrotational everywhere except on this disturbed sheet; the perturbation is thus "isovortical" in the sense that the disturbed vorticity is obtained by a virtual flux-conserving displacement of the undisturbed vorticity field. The velocity above and below the interface then takes the form

$$\mathbf{u} = (-U/2, 0, 0) + \nabla\phi_1 \quad \text{for } y > \eta\,, \tag{5.4}$$

$$\mathbf{u} = (+U/2, 0, 0) + \nabla\phi_2 \quad \text{for } y < \eta\,, \tag{5.5}$$

where, by virtue of incompressibility,

$$\nabla^2\phi_1 = 0 \quad \text{and} \quad \nabla^2\phi_2 = 0. \tag{5.6}$$

Since moreover the perturbation velocity must vanish as $y \to \pm\infty$, it follows that

$$\phi_1 = \Phi_1(t)e^{-ky+ikx}, \quad \phi_2 = \Phi_2(t)e^{ky+ikx}, \tag{5.7}$$

where $\Phi_1(t)$ and $\Phi_2(t)$ are to be found.

There are now two important conditions that must be satisfied on the vortex sheet $y = \eta(x,t)$. First, since this sheet moves with the fluid, its Lagrangian derivative must vanish, i.e.,

$$\frac{D}{Dt}(y - \eta(x,t)) \equiv \left(\frac{\partial}{\partial t} + \mathbf{u}\cdot\nabla\right)(y - \eta(x,t)) = 0 \quad \text{on } y = \eta\,. \tag{5.8}$$

Now $Dy/Dt \equiv \mathbf{u}\cdot\nabla y = \partial\phi_{1,2}/\partial y$ according as we approach the sheet from above or below. Also, for so long as the disturbance remains small, the problem may be linearized, i.e., squares and products of the small quantities η, Φ_1 and Φ_2 may be neglected and the jump conditions may be applied on $y = 0$ instead of $y = \eta$. It follows that

$$\frac{\partial\phi_1}{\partial y} = \frac{\partial\eta}{\partial t} - \frac{1}{2}U\frac{\partial\eta}{\partial x} \quad \text{and} \quad \frac{\partial\phi_2}{\partial y} = \frac{\partial\eta}{\partial t} + \frac{1}{2}U\frac{\partial\eta}{\partial x} \quad \text{on } y = 0\,. \tag{5.9}$$

Second, the pressure $p = \text{cst} - \rho\partial\phi/\partial t + \rho\mathbf{u}^2/2$ must be continuous across $y = \eta$, so that on linearizing,

$$\frac{\partial\phi_2}{\partial t} - \frac{\partial\phi_1}{\partial t} + \frac{1}{2}U\left(\frac{\partial\phi_2}{\partial x} + \frac{\partial\phi_1}{\partial x}\right) = 0 \quad \text{on } y = 0\,. \tag{5.10}$$

Equations (5.9) and (5.10) may now be combined to give, after some simple algebra, the amplitude equation

$$\frac{\partial^2\eta}{\partial t^2} = \frac{1}{4}k^2U^2\eta\,, \tag{5.11}$$

with exponential solutions $\eta \propto e^{\sigma t}$ where $\sigma = \pm kU/2$. Thus the mode for which

$$\sigma = +kU/2 \tag{5.12}$$

grows exponentially until the linearized theory ceases to be valid. These modes (for varying wavenumber k) are unstable, and the growth rate is proportional to k, increasing as the wavelength $2\pi/k$ of the disturbance decreases.

The physical mechanism of this instability is that the local strength of the perturbed vortex sheet, given for the unstable mode by

$$\Gamma(x,t) = U + \frac{\partial\phi_2}{\partial x} - \frac{\partial\phi_1}{\partial x} = U + 2i\frac{\partial\eta}{\partial t} = U + ikU\eta\,, \tag{5.13}$$

is $\pi/2$ out of phase with η; the perturbation vorticity is maximal at the points of inflexion where the slope of η is positive, and the induced velocity is such as to amplify the perturbation (Fig. 4).

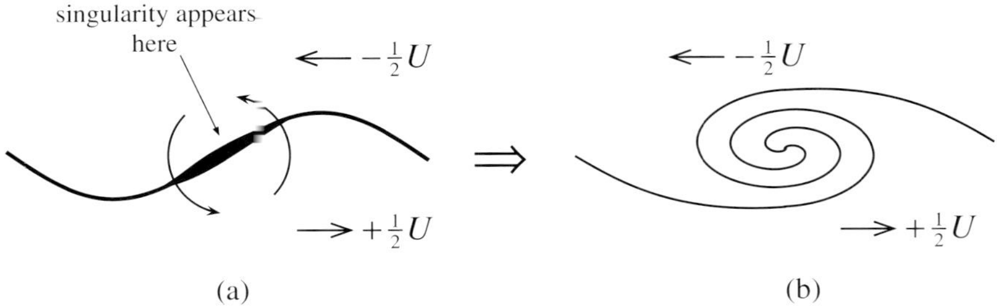

Fig. 4. The Kelvin–Helmholtz instability of a vortex sheet: (a) Vorticity accumulates in the sheet at the upward sloping inflexion points; (b) Spiral wind-up after the Moore singularity.

This interpretation of the instability mechanism actually continues into the nonlinear regime, investigated by Moore (1979). Moore noted first that, even on linear theory, some kind of singular behavior is to be expected after a finite time. For, by way of example, suppose that the initial disturbance is periodic in x with period λ, with convergent Fourier series of the form

$$\eta(x,0) = \sum_{n=1}^{\infty} A_n \sin \frac{n\pi x}{\lambda}\,, \tag{5.14}$$

where

$$A_n = e^{-n} n^{-p}\,, \tag{5.15}$$

with $p > 0$. Thus $\eta(x,t)$ and all its x-derivatives exist at time $t = 0$. However, by virtue of Eq. (5.12), selecting only the unstable modes, the disturbance at time t is given by

$$\eta(x,t) = \sum_{n=1}^{\infty} A_n \exp \frac{n\pi U t}{2\lambda} \sin \frac{n\pi x}{\lambda}\,, \tag{5.16}$$

and this series diverges for $t > t_c = 2\lambda/\pi U$, because the exponential growth of the coefficients then defeats the power-law decay for large n.

Now nonlinear effects generate harmonics of the initial disturbance even when this consists of a single Fourier mode, so that a series of the form Eq. (5.14) is soon established. Moore's achievement was to show that the exact nonlinear solution for $\eta(x,t)$ becomes singular at a finite time of order λ/U at the upward-sloping inflexion points where, as indicated above, the accumulation of vorticity becomes more and more concentrated.

This singularity appears as a discontinuity of curvature, and the vortex sheet strength is cuspidal in form. Beyond the singularity time, observation suggests that the sheet rolls up in a periodic sequence of spiral vortices (Fig. 4(b)), although no analytical solution is as yet available to describe this behavior.

What is important here is that any vortex sheet is absolutely unstable, with a tendency to break up into a series of concentrations of vorticity, more like vortex tubes than a vortex sheet. The vortex tube appears in general to be a much more robust structure than the vortex sheet which has at best a transitory existence, even in turbulent flows.

The Kelvin–Helmholtz instability, as described above, occurs not only for vortex sheets, but also for parallel shear flows having an inflexion point in the velocity profile; the "tanh" profile

$$\mathbf{U} = (-U/2 \tanh y/\delta, 0, 0)\,, \tag{5.17}$$

for which vorticity is distributed in a layer of thickness $O(\delta)$, is a useful prototype. Such a velocity field is unstable to sinusoidal perturbations of wavelength large compared with δ; on such scales, the velocity profile "looks like" the discontinuous profile (5.1), so it is not surprising that it exhibits the same type of instability leading to spiral wind-up of the whole vortical layer.

In fact, the existence of at least one inflexion point in the profile of a parallel shear flow of an inviscid fluid is known to be a necessary condition for (linearized) instability of the flow (see, for example (Drazin and Reid, 2005)). Plane Poiseuille flow, with its parabolic profile, is therefore stable in the limit of infinite Reynolds number ($\nu = 0$). The source of the (slow) instability of this and similar flows must therefore be sought in the dual role of viscosity, usually thought to be merely stabilizing!

6. Transient Instability and Streamwise Vortices

There is however another, potentially more potent, mechanism by which plane parallel non-inflexional flows may be destabilized; this arises through consideration of the shearing of disturbances of finite (rather than infinitesimal) amplitude. Such disturbances, as might be anticipated, can be drawn out into long structures parallel to the flow (or "streamwise vortices") which, when superposed on the underlying shear flow, provide locally inflexional profiles, which are then subject to the Kelvin–Helmholtz instability.

We shall illustrate this behavior by considering the simplest case of uniform shear flow

$$\mathbf{U} = (\alpha y, 0, 0)\,, \tag{6.1}$$

on which, at time $t = 0$, we superpose a sinusoidal disturbance of the form

$$\mathbf{u}(\mathbf{x}, t) = \mathbf{A}_0 \exp(i\mathbf{k}_0 \cdot \mathbf{x})\,, \tag{6.2}$$

with $\mathbf{k}_0 \cdot \mathbf{A}_0 = 0$ (by incompressibility). For the moment, we retain the effects of viscosity. The analysis that follows was presented by Moffatt (1967), and developed in the context of turbulent shear flow by Townsend (1976).

We suppose that the perturbation, although finite, is still sufficiently weak to allow linearization of the Navier–Stokes equation:

$$\frac{\partial \mathbf{u}}{\partial t} + \mathbf{U} \cdot \nabla \mathbf{u} + \mathbf{u} \cdot \nabla \mathbf{U} = -\frac{1}{\rho}\nabla p + \nu \nabla^2 \mathbf{u}\,, \tag{6.3}$$

where p is the perturbation pressure associated with the disturbance. This equation admits a solution of the form

$$\mathbf{u} = \mathbf{A}(t) \exp(i\mathbf{k}(t) \cdot \mathbf{x})\,, \quad p/\rho = P(t) \exp(i\mathbf{k}(t) \cdot \mathbf{x})\,, \tag{6.4}$$

in which both wavevector $\mathbf{k}(t)$ and amplitudes $\mathbf{A}(t)$ and $P(t)$ are allowed to vary with time. Such disturbances, first recognized by Lord Kelvin (1887), are known as "Kelvin modes." We may note that for a single mode of this kind, the omitted nonlinear term $\mathbf{u} \cdot \nabla \mathbf{u}$ in Eq. (6.3) is in fact identically zero, so that Eqs. (6.4) can provide an exact solution of the Navier–Stokes equation. However, a superposition of modes of different wave vectors do involve significant nonlinear interactions, which we do not consider here.

Substituting Eqs. (6.4) in Eq. (6.3) gives

$$\dot{\mathbf{A}} + i(\dot{\mathbf{k}} \cdot \mathbf{x})\mathbf{A} + \alpha A_2(1, 0, 0) + i\alpha y k_1 \mathbf{A} = -i\mathbf{k}P - \nu k^2 \mathbf{A}\,, \tag{6.5}$$

and we have also, by incompressibility,

$$\mathbf{k}(t) \cdot \mathbf{A}(t) = 0\,. \tag{6.6}$$

The coefficients of x, y and z in Eq. (6.5) must vanish; hence $\dot{k}_1 = 0$, $\dot{k}_2 = -\alpha k_1, \dot{k}_3 = 0$, so that

$$k_1 = k_{01}\,, \quad k_2(t) = k_{02} - \alpha k_1 t\,, \quad k_3 = k_{03}\,. \tag{6.7}$$

This simply describes the shearing of the wave fronts, which become more and more aligned parallel to the plane $y = 0$. If $k_1 = 0$, then the wave vector $(0, k_2, k_3)$ remains constant, whereas if $k_1 \neq 0$, then the effect of the shear is asymptotically to align the wave vector in the (0, 1, 0) direction and to increase its magnitude linearly with time.

Here we may note immediately that the effect of the viscous term is simply to introduce a factor

$$\exp\left[-\nu\int_0^t (\mathbf{k}(t))^2 dt\right] = \exp\left[-\nu(k_0^2 t - k_1 k_{02}\alpha t^2 + k_1^2\alpha^2 t^3/3)\right], \qquad (6.8)$$

where $k_0 = |\mathbf{k}_0|$, so that, provided $k_1 \neq 0$, this Kelvin mode experiences "accelerated decay" on a timescale

$$\alpha t = O(\alpha/\nu k_1^2)^{1/3}. \qquad (6.9)$$

Modes for which k_1/k_0 is small survive for a long time (when ν is small); the exceptional modes for which $k_1 = 0$ survive for the much longer timescale $O(1/\nu k_0^2)$, unaffected by the shear. It is the decay of all modes as described by Eq. (6.8) that accounts for the stability of the flow $\mathbf{U}$ on linearized analysis. However, before this ultimate decay sets in, the amplitude $|\mathbf{A}(t)|$ may increase by an arbitrarily large factor, as we shall now show.

Noting first, from Eq. (6.6), that $\dot{\mathbf{k}}\cdot\mathbf{A} + \dot{\mathbf{A}}\cdot\mathbf{k} = 0$, we have, from Eq. (6.5),

$$-ik^2 P = -\dot{\mathbf{k}}\cdot\mathbf{A} + \alpha A_2 k_1 = 2\alpha A_2 k_1, \qquad (6.10)$$

and the part of Eq. (6.5) not involving x, y and z is then satisfied provided

$$\dot{\mathbf{A}} + \alpha A_2(1,0,0) = -i\mathbf{k}P = 2\alpha A_2 k_1 \mathbf{k}/k^2. \qquad (6.11)$$

Integration of the second component of this equation, then of the first and third components, is straightforward; with the notation

$$l^2 = k_1^2 + k_3^2, \quad \tan\theta = l/k_2(t), \quad [\psi] = \psi(t) - \psi(0), \qquad (6.12)$$

the solution is

$$A_1(t) = A_{01} - A_{02}\left\{\frac{k_0^2 k_3^2}{k_1 l^3}[\theta] + \frac{k_1 k_0^2}{l^2}\left[\frac{k_2}{k^2}\right]\right\}, \qquad (6.13)$$

$$A_2(t) = A_{02}k_0^2/k^2, \qquad (6.14)$$

$$A_3(t) = A_{03} + A_{02}\frac{k_3 k_0^2}{l^3}\left\{[\theta] + l\left[\frac{k_2}{k^2}\right]\right\}. \qquad (6.15)$$

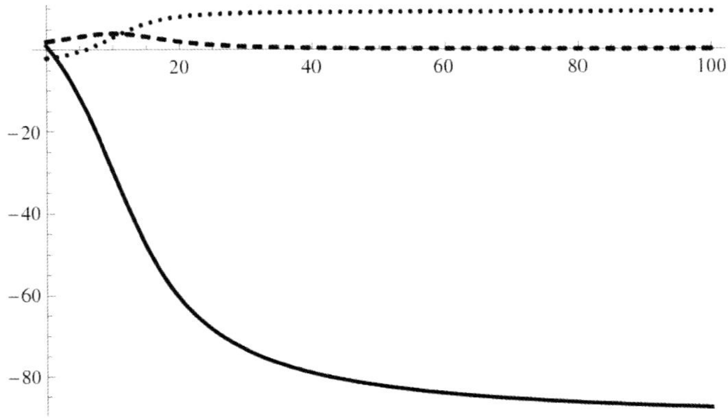

Fig. 5. Evolution of $A_1(t)$ (solid curve), $A_2(t)$ (dashed), and $A_3(t)$ (dotted), as given by Eqs. (6.13)–(6.15), with initial conditions $\mathbf{k}_0 = (0.1, 1, 1)$ and $\mathbf{A}_0 = (1, 1, -1.1)$ (so $\mathbf{k}_0 \cdot \mathbf{A}_0 = 0$); note the relatively long period of linear growth of $A_1(t)$, a symptom of transient instability.

These three components are plotted in Fig. 5 for the initial conditions $\mathbf{k}_0 = (0.1, 1, 1)$ and $\mathbf{A}_0 = (1, 1, -1.1)$, for which $k_1/k_0 \approx 0.07$, small enough for there to be a relatively long period of approximately linear growth of $|A_1(t)|$. This period of linear growth increases as k_1/k_0 decreases. The linear growth, or "transient instability," results from the $(\mathbf{u} \cdot \nabla)\mathbf{U} = u_2 \partial \mathbf{U}/\partial y$ term in Eq. (6.3), which corresponds to persistent transport of mean-flow x-momentum in the y-direction.

If a random superposition of modes with isotropically distributed initial wavevectors $\mathbf{k}_0$ is subjected to the above shearing, then the dominant contribution to the disturbance energy will ultimately come from modes with wavevectors in an increasingly narrow neighborhood of the plane (in wavenumber space) $k_1 = 0$, i.e., from modes for which $\mathbf{k}_0 \cdot \mathbf{U} \approx 0$. Physically this corresponds to the emergence of structures having little or no variation in the streamwise direction. Such structures are known, for obvious reasons, as "streamwise vortices;" they grow in strength, under the action of the mean shear, until the appearance of inflexion points in the profile of the total x-component of velocity is inevitable. At that stage the flow is prone to "secondary instability" of Kelvin–Helmholtz (K–H) origin; the flow becomes fully three-dimensional, and the transition to turbulence is well underway. All this applies of course only if the viscosity parameter ν is sufficiently weak.

The theory described above is a particular case of what is known as "Rapid Distortion Theory" (RDT), which more generally describes the linearized uniform distortion of a field of turbulence by a mean velocity field of the form

$$U_i(\mathbf{x}) = c_{ij}x_j, \tag{6.16}$$

of which Eq. (6.1) is obviously a special case. Such flows may be either elliptic or hyperbolic in character. It is possible to incorporate additional effects relevant in geophysical applications, e.g., uniform density stratification and/or Coriolis effects associated with the Earth's rotation. Such effects have been explored in detail by Sagaut and Cambon (2008), where extensive references to previous work on RDT may be found.

It is also worth noting that transient instabilities, as described above, and as greatly developed by Schmid and Henningson (1994), play an important part in more recent work in which new steady and traveling-wave solutions of the classical problems of Couette flow and Poiseuille flow in a pipe have been found. The essential idea (see, for example (Waleffe, 2003; Pringle and Kerswell, 2007)) is that coherent structures formed by transient instability are unstable to K–H-type instability, and that these (secondary) instabilities interact coherently in such a way as to regenerate the original finite-amplitude perturbations to the flow. The highly original new ideas and results in this area, which have a bearing on the important problem of transition to turbulence, are among the most exciting to emerge in recent years.

7. Turbulence, Viewed as a Random Field of Vorticity

Over the last 20 years, turbulence has been increasingly subjected to Direct Numerical Simulation (DNS), i.e., computational treatment of the Navier–Stokes equations without approximation, by either finite-difference or spectral techniques, and "post-processing" of the numerical output. Figure 6 shows the vorticity distribution in high vorticity regions of a field of turbulence, from a "state-of-the-art" simulation on the Earth Simulator (Yokokawa *et al.*, 2002); what is important to note here is the apparent "tube-like" structure of this random field. We referred in the introduction to the persistent stretching of vortex lines in a turbulent flow. Figure 6 gives some substance to this description: Each vortex tube is subject to stretching associated with the induced velocity of the whole vorticity distribution (possibly dominated by that of neighboring

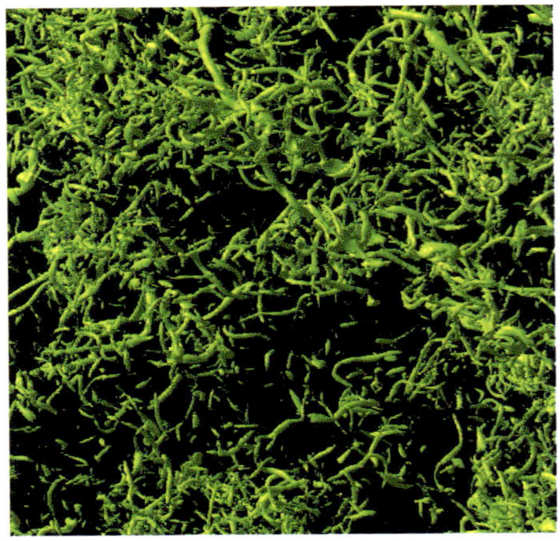

Fig. 6. Intense-vorticity iso-surfaces ($|\omega| > \langle\omega\rangle + 4\sigma$, where σ is the standard deviation of $|\omega|$), in a direct numerical simulation of homogeneous turbulence (from Yokokawa *et al.* (2002), by permission); this simulation was carried out in a periodic box with 4096^3 grid points, and at a Reynolds number $Re_\lambda = 732$; this Reynolds number is $O(Re^{1/2})$, where $Re = u_0 L/\nu$. This figure shows a "zoomed-in" high vorticity region of size $(748^2 \times 1496)l_v$, where l_v is the "inner" Kolmogorov scale. Vorticity fluctuations down to this scale are reasonably well resolved.

vortices), in a manner reminiscent of Burgers' vortex model discussed in Sec. 4.

Of course, such a description presupposes that there is indeed a systematic stretching effect (rather than the opposite — a systematic contraction). This stretching arises from a natural tendency for any two fluid particles, initially close together, to move apart under the action of a random incompressible velocity field. Indeed, if $\delta\mathbf{x}(\mathbf{t})$ is the separation of two particles, with $\delta\mathbf{x}(0) = \delta\mathbf{a}$ assumed infinitesimally small and non-random, then it can be shown (Orszag, 1977) that in homogeneous, isotropic turbulence (i.e., turbulence whose statistical properties are invariant under translation and rotation)

$$\left\langle \delta\mathbf{x}^2 \right\rangle \geq \delta\mathbf{a}^2 . \tag{7.1}$$

When coupled with an assumption concerning the "finite-memory" of turbulence (which amounts to assuming that the turbulence field for times greater than $t + t_c$ is uncorrelated with that at time t), this is sufficient to establish that $\left\langle \delta\mathbf{x}^2 \right\rangle$ increases systematically in time (Davidson, 2004). In particular, if $\delta\mathbf{x}$ is aligned with a vortex line, this element of the vortex line will be systematically stretched by the flow (and this applies to every element of every vortex line!).

The essential ingredients of the dynamics of turbulence may thus be thought of as a combination of three elements: Formation of sheet-like structures by shearing of random vorticity (the transient instability mechanism); all-pervasive Kelvin–Helmholtz instability of such structures leading to tube-like structures with possibly some remnants of spiral wind-up; and persistent stretching of such vortices by the strain induced by the surrounding vorticity field. Each of these ingredients has a tendency to decrease the scale of the velocity field, i.e., to contribute to the energy cascade towards the smallest scales of the turbulence, a fundamental aspect of the problem to which we now turn.

8. The Kolmogorov–Obukhov Energy-Cascade Theory

The random character of a turbulent velocity field necessitates a statistical treatment in which an "ensemble average" $\langle \cdots \rangle$ can be defined. By "homogeneous" turbulence, as indicated above, we mean turbulence for which all such averages are invariant under translation, i.e., independent of the origin of the coordinate system adopted. By "isotropic" turbulence, we mean turbulence that is homogeneous and, in addition, invariant under rotation of the frame of reference, i.e., statistically "the same in every direction." We note that, if homogeneous turbulence is subjected to uniform strain of the form Eq. (6.16), then it remains homogeneous, but develops increasingly marked anisotropy, even if isotropic initially. Homogeneous turbulence has been intensively studied since the pioneering investigations recorded by Batchelor (1953). A modern treatment of the subject, with emphasis on the Kolmogorov (1941) theory and its later modifications, is provided by Frisch (1995).

We restrict attention here to the situation when the mean velocity vanishes: $\langle \mathbf{u} \rangle = 0$. Then attention must be focused on correlations such as

$$R_{ij}(\mathbf{r}) = \langle u_i(\mathbf{x}) u_j(\mathbf{x}+\mathbf{r}) \rangle \,, \quad S_{ijk}(\mathbf{r}) = \langle u_i(\mathbf{x}) u_j(\mathbf{x}) u_k(\mathbf{x}+\mathbf{r}) \rangle \,, \ldots \,, \tag{8.1}$$

in standard suffix notation. Equations for such correlation tensors can be obtained from the Navier–Stokes equations in a straightforward way; the trouble is that, due to the nonlinearity of these equations, the equation for $\partial R_{ij}/\partial t$ involves terms like $S_{ijk}(\mathbf{r})$; more generally, the time derivative of any nth-order correlation inevitably involves the current value of $(n+1)$th-order correlations. This is the famous "closure problem" that bedevils the subject. No completely satisfactory "closure" hypothesis (providing an instantaneous relationship between nth-order correlations and those of lower order) has yet been found.

There is however one equation for a second-order quantity that does not involve higher-order quantities[a]: This is the energy equation, easily derived from Eqs. (3.1):

$$\frac{d}{dt}\frac{1}{2}\left\langle \mathbf{u}^2 \right\rangle = -\nu \left\langle \boldsymbol{\omega}^2 \right\rangle + \epsilon \,. \tag{8.2}$$

The nonlinear term of Eqs. (3.1) makes no contribution to this energy equation, because it simply redistributes energy over an ever-increasing range of length-scales (as if through the generation of harmonics and sub-harmonics). We include a term ϵ in Eq. (8.2), representing the rate of input of energy to the turbulence by some stirring mechanism on a scale L; on dimensional grounds, the level of turbulent energy generated is then of order

$$u_0^2 \equiv \left\langle \mathbf{u}^2 \right\rangle \sim (\epsilon L)^{2/3} \,, \tag{8.3}$$

and we assume that

$$Re = u_0 L/\nu \gg 1 \,. \tag{8.4}$$

Under statistically steady conditions, from Eq. (8.2),

$$\left\langle \boldsymbol{\omega}^2 \right\rangle = \epsilon/\nu \,, \tag{8.5}$$

from which we note immediately that the enstrophy $\left\langle \boldsymbol{\omega}^2 \right\rangle \to \infty$ as $\nu \to 0$.

The picture then, as conceived by Richardson (1926) and formalized by Kolmogorov (1941), is that energy cascades at a rate ϵ from scales of order L down to scales of order $l_v (\ll L)$ at which viscous effects can dissipate the energy (to heat). The only dimensional parameters on which the scale l_v can depend are ϵ and ν, and it therefore follows on dimensional grounds that

$$l_v \sim (\nu^3/\epsilon)^{1/4} \,. \tag{8.6}$$

It then follows that

$$l_v/L \sim Re^{-3/4} \,, \tag{8.7}$$

[a] There is also a similar equation for the mean helicity which involves a dissipative term $-\nu\langle \boldsymbol{\omega} \cdot \nabla \times \boldsymbol{\omega} \rangle$; however, since helicity is not sign-definite, positive helicity generation at one scale can be compensated by negative helicity generation at another, even neglecting the effect of viscosity. This means that the concept of a "helicity cascade" must be treated with caution.

so that there is indeed a wide range of scales between the "energy injection scale" L and the "dissipation scale" l_v. It is over this range that the energy cascade can proceed.

Kolmogorov's (1941) theory is concerned with the statistical properties of turbulence on scales small compared with L, and he assumed that on such scales, these statistical properties are isotropic and depend only on the parameters ϵ and ν, as well as on the separation variable r. Moreover, if $L \gg r \gg l_v$ (the "inertial range" of scales), then the statistical properties do not depend on ν. Thus, for example, the "second-order structure function" $\langle(\mathbf{u}(\mathbf{x}+\mathbf{r})-\mathbf{u}(\mathbf{x}))^2\rangle$ must, on dimensional grounds, have the behavior

$$\langle(\mathbf{u}(\mathbf{x}+\mathbf{r})-\mathbf{u}(\mathbf{x}))^2\rangle \sim (\epsilon\, r)^{2/3}\,. \tag{8.8}$$

Similarly, the mean-square separation of two fluid particles $\langle(\Delta\mathbf{x})^2\rangle$ must increase like

$$\langle(\Delta\mathbf{x})^2\rangle \sim \epsilon\, t^3\,, \tag{8.9}$$

for so long as this quantity remains within the inertial range, a result foreshadowed by Richardson (1926) in an early study of atmospheric diffusion. This is more rapid than conventional diffusion in three dimensions with diffusivity D, namely $\langle(\Delta\mathbf{x})^2\rangle \sim 6Dt$, because, as the particles separate, eddies on progressively larger scales contribute to the diffusive process.

An equivalent formulation of the energy cascade in wavenumber space (Obukhov, 1941) gives a result for the energy spectrum function $E(k)$ equivalent to (8.8), namely

$$E(k) = C\epsilon^{2/3}k^{-5/3} \quad (L^{-1} \ll k \ll k_v = l_v^{-1})\,. \tag{8.10}$$

This function $E(k)$ is defined in such a way that

$$\langle(\mathbf{u}(\mathbf{x})^2\rangle = 2\int_0^\infty E(k)\,dk\,, \tag{8.11}$$

so that $E(k)\,dk$ is the contribution to the mean kinetic energy from wavenumbers in the spherical shell $\{k, k+dk\}$ in wavenumber space. According to the theory, the dimensionless constant C should be the same in all fields of turbulence, irrespective of the nature of the source of energy on scales of order L, and irrespective of the context, whether environmental, meteorological, astrophysical, or whatever. The first convincing evidence for a $k^{-5/3}$ spectral range came from measurements of turbulence at a Reynolds number of order 10^8 in the tidal channel to the east of Vancouver Island by Grant *et al.* (1962). Since then, the Kolmogorov theory (sketched

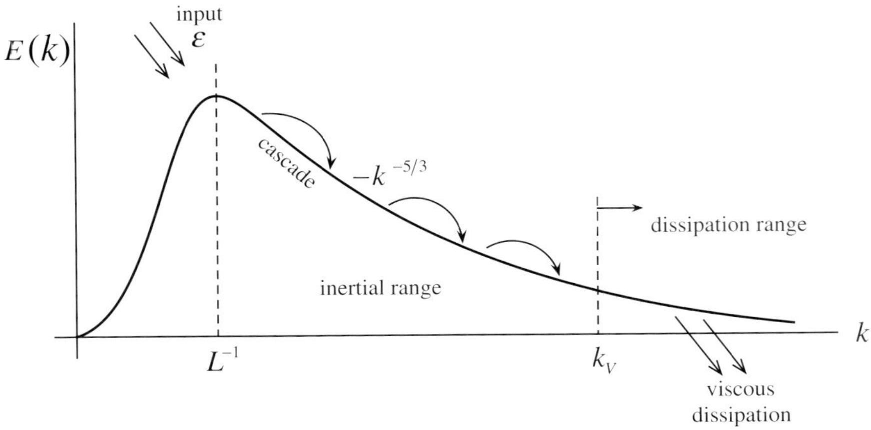

Fig. 7. Energy cascade according to the Kolmogorov–Obukhov scenario. Energy is supplied to the turbulence at a rate ϵ on scales of order L, and is dissipated at wavenumbers of order $k_v = (\epsilon/\nu^3)^{1/4}$; for wavenumbers in the inertial range $L^{-1} \ll k \ll k_v$, the energy spectrum function follows a $k^{-5/3}$ power law.

schematically in Fig. 7) has provided the bedrock of our understanding of turbulence.

Yet all was not well with the theory, as Kolmogorov (1962) himself recognized; for the rate of dissipation of energy is itself a function of position and time: $\epsilon = \epsilon(\mathbf{x}, t)$, and in regions where $\epsilon > \langle\epsilon\rangle$, the energy cascade presumably proceeds more vigorously, a runaway effect that is now known to generate "intermittency" in a field of turbulence, i.e., regions of relatively intense vorticity embedded in more quiescent regions, very much as revealed by DNS. Although intermittency has at most a weak effect on the second-order structure function and on the energy spectrum function (the $k^{-5/3}$-law being apparently quite robust), higher-order statistics are more seriously affected, and the conceptual basis for the Kolmogorov theory is seriously undermined. Huge research effort has been devoted to the problem of intermittency (see, for example, (Frisch, 1995)), but it seems fair to say that the phenomenon still poses a great challenge to theoreticians.

A further great challenge that remains concerns the behavior in the "dissipation range" of wavenumbers $k \sim k_v$ and greater, where $k_v = l_v^{-1} = (\epsilon/\nu^3)^{1/4}$. Here the experimental evidence is that $E(k)$ decays exponentially for $k > k_v$, implying smoothness of the velocity field at the smallest scales (always of course within the limits of a continuum description). On the other hand, we have the result Eq. (8.5) implying the divergence of enstrophy as $\nu \to 0$. This brings us back to the problem posed at the outset of precisely how the energy of turbulence is dissipated at the smallest scales. The Burgers model of Sec. 4 provides an important clue and starting point, but the crucial problem of the *interaction of skewed vortices*, as detected in DNS, remains of central importance at these smallest scales. We may note that, at a Reynolds number of order 10^8 as in the Vancouver tidal channel, if $L \sim 1\,\text{km}$, then $l_v \sim Re^{-3/4}L \sim 1\,\text{mm}$; this range of scales from kilometers down to millimeters in a $3D$ field of turbulence is far beyond what can be simulated in even the most powerful supercomputers of the current era; hence the continuing need for theoretical analysis of turbulence in parallel with experimental observation and carefully crafted numerical simulation.

In this brief introduction to the huge subject of vortex dynamics and turbulence, we have only been able to scrape the surface. Many books are now available for students wishing to pursue the subject in depth. Notable among these is the two-volume encyclopedic work of Monin and Yaglom (1975). The more recent volumes of Davidson (2004) and Sagaut and Cambon (2008) bear testimony to the continuing vitality of the subject. These and other notable books are distinguished by two asterisks (**) in the list of references that follows.

Acknowledgments

I thank Mark Hallworth for help with preparation of the figures.

References

Aref, H. (Ed.) (2010). *150 years of Vortex Pynamics*, P∼C IUTAM Book series, Vol. 20 (Springer, Dorctrechz).

Arnol'd, V. I. (1974). The asymptotic Hopf invariant and its applications (in Russian). In Proc. Summer School in Differential Equations, Erevan, Armenian SSR. Acad. Sci. [English translation: *Sel. Math. Sov.* **5**, pp. 327–345 (1986)].

Batchelor, G. K. (1953). *Homogeneous Turbulence* (Cambridge University Press, Cambridge, UK**).

Bazant, M. Z. and Moffatt, H. K. (2005). Exact solutions of the Navier–Stokes equations having steady vortex structures, *J. Fluid Mech.* **541**, 55, pp. 226–264.

Beale, J. I., Kato, T. and Majda, A. (1984). Remarks on the breakdown of smooth solutions for the 3-D Euler equations, *Comm. Math. Phy.* **94**, pp. 61–66.

Burgers, J. M. (1948). A mathematical model illustrating the theory of turbulence, *Adv. Appl. Mech.* **1**, pp. 171–199.

Davidson, P. A. (2004). *Turbulence: An Introduction for Scientists and Engineers* (Oxford University Press, Cambridge, Mass).

Dombre, T., Frisch, U., Greene, J. *et al.* (1986). Chaotic streamlines in the ABC flow, *J. Fluid Mech.* **167**, pp. 353–391.

Drazin, P. G. and Reid, W. H. (2005). *Hydrodynamic Stability*, 2nd edn. (Cambridge University Press, Cambridge, UK).

Euler, L. (1755). Principes généraux du mouvement des fluides, *Opera Omnia, ser. 2* **12**, pp. 54–91, [Reproduced in English translation in: *Physica D* **237** (2008), 1825–1839].

Eyink, G., Frisch, U., Moreau, R. *et al.* (2008). Euler equations: 250 years on, *Physica D* **237**.

Frisch, U. (1995). *Turbulence — The Legacy of A. N. Kolmogorov* (Cambridge University Press**, Cambridge, UK).

Grant, H. C., Stewart, R. W. and Moilliet, A. (1962). Turbulence spectra from a tidal channel, *J. Fluid Mech.* **12**, pp. 241–268.

Helmholtz, H. (1858). Uber integrale der hydrodynamischen Gleichungen, welche den Wirbelbewegungen entsprechen, *Crelle's Journal* **55**, pp. 25–55, [Trans. P. Tait (1867), On integrals of the hydrodynamic equations, which express vortex motion] *Phil. Mag.* **33**, pp. 485–510.

Kelvin, Lord (William Thomson) (1867). On vortex atoms, *Phil. Mag.* **34**, pp. 15–24.

Kelvin, Lord (William Thomson) (1869). On vortex motion, *Trans. Roy. Soc. Edin.* **25**, pp. 217–260.

Kelvin, Lord (William Thomson) (1887). Stability of fluid motion: rectilinear motion of viscous fluid between two parallel plates, *Phil. Mag.* **24**, 5, pp. 188–196.

Kolmogorov, A. N. (1941). The local structure of turbulence in incompressible viscous fluid for very large Reynolds number, *Dokl. Akad. Nauk. SSSR* **30**, pp. 9–13.

Kolmogorov, A. N. (1962). A refinement of previous hypotheses concerning the local structure of turbulence in a viscous incompressible fluid at high Reynolds number, *J. Fluid Mech.* **13**, pp. 82–85.

Moffatt, H. K. (1967). "Interaction of turbulence with strong wind shear," in A. Yaglom and V. Tatarski (eds), *Atmosphere Turbulence and Radio Wave Propagation* (Nauka, Moscow), pp. 139–156.

Moffatt, H. K. (1969). The degree of knottedness of tangled vortex lines, *J. Fluid Mech.* **36**, pp. 117–129.

Moffatt, H. K., Kida, S. and Ohkitani, K. (1994). Stretched vortices — the sinews of turbulence; high Reynolds number asymptotics, *J. Fluid Mech.* **259**, pp. 241–264.

Monin, A. C. and Yaglom, A. M. (1975). *Statistical Fluid Mechanics, I and II* (MIT Press, US).

Moore, D. W. (1979). The spontaneous appearance of a singularity in the shape of an evolving vortex sheet, *Proc. Roy. Soc. London. A* **365**, pp. 105–119.

Moreau, J.-J. (1961). Constants d'un ilot tourbillonnaire en fluide parfait barotrope, *CR Acad. Sci. Paris* **252**, pp. 2810–2812.

Obukhov, A. M. (1941). On the distribution of energy in the spectrum of turbulent flow, *Dokl. Akad. Nauk. SSSR* **32**, pp. 22–24.

Orszag, S. A. (1977). "Lectures on the statistical theory of turbulence," in R. Balian and J. L. Peube (eds), *Fluid Dynamics* (Gordon and Breach, New York), pp. 237–374.

Pringle, C. and Kerswell, R. R. (2007). Asymmetric, helical and mirror-symmetric travelling waves in pipe flow, *Phys. Rev. Lett.* **99**, 074502 [4 pages].

Richardson, L. F. (1926). Atmospheric diffusion shown on a distance-neighbour graph, *Proc. Roy. Soc. London A* **110**, pp. 709–737.

Saffman, P. G. (1995). *Vortex Dynamics* (Cambridge Univ. Press, Cambridge, UK).

Sagaut, P. and Cambon, C. (2008). *Homogeneous Turbulence Dynamics* (Cambridge Univ. Press, Cambridge, UK).

Schmid, P. J. and Henningson, D. S. (1994). Optimal energy density growth in Hagen–Poiseuille flow, *J. Fluid Mech.* **277**, pp. 197–225.

Tait, P. G. (1867). On integrals of the hydrodynamic equations, which express vortex motion, *Phil. Mag.* **33**, pp. 485–510.

Townsend, A. A. (1976). *The Structure of Turbulent Shear Flow*, 2nd edn. (Cambridge Univ. Press, Cambridge, UK).

Waleffe, F. (2003). Homotopy of exact coherent structures in plane shear flows, *Phys. Fluids* **15**, pp. 1517–1534.

Yokokawa, M., Itakura, K., Uno, A. *et al.* (2002). 16.4-Tflops direct numerical simulation of turbulence by a Fourier spectral method on the Earth Simulator. Available at: http://www.sc-2002.org/paperpdfs/pap273.pdf.

Chapter 2

Geophysical and Environmental Fluid Dynamics

Tieh-Yong Koh[1] and Paul F. Linden[2]

[1]*School of Physical and Mathematical Sciences*
Nanyang Technological University
21 Nanyang Link, SPMS-04-01
Singapore 637371, Singapore
kohty@ntu.edu.sg

[2]*Department of Mechanical and Aerospace Engineering*
University of California, San Diego
9500 Gilman Drive
La Jolla, CA 92130, USA
pflinden@ucsd.edu

In this chapter, the basic mechanics of stratified, rotating fluids as the background to geophysical and environmental flows are discussed. The following topics are included: Stable stratification and internal waves; gravity currents; plumes and convective flows; similarity theory of the atmospheric boundary layer; geostrophic motion and inertial waves; geostrophic adjustment.

1. Introduction

The Earth is enveloped by two important fluids: The atmosphere and the oceans. Both fluids are in a constant state of motion. A visit to the coast immediately reveals the restlessness of these fluids: There are winds buffeting the coast and waves crashing against the shore. Geophysical fluid dynamics is the study of the motion of the atmosphere and oceans according to the principles of dynamics and thermodynamics.

2. Stratified Flows

2.1. *Surface Gravity Waves*

Ocean waves seen crashing onshore are examples of surface waves. Such wave motion results when the water surface is displaced above its equilibrium level and gravity acts to pull it downwards. As the water falls, it does not stop at the equilibrium level but continues beyond to form a depression in the surface. Pressure from the surrounding water mass forces the surface depression to rise and the rising motion again overshoots the equilibrium level. Gravity acts once more to restore the water surface. The consequent oscillatory motion spreads horizontally, creating surface gravity waves.

2.1.1. *Dimensional analysis*

A plane surface wave may be represented as a disturbance to the surface height h

$$h = H + h_0 \exp[i(\omega t - kx)] \tag{2.1}$$

where H is the equilibrium height of the water measured from the bottom of the ocean, and h_0, ω and k are the amplitude, angular frequency and wavenumber of the wave, respectively. t denotes time and x denotes the displacement along the direction of wave propagation. The phase speed c of the wave is defined as

$$c = \frac{\omega}{k}, \tag{2.2}$$

and c depends on the gravitational field strength g, the equilibrium depth H and the wavenumber k.

Using dimensional analysis, it is possible to derive the functional form of phase velocity c with dimensions $[c] = \mathrm{LT}^{-1}$, in terms of the above factors, with dimensions $[g] = \mathrm{LT}^{-2}$, $[H] = \mathrm{L}$, $[k] = \mathrm{L}^{-1}$. Let $c = g^l H^m k^n$. Then,

$$LT^{-1} = L^{l+m-n}T^{-2l}$$

$$\Rightarrow \begin{cases} l = \dfrac{1}{2} \\ m = n + \dfrac{1}{2} \quad or \quad n = m - \dfrac{1}{2} \end{cases}$$

$$\therefore c = \sqrt{gH}(kH)^n \quad or \quad \sqrt{\frac{g}{k}}(kH)^m.$$

Since m and n can be any number, it is conceivable that

$$c = \sqrt{gH}\ f_1(kH) \quad or \quad \sqrt{\frac{g}{k}}\ f_2(kH) \tag{2.3}$$

for some dimensionless functions f_1 or f_2.

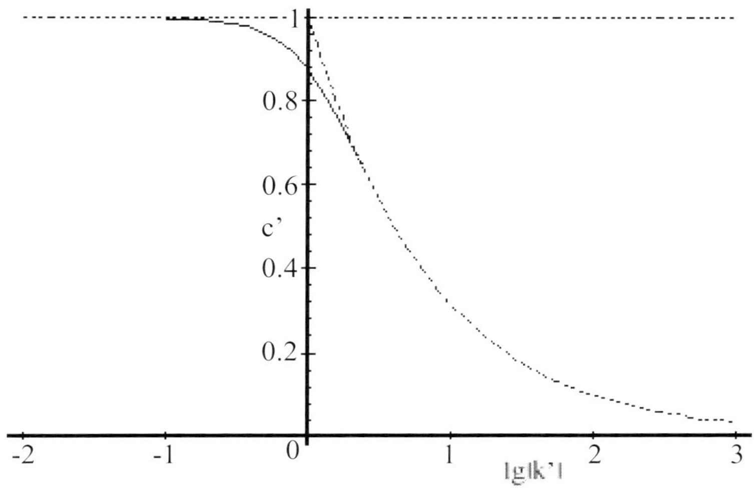

Fig. 1. The exact dispersion relation of surface waves, where $c' = c/\sqrt{gH}$ and $k' = kH$ are nondimensionalized phase velocity and wavenumber, respectively. Note that the logarithm of k' is plotted on the abscissa. Dashed lines denote the two asymptotic functions, $c' = 1$ and $c' = 1/\sqrt{|k'|}$, for small and large k', respectively.

2.1.2. *Exact dispersion relation*

The dependence of phase velocity c on the wavenumber k can be derived exactly from fluid dynamical considerations (see (Secs. 5.2 and 5.3 of Gill, 1982)). In particular, when the wave amplitude is small (i.e., $h_0 k \ll 1$) and nonlinear effects can be neglected,

$$c = \sqrt{\frac{g}{k}\tanh(kH)} \tag{2.4}$$

This relationship is shown in Fig. 1.

In the shallow-water or long-wave limit, i.e., when $kH \ll 1$, $\tanh(kH) \to kH$. Thus,

$$c = \sqrt{gH}. \tag{2.5}$$

This result is consistent with dimensional analysis (2.3) when assuming that the speed is independent of the wavelength, which is the physically relevant limit for shallow water.

The phase velocity is independent of wavenumber and all long waves travel at the same speed. For example, earthquakes on the sea floor can excite tidal waves or tsunamis, with wavelengths up to hundreds of kilometers, while the ocean is at most a few kilometers deep. Thus, a tsunami propagates at speed $\sqrt{gH} \sim 200\,\text{ms}^{-1} \sim 720\,\text{kmh}^{-1}$ without dispersion, allowing its energy density to be maintained as it crosses a vast expanse of ocean. As a tsunami approaches the shore, the water depth decreases and slows it down. Since $\lambda = 2\pi c/\omega$, and frequency ω is constant, the wavelength λ diminishes. For the same energy density, the amplitude h_0 increases until nonlinear effects become important: The water depth at the wave crest would be significantly larger than that at the wave trough. This causes the wave crest to move faster than the wave trough and eventually the wave rolls over and breaks. In fact, h_0 can grow to as big as tens of meters. The large wave amplitude allows the tsunami to propagate inland for kilometers, causing much harm to life and property along the coast.

In the deep-water or short-wave limit, i.e., when $kH \gg 1$, $\tanh(kH) \to 1$. Thus,

$$c = \sqrt{\frac{g}{k}}. \tag{2.6}$$

Again the above equation is consistent with dimensional analysis (2.3), but in this case applying the reasonable assumption that in deep water the phase velocity is independent of depth. Long waves travel faster than the short waves so that deep-water waves are dispersive. For example, for waves of wavelength up to tens of meters, the water depth away from the shore is much deeper and these waves propagate dispersively.

On the other hand, as the waves approach the shore they start to propagate into shallow water and their speed then depends on the local depth H. As a wave approaches the shore obliquely (Fig. 2), the portion of its wave crest that is nearer the shore is in shallower water and so, from Eq. (2.5), propagates more slowly and so the wave crests turn parallel to the beach. This is why surfing towards the shore is possible no matter which direction the arriving swell comes from!

2.2. *Froude Number*

Consider a river of depth H flowing with uniform velocity U. A stone is thrown into the river and excites waves. In the river's moving frame of reference, the waves travel with speed c outwards in all directions from the point of entry of the stone. But in the stationary frame, a uniform velocity U will be superimposed on the propagation of the waves. Thus, when $U > c$, all the waves will be carried downstream with a net velocity between $U - c$ and $U + c$. Conversely, for the disturbance introduced by the

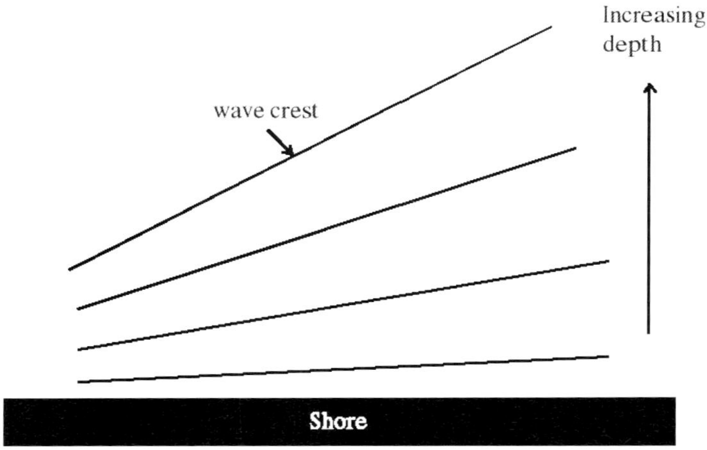

Fig. 2. Constant-phase lines such as wave crests of an approaching plane wave refract toward the shore due to changing depth of the water and dispersion effects.

stone to propagate upstream in the stationary frame, it is necessary that $U < c$.

This different qualitative behavior between $U < c$ and $U > c$ is characterized by a Froude number, defined as

$$F \equiv \frac{U}{c}. \tag{2.7}$$

Flows for which $F > 1$ are said to be *supercritical*; those for which $F < 1$ are said to be *subcritical*. When $F = 1$, the flow is said to be *critical*. Since the phase speed of surface waves has an upper bound of $\sqrt{gH}$ (cf. Fig. 1), the Froude number for surface waves is

$$F = \frac{U}{\sqrt{gH}}. \tag{2.8}$$

2.3. *Stratification and Buoyancy Frequency*

Waves, called *internal gravity waves* can also occur in the interior of the atmosphere and the oceans. In both surface and internal gravity waves, vertical displacement leads to restoring forces. In the interior of a fluid, these restoring forces result from gravity acting on density differences caused by displacement of fluid parcels from their equilibrium positions.

Figure 3 shows salinity and temperature distributions in the Pacific. With the exception of the surface near the tropics, the salinity increases

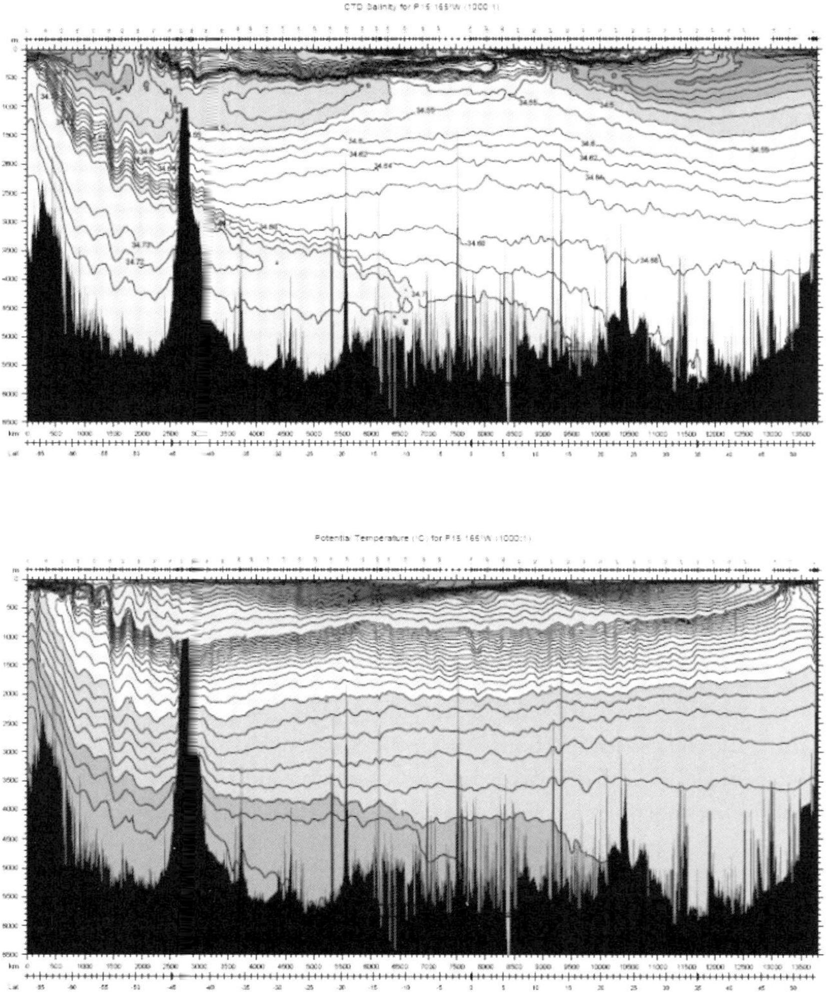

Fig. 3. Distribution of salinity (top) and potential temperature in a north–south transect in the Pacific Ocean. Salinity is in practical salinity units (or roughly parts per thousand by mass) and potential temperature is in degrees Celsius. Potential temperature is the temperature of the sea water if it is brought adiabatically to standard sea level pressure and it indicates the temperature of the sea water without the warming effect of adiabatic compression in the depths of the ocean. Red and yellow denote higher values. [*Source*: Woce Pacific Ocean Atlas (Talley, 2007).]

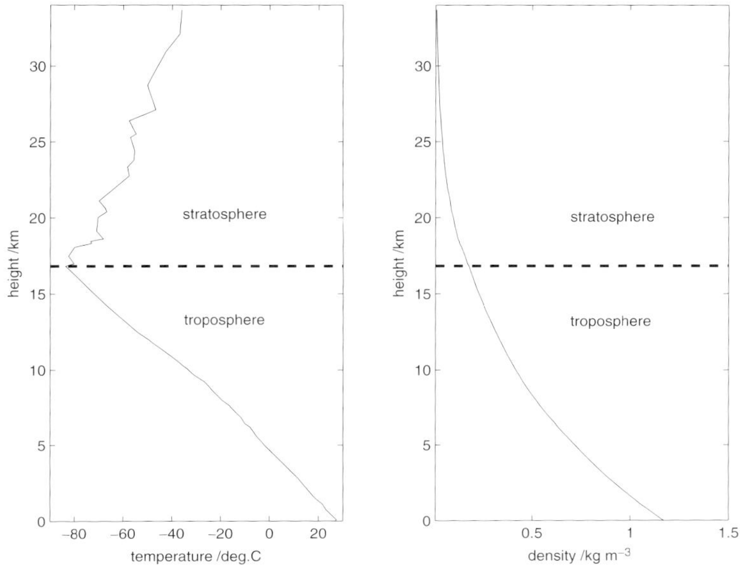

Fig. 4. Distribution of temperature (left) and density (right) of the atmosphere above Singapore at 00 UTC, 20 April 2009, as measured by a balloon radiosonde launched from Changi Meteorological Station. The reason the density of the air decreases with height in the troposphere, despite the fact that the temperature is also decreasing, is due to the effects of pressure. As the pressure drops with height the air expands and so its density decreases — and this effect exceeds the influence of temperature.

and the temperature decreases with depth. Similarly, Fig. 4 illustrates that the atmosphere gets colder with height in the troposphere and warmer with height in the stratosphere. As a result the density of air decreases with height throughout the atmosphere. Therefore, at large scales, both the atmosphere and ocean are *stably stratified* fluids with less dense fluid lying above denser fluid.

Consider a small parcel of fluid of volume V being raised to a height of δz above its equilibrium level in a stably stratified fluid (Fig. 5). The buoyancy force acting on the parcel is $gV\ \delta\rho$, where $\delta\rho$ is the difference between the parcel density and that of the environment. This force is directed in the opposite direction to the displacement providing a restoring force. Newton's second law implies that

$$\rho V\ \frac{d^2}{dt^2}\delta z = -gV\ \delta\rho.$$

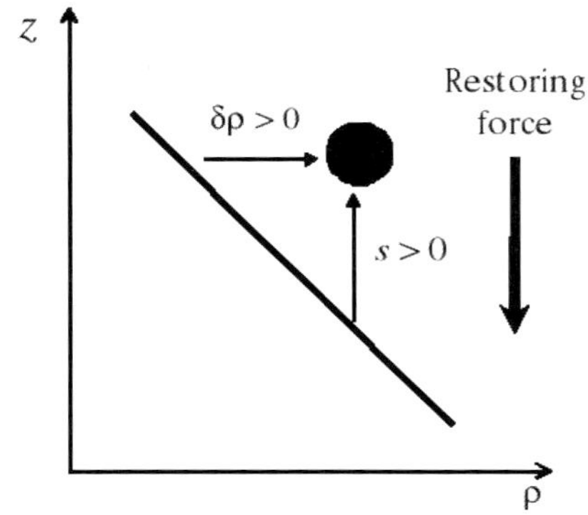

Fig. 5. In a stably stratified fluid, i.e., where $d\rho/dz < 0$, upward displacement of a fluid parcel leads to positive density difference $\delta\rho$ from the environment. The converse is true for downward displacement (not shown). The result is always a restoring buoyancy force towards the equilibrium level of the parcel.

For small displacements, $\delta\rho \approx -\frac{\mathrm{d}\rho}{\mathrm{d}z}\delta z$. Hence,

$$\frac{d^2}{dt^2}\delta z + N^2\delta z = 0, \tag{2.9}$$

where N is the buoyancy frequency

$$N^2 \equiv -\frac{g}{\rho}\frac{d\rho}{dz}. \tag{2.10}$$

For $N^2 > 0$, which corresponds to the density decreasing with height, the motion is an oscillation with frequency N,

$$\delta z = A\cos Nt + B\sin Nt, \tag{2.11}$$

where A and B are constants. For both the oceans and the atmosphere, $N \sim 10^{-2}\,\mathrm{s}^{-1}$, and so typical wave periods are about 10–20 minutes long.

2.4. *Internal Gravity Waves*

In the presence of continuous stable stratification, internal gravity waves propagate both horizontally and vertically. Suppose an internal gravity wave with wave vector $\mathbf{k} = (k, l, m)$ propagates at an angle θ to the vertical (Fig. 6). For an incompressible fluid, i.e., $\nabla \cdot \mathbf{u} = \mathbf{k} \cdot \mathbf{u} = 0$, the displacement is in the plane normal to the wave vector. The component δs of this

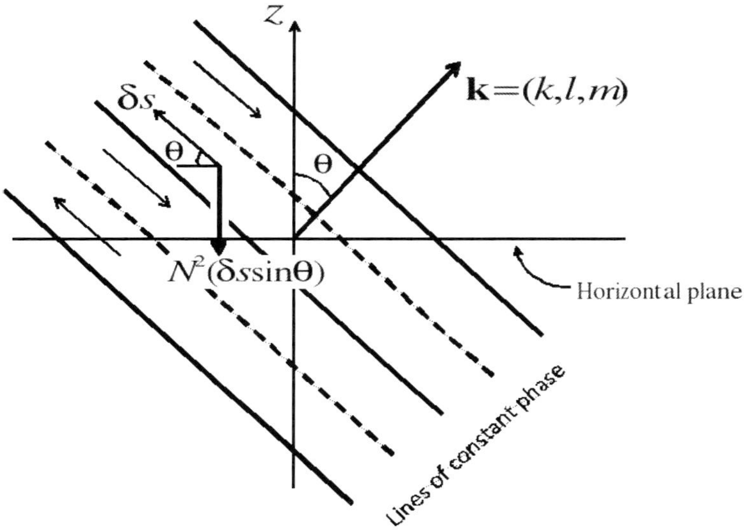

Fig. 6. An internal gravity wave with wave vector (k, l, m) propagates at an angle θ to the vertical and the resultant displacement δs and buoyancy force per unit mass $N^2 \delta s \sin\theta$.

displacement that is coplanar with the wave vector and the vertical has a vertical component $\delta s \sin\theta$ and results in a buoyancy force per unit mass $b = N^2 \delta s \sin\theta$. Only the component of buoyancy force $\delta b \sin\theta$ normal to the wave vector results in a restoring acceleration. Thus,

$$\frac{d^2}{dt^2}\delta s = -\delta b \sin\theta,$$

$$\frac{d^2}{dt^2}\delta s + (N^2 \sin^2\theta)\ \delta s = 0.$$

Therefore, the frequency of the oscillations and, hence, of the wave is

$$\begin{aligned} \omega &= N \sin\theta, \\ &= N\ \frac{|\mathbf{k} \times \hat{z}|}{|\mathbf{k}|}, \\ \omega^2 &= N^2 \frac{k^2 + l^2}{k^2 + l^2 + m^2}. \end{aligned} \tag{2.12}$$

The same dispersion relation can be derived from the linearized fluid dynamical equations for small wave amplitudes (see Secs. 6.4 and 6.5 of Gill

1982). The phase velocity $\mathbf{c_p}$ and the group velocity $\mathbf{c_g}$ are defined as

$$\mathbf{c_p} \equiv (c_{px}, c_{py}, c_{pz}) \equiv \frac{\omega \mathbf{k}}{|\mathbf{k}|^2} = \frac{\omega}{k^2 + l^2 + m^2}(k, l, m), \tag{2.13}$$

$$\mathbf{c_g} \equiv (c_{gx}, c_{gy}, c_{gz}) \equiv \left(\frac{d\omega}{dk}, \frac{d\omega}{dl}, \frac{d\omega}{dm}\right). \tag{2.14}$$

The phase velocity is

$$\begin{aligned} |\mathbf{c_p}| &= \frac{\omega \mathbf{k}}{|\mathbf{k}|^2}, \\ &= \frac{N}{|\mathbf{k}|^2}\sqrt{k^2 + l^2} < \frac{N}{|\mathbf{k}|}. \end{aligned}$$

Short waves propagate more slowly than long waves and so internal gravity waves are dispersive. Using Eqs. (2.12), (2.13) and (2.14),

$$\begin{aligned} c_{gx}c_{px} &= \frac{k^2}{|\mathbf{k}|^2}\frac{d(\omega^2)}{d(k^2)}, \\ &= \frac{k^2}{|\mathbf{k}|^2}\frac{N^2}{|\mathbf{k}|^2}\frac{m^2}{|\mathbf{k}|^2} > 0. \end{aligned} \tag{2.15}$$

$$\begin{aligned} c_{gy}c_{py} &= \frac{l^2}{|\mathbf{k}|^2}\frac{d(\omega^2)}{d(l^2)}, \\ &= \frac{l^2}{|\mathbf{k}|^2}\frac{N^2}{|\mathbf{k}|^2}\frac{m^2}{|\mathbf{k}|^2} > 0. \end{aligned} \tag{2.16}$$

$$\begin{aligned} c_{gy}c_{py} &= \frac{m^2}{|\mathbf{k}|^2}\frac{d(\omega^2)}{d(m^2)}, \\ &= -\frac{m^2}{|\mathbf{k}|^2}\frac{\omega^2}{|\mathbf{k}|^2} < 0. \end{aligned} \tag{2.17}$$

Thus, the horizontal group and phase velocities always point in the same direction, while the vertical group and phase velocities always point in opposite directions. It can also be verified that

$$\mathbf{c_p} \cdot \mathbf{c_g} = 0. \tag{2.18}$$

Thus, the group velocity is normal to the phase velocity and the wave vector. This relation results from the fact that the fluid motion is perpendicular to the wavenumber vector (Fig. 6).

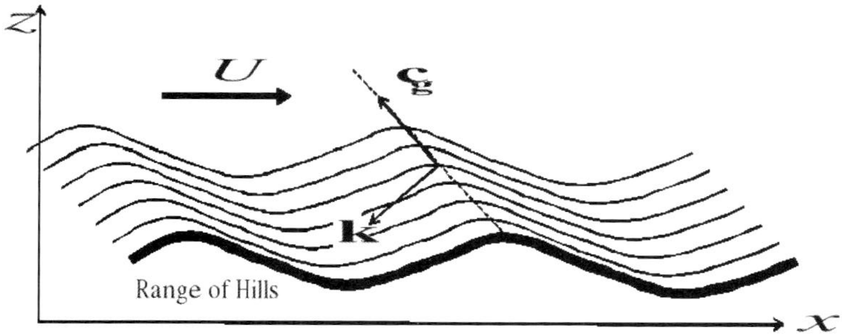

Fig. 7. An infinite range of hills excite internal gravity waves when the mean wind above them, U, is subcritical. Note that the line of constant phase is tilted upwind with height, so that the wave vector $\mathbf{k}$ and hence the phase velocity points upwind and downward, while the group velocity $\mathbf{c}_g$ points upwind and upward.

2.5. *Mountain Waves*

Mountain waves illustrate a particular application of the theory of internal gravity waves. Consider an infinite range of hills of height $z = h \cos kx$. Suppose the air above the hills is stably stratified with buoyancy frequency N and is flowing at uniform velocity of U (Fig. 7). As the air flows past the hills, internal gravity waves are excited. Assuming that the maximum slope of the hills $kh \ll 1$, the amplitude of the waves is small and so the dispersion relation (2.12) is applicable.

In the stationary frame (as in Fig. 7), there is a mean wind blowing to the right and the internal gravity waves are not moving. But in the frame that moves with the mean wind, the basic state of the atmosphere is at rest while the hills and the waves are propagating to the left. Thus, it is deduced that the horizontal component of the wave vector points to the left. On the other hand, energy is radiated by the wave in the direction of the group velocity. This means that the vertical group velocity must be pointing upward as energy from the surface radiates upward into the atmosphere. Equation (2.17) shows that the vertical phase velocity and hence the vertical component of the wave vector points in the opposite direction, i.e., downward. Therefore, the wave vector points downward and to the left and the line of constant phase accordingly tilts upwind with height. Since rising air expands and cools, condensation of water vapor can occur and form lenticular clouds upwind of hilltops.

The stationary wave pattern in Fig. 7 can only form if the mean wind U is subcritical, i.e.,

$$F = \frac{U}{N/|\mathbf{k}|} \le 1,$$
$$\therefore |\mathbf{k}|U \le N.$$

When U is supercritical, i.e., $F > 1$, all the waves are swept downwind and at these high windspeeds the effects of stratification are relatively unimportant.

2.6. *Mass, Momentum and Energy Fluxes*

The continuity, horizontal momentum and buoyancy equations in a two-dimensional, linear internal gravity wave propagating in the (x, z)-plane are

$$\frac{\partial w'}{\partial z} = -\frac{\partial u'}{\partial x}, \tag{2.19}$$
$$\frac{\partial u'}{\partial t} = -\frac{1}{\rho_0}\frac{\partial p'}{\partial x}, \tag{2.20}$$
$$\frac{\partial \rho'}{\partial t} = -w'\frac{d\rho_0}{dz}, \tag{2.21}$$

where $\rho_0(z)$ is the ambient density distribution and u', w', p' and ρ' denote the small perturbations in horizontal and vertical velocity, pressure and density, respectively.

These equations show that for an internal gravity wave of the form $\exp[i(\omega t - kx - mz)]$ (i.e., plane sinusoidal waves): w' is either in phase $(km < 0)$ or in anti-phase $(km > 0)$ with u'; u' is either in phase $(\omega k > 0)$ or in anti-phase $(\omega k < 0)$ with p'; and w' is $\pi/4$ radians out of phase with ρ'. Therefore,

$$\overline{w'u'} \ne 0, \tag{2.22}$$
$$\overline{w'p'} \ne 0, \tag{2.23}$$
$$\overline{w'\rho'} = 0. \tag{2.24}$$

Using relations (2.22), (2.23) and (2.24) further implies that for plane sinusoidal waves,

$$\overline{u'p'} \ne 0, \tag{2.25}$$
$$\overline{u'\rho'} = 0. \tag{2.26}$$

Equation (2.22) shows that the vertical flux of horizontal momentum and horizontal flux of vertical momentum are non-zero, i.e., momentum is radiated by the waves. The covariance between velocity and pressure perturbations is the rate at which work is done by the fluid against the pressure force (per unit area normal to the velocity) in the wave (see Sec. 1.10.1 of Vallis, 2006). Hence, Eqs. (2.24) and (2.26) signify that work is done by the waves in the atmosphere and hence energy is radiated away from the surface by the wave. The zero mass flux in Eqs. (2.24) and (2.26) means that no net transport of mass occurs in a linear internal gravity wave.

3. Convection

3.1. *Unstable Stratification*

Hot air rises because gas expands when heated thereby reducing the density (conserving mass). Air is well described by the perfect gas equation

$$\rho = \frac{p}{RT}, \tag{3.1}$$

where R is the specific gas constant. Thus an increase in temperature leads to decrease in density. Differentiation of this equation shows that the coefficient of expansion at constant pressure is

$$\alpha \equiv -\frac{1}{\rho}\frac{\partial \rho}{\partial T} = \frac{1}{T}. \tag{3.2}$$

The same is true for water, where the equation of state is approximated about (ρ_0, T_0) in the form

$$\rho = \rho_0(1 - \alpha(T - T_0)). \tag{3.3}$$

Thus in a gravitational field, liquid rises when it is heated from below — as in cooking!

3.2. *Parcel Argument*

We repeat the argument given in Sec. 2.3 considering a fluid parcel moved from its equilibrium position, but now in a situation where the density *increases* with height — an *unstably stratified fluid* (Fig. 8). The governing equation remains as Eq. (2.9) where now the buoyancy frequency is imaginary as

$$N^2 \equiv -\frac{g}{\rho}\frac{d\rho}{dz} < 0.$$

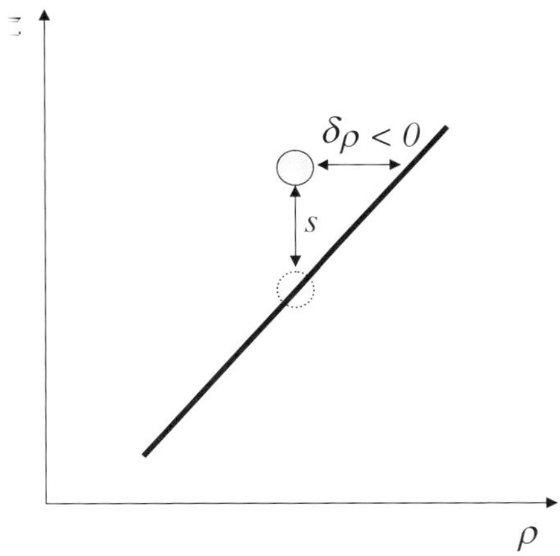

Fig. 8. A schematic showing the density perturbation for a small parcel raised a distance s from its equilibrium position in a statically unstable stratification.

Write $M^2 = -N^2 > 0$, then

$$s(t) = Ae^{Mt} + Be^{-Mt}.$$

In this case, the first term on the right-hand side dominates and $s(t)$ *increases exponentially with time* so that the parcel accelerates away from its equilibrium position and the stratification is said to be *statically unstable*. The motion that results is called *convection*.

3.3. *Dimensional Analysis*

In a layer of depth H with an imposed temperature difference ΔT in a fluid with viscosity ν and thermal diffusivity κ under gravity, as sketched in Fig. 9, there are two governing dimensionless parameters:

- Rayleigh number

$$Ra = \frac{g(\Delta T/T)H^3}{\nu\kappa}$$

- Prandtl number

$$\sigma = \frac{\nu}{\kappa}$$

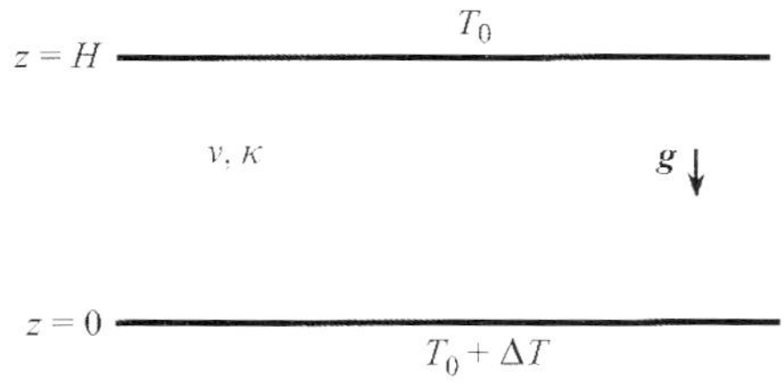

Fig. 9. A layer of fluid of depth H heated from below by a temperature difference ΔT.

The Prandtl number σ depends on the physical properties of the fluid and takes the values of $\sigma \approx 0.7$ for air and $\sigma \approx 7$ for water.

3.3.1. *Rayleigh number*

As can be seen from the approximate equation. Otherwise, pressure variations must come into the equation below, on potential temperature replaces temperature. Eq. (3.3), the quantity

$$g\frac{\Delta T}{T} = g\frac{\Delta \rho}{\rho} \equiv g'$$

is the *reduced gravity* (or buoyancy) associated with a temperature difference ΔT, and is a measure of the driving force of the convection. In order to see the physical relevance of the Rayleigh number we consider the force balance on a heated parcel of fluid.

A parcel of fluid of size a has a (positive) buoyancy force proportional to $g\Delta\rho a^3$. The motion of this parcel is retarded by viscosity giving a force balance

$$g\Delta\rho a^3 \sim \rho\nu a w, \tag{3.4}$$

where $w = \frac{\mathrm{d}z}{\mathrm{d}t}$ is its vertical velocity, and hence

$$\frac{\mathrm{d}z}{\mathrm{d}t} \sim \frac{g' a^2}{\nu}. \tag{3.5}$$

The buoyancy of the parcel is reduced by loss of heat by conduction

$$\frac{\mathrm{d}g'}{\mathrm{d}t} \sim \frac{\kappa g'}{a^2}. \tag{3.6}$$

Hence

$$g' \sim g_0' e^{-\frac{\kappa t}{a^2}}. \tag{3.7}$$

where g_0' is the initial buoyancy of the parcel. So if $z(0) = 0$,

$$z = \frac{g_0' a^4}{\nu\kappa}(1 - e^{-\frac{\kappa t}{a^2}}). \tag{3.8}$$

Thus the maximum height of rise of the parcel as $t \to \infty$ is

$$z_{\max} \sim \frac{g_0' a^4}{\nu\kappa}. \tag{3.9}$$

For convection to occur the parcel needs to rise at least across the depth of the fluid layer ($z_{\max} > H$) and, since $a \le H$, the largest value of $z_{\max}$ is achieved when $a = H$. Further, since the maximum value of $g_0' = g\frac{\Delta T}{T}$, this implies otherwise, according to the inequalities put fort it would appear that the critical value is 1 there is a critical value of the Rayleigh number $Ra = \frac{g\frac{\Delta T}{T}H^3}{\nu\kappa}$ that must be exceeded for convection to occur.

3.4. *Convection Strength*

The strength of the convection is measured by the heat flux H across the fluid layer. A nondimensional measure of the heat flux is the Nusselt number

$$Nu \equiv \frac{\mathsf{H}}{\frac{\kappa\Delta T}{H}}, \tag{3.10}$$

which is the ratio of the heat flux to the conductive flux that would occur across the fluid layer without fluid motion.

Dimensionally, the Nusselt number is a function of the two dimensionless parameters of the Rayleigh and Prandtl numbers:

$$Nu = f(Ra, \sigma), \tag{3.11}$$

and for given fluid properties

$$Nu = f(Ra). \tag{3.12}$$

3.5. *High Rayleigh Number*

Figure 10 shows average temperature profiles across the fluid layer at different values of the Rayleigh number. At the horizontal boundaries, where there is no fluid velocity as a result of the "no-slip" boundary condition, heat enters the fluid at the bottom and leaves at the top by conduction alone. As ΔT, and hence Ra, increases it is reasonable to suppose that more heat is transferred across the fluid, and so the conduction at the top

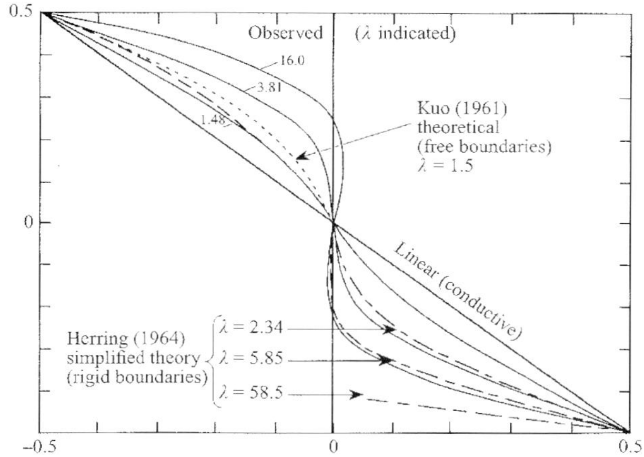

Fig. 10. Profiles of temperature with increasing Rayleigh number (λ in the figure) — note that the boundary layers at the top and bottom become thinner in order to provide the flux by conduction from the solid boundaries. Buoyant elements break off from these layers and drive convection in the interior. Note also the "counter-gradient" profile with a slightly stable interior at high Ra. Otherwise the axes are not labelled. [*Source*: Gille (1967).]

and bottom boundaries must also increase. This means that the vertical temperature gradients at these boundaries must increase, and a smaller fraction of the temperature drop across the fluid layer occurs in the center of the layer. This change in the temperature profiles is seen in Fig. 10.

At very high values of Ra, almost all the temperature drop occurs at the two boundaries in these two conductive boundary layers and it is reasonable to suppose that changing the vertical separation between these two layers has little effect on the overall heat flux. Mathematically, this is equivalent to assuming that, as $Ra \to \infty$, the heat flux across the fluid layer is independent of the depth of the layer. Consequently,

$$Nu \propto Ra^{1/3}, \tag{3.13}$$

or that the heat flux,

$$\mathsf{H} \propto \Delta T^{4/3}: \tag{3.14}$$

— the so-called "four-thirds" law.

The assumption that the heat flux is independent of the depth is found not to be generally the case. As can be seen in the scaling argument, the parcel that satisfies the inequality ($z_{\max} > H$) has a size $a \sim H$.

This is observed in practice — hot fluid leaves the lower boundary and flows across the layer in motions with scales comparable with the layer depth. This dependence on depth is also linked to nonlinear effects such as the generation of mean flows within the fluid. As a result of this dependence on the layer depth, the "four-thirds" law is only an approximation. A recent discussion of these complex and subtle issues can be found in Herring (1963).

3.6. *Very High Rayleigh Number*

As discussed above, heat leaves the boundaries by conduction. Across a boundary layer, thickness δ the conductive heat flux is

$$\mathsf{H} = \frac{\kappa \Delta T}{\delta}. \tag{3.15}$$

Thus

$$Nu = \frac{H}{\delta} > 1. \tag{3.16}$$

A "boundary-layer" Rayleigh number can be introduced as $Ra^{\delta} \equiv \frac{g\Delta T \delta^3}{T\nu\kappa}$ so that Eq. (3.16) becomes

$$Nu = \left(\frac{Ra}{Ra^{\delta}}\right)^{1/3} > 1, \tag{3.17}$$

which is in accordance with the "four-thirds" law. When Ra^{δ} exceeds a critical value $Ra_c^{\delta} \sim O(10^3)$, the *boundary layer* itself becomes unstable.

From Eq. (3.17), the Rayleigh number Ra for the convecting flow is

$$Ra > Ra_c^{\delta} \sim O(10^3). \tag{3.18}$$

At these high Ra, buoyant fluid is ejected from the thin boundary layers and the horizontal scales of these buoyant elements are comparable to $\delta \ll H$. Consequently, when viewed on the scale of the fluid layer they arise from small sources — such buoyant elements take the form of plumes if the source is maintained over time or thermals if the source is transient. These are discussed in the next section.

4. Plumes

Plumes arise from sustained and localized sources of buoyancy. Common examples are hot gases emitted from a chimney stack, air rising above a

Fig. 11. A plume from which Buncefield fire near London in 2005. Note the increase in the width of the plume with height — as a result of entrainment.

person or piece of electrical equipment, air rising from a thin boundary layer at high Ra, and smoke rising above a fire (as in Fig. 11). Dynamically, plumes are characterized by the buoyancy flux B, which for air is proportional to the heat flux H

$$B = \frac{g\mathsf{H}}{T\rho C_p}, \tag{4.1}$$

where C_p is the specific heat at constant pressure.

4.1. *Plumes–Dimensional Analysis*

Plumes arising from a point source are observed to be self-similar and on average conical in shape, as can be seen in the laboratory images in Figs. 12 and 13.

A plume is characterized by its mean radius b, vertical velocity w and buoyancy g' which depend only on B and height z. The buoyancy flux has dimensions $[B] = \mathrm{L}^4\ \mathrm{T}^{-3}$. Hence, dimensional analysis implies that the width-average vertical velocity and buoyancy, and the plume radius are given by, respectively

$$w = c_1 B^{1/3} z^{-1/3}, \tag{4.2}$$

$$g' = c_2 B^{2/3} z^{-5/3}, \tag{4.3}$$

$$b = \beta z. \tag{4.4}$$

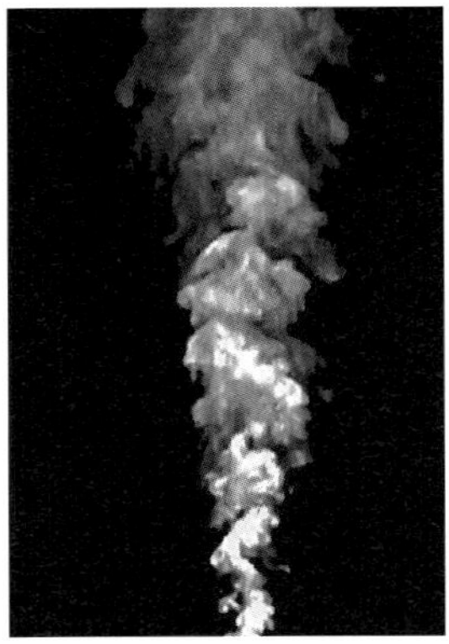

Fig. 12. A laser image of a plume rising in an unstratified fluid. [*Source*: Jens Huber (2006, private communication)].

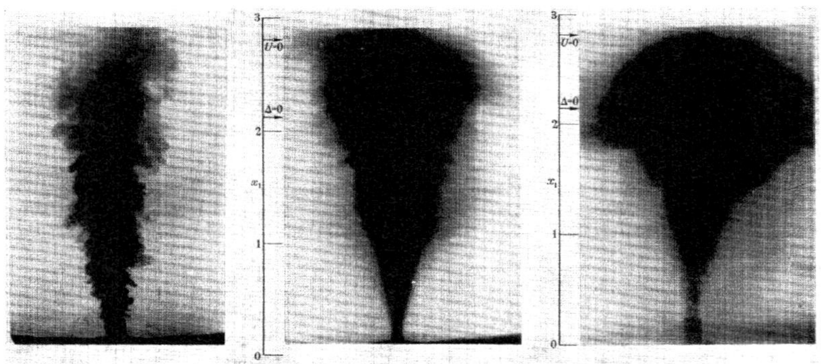

FIGURE 2 FIGURE 3 FIGURE 4

FIGURE 2. The maintained plume in a uniform ambient fluid.

FIGURE 3. The maintained plume in a stably stratified fluid; a time exposure during an early stage of the release of dyed fluid so that the edges of the plume are not obscured by the top layer.

FIGURE 4. The maintained plume in a stably stratified fluid; a time exposure during a later stage of the release when a layer of dyed fluid is growing at the top.

Fig. 13. A plume rising in a stably stratified fluid. [*Source*: Morton *et al.* (1956).]

where c_1 and c_2 are dimensionless contants.

From dimensional analysis, the volume flux

$$Q \equiv 2\pi \int_0^\infty rw\mathrm{d}r \propto B^{1/3} z^{5/3}. \tag{4.5}$$

The volume flux increases with height due to *entrainment* of ambient fluid, which also causes the plume buoyancy to decrease with height.

4.2. *Entrainment*

The continuity equation in polar coordinates is

$$\frac{1}{r}\frac{\partial(ru)}{\partial r}+\frac{\partial w}{\partial z}=0. \tag{4.6}$$

Integrate Eq. (4.6) across the plume

$$\int_0^\infty r\frac{\partial w}{\partial z}\mathrm{d}r=-\int_0^\infty \frac{\partial(ru)}{\partial r}\mathrm{d}r, \tag{4.7}$$

$$\frac{\mathrm{d}}{\mathrm{d}z}\int_0^\infty rw\mathrm{d}r=-[ru]_0^\infty, \tag{4.8}$$

$$\frac{1}{2\pi}\frac{\mathrm{d}Q}{\mathrm{d}z}=-ru|_\infty. \tag{4.9}$$

showing that the increase in volume flux is compensated by inflow from the ambient fluid — by *entrainment*.

4.2.1. *Entrainment assumption*

This inflow can be written in terms of an entrainment velocity u_e at the plume edge b:

$$bu_e=-ru|_\infty. \tag{4.10}$$

Since entrainment is a turbulent process, the calculation of this inflow velocity cannot be made without recourse to some assumption. The *entrainment assumption* is that the inflow velocity is proportional to the mean vertical velocity of plume:

$$u_e=\alpha w. \tag{4.11}$$

It is found from experiment that the entrainment constant $\alpha \sim 0.1$.

4.3. *Self-similarity*

Entrainment increases the volume flux in the plume at a rate

$$\frac{\mathrm{d}Q}{\mathrm{d}z}\propto bu_e=\alpha bw. \tag{4.12}$$

Since $w \propto z^{-1/3}$ and $b \propto z$ then it follows that $Q \propto z^{5/3}$, the same dependence on height as (4.5) obtained by dimensional analysis, which assumed that the plume is self-similar. Thus the entrainment assumption is equivalent to self-similarity in an unstratified fluid. This relation provides a strong argument in support of the entrainment assumption.

4.4. *Plume Rise in a Stratified Fluid*

As a plume rises through a stably stratified fluid, the density of the fluid in the plume increases with height and the density of the surrounding fluid decreases with height. Generally, there is a height — the level of neutral buoyancy — at which the density of the fluid within the plume equals that of the fluid outside the plume. As the plume is carried upwards past the level of neutral buoyancy by its upward momentum it is then denser than the surrounding fluid and the buoyancy forces now act downward and bring the plume to rest. It then spreads out horizontally into the surroundings, at a height z_N.

4.4.1. *Dimensional analysis*

The maximum height of rise z_N depends on the strength of plume B, dimensions $[B] = L^4\mathrm{T}^{-3}$ and strength of stratification characterized by the buoyancy frequency N, dimensions $[N] = T^{-1}$. Dimensional analysis then implies that

$$z_N = cB^{1/4}N^{-3/4}, \tag{4.13}$$

where c is a dimensionless constant that needs to be determined by theory, computation, or experiment.

Figure 14 shows a compilation of plume rise data from laboratory experiments to a large oil fire — and the relation holds over five decades in plume rise height.

4.4.2. *Impact on the external environment — the "filling box"*

In an enclosed environment the plume eventually heats all the fluid. Figure 15 shows the flow generated in a closed box. When the plume is initiated it rises to the top of the box and spreads out along the upper boundary. Subsequently, the plume rises through some of this preheated fluid and so it arrives at the top of the box warmer than the previous fluid. It displaces the fluid near the top and generates a downward motion in the environment, and creates a stable stratification outside the plume.

The time, τ, taken for the box to fill is the time it takes for all the fluid in the box to pass through the plume, and this is given by the volume, V,

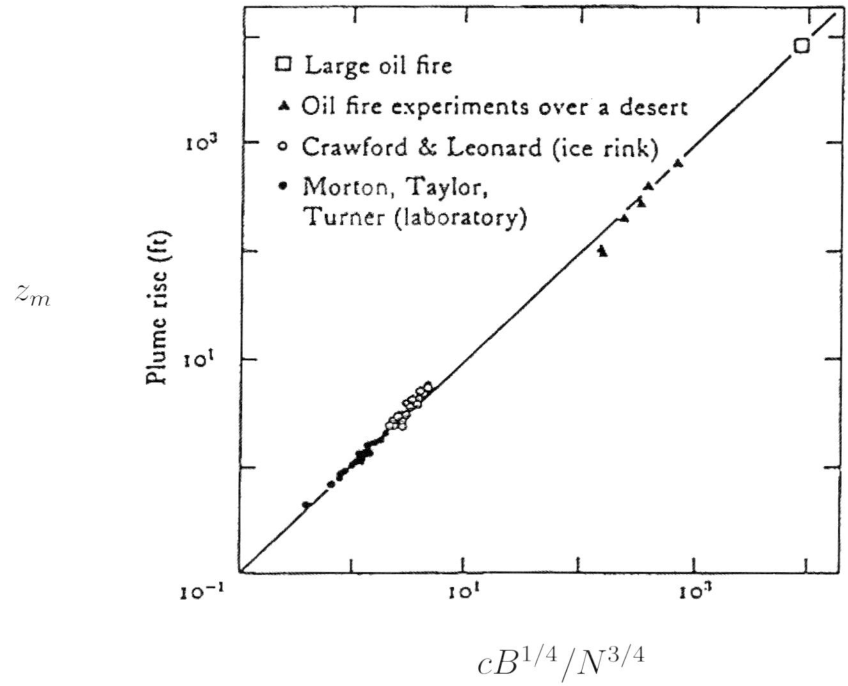

Fig. 14. Height of rise in a stratified fluid — this is the second largest extrapolation of scales in physics. [*Source*: Morton *et al.* (1956).]

of the box divided by the maximum volume flux in the plume which occurs at the top of the box. Thus, the "filling box" time — the time for all the fluid in the box to pass through the plume is

$$\tau \propto \frac{V}{B^{1/3}H^{5/3}}. \tag{4.14}$$

If there are many plumes — such as in an industrial or urban setting (Fig. 16) — they "compete" for air to entrain, providing the same effect as being in a confined box.

4.5. *Fires*

A fire requires a mixture of fuel and oxygen — expressed as the stoichiometric fuel-air mass ratio f, which is the reciprocal of the mass of air required to burn a unit mass of fuel. Typically, $f < 0.2$ for fuels of interest, primarily

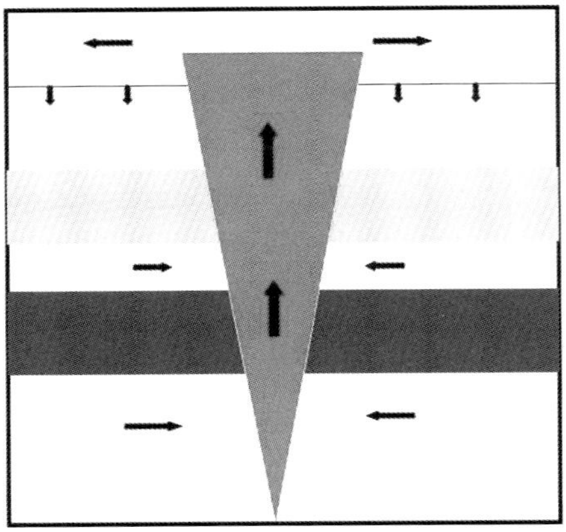

Fig. 15. Plume in an enclosed box. [*Source*: (Baines & Turner (1969).]

Fig. 16. Smog in LA caused by the filling box process.

because of the low molecular weight of hydrogen, as well as the excess nitrogen in air. This small value of f, requiring a large mass of air to burn a unit mass of fuel, means that entrainment is important to sustain a fire.

5. Gravity Currents

5.1. *Horizontal Stratification*

Fluid of density $\rho(\mathbf{x})$ at rest under gravity $\mathbf{g}$ satisfies the hydrostatic relation

$$\frac{\partial p}{\partial z} = -g\rho. \tag{5.1}$$

Unless $\rho = \rho(z)$ only, the pressure p will, in general, be different at different horizontal locations resulting in motion. Thus a fluid can only be at rest

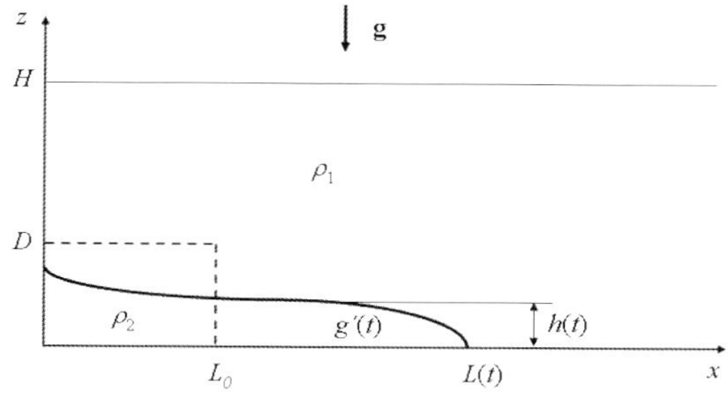

Fig. 17. Schematic of a finite volume release.

if it is vertically stratified; *horizontal stratification always drives a flow.* This is the result of *baroclinic generation of vorticity* which occurs when $\nabla\rho \times \nabla p \neq 0$.

5.2. *Gravity Currents*

5.2.1. *Dimensional analysis*

Consider the release of a constant volume of dense fluid in a channel as shown in Figs. 17 and 18. One example in nature of such a gravity current occurs in an avalanche (Fig. 19). The speed U depends on g' and time t and the volume of the release characterized by the height D and horizontal extent L_0. The gravity current is two-dimensional and spans the width of the channel.

5.2.1.1. Constant velocity phase

Initially U is observed to be constant and independent of the horizontal scale L_0. In that case dimensionally

$$U = F\sqrt{g'D}, \tag{5.2}$$

where F is a dimensionless constant conventionally called the Froude number.

5.2.1.2. Similarity phase

At later times the finite release size, characterized by L_0, becomes important. The flow is then determined by the total buoyancy per unit width $'B = g'HL = g'DL_0$, which is constant by conservation of mass. The

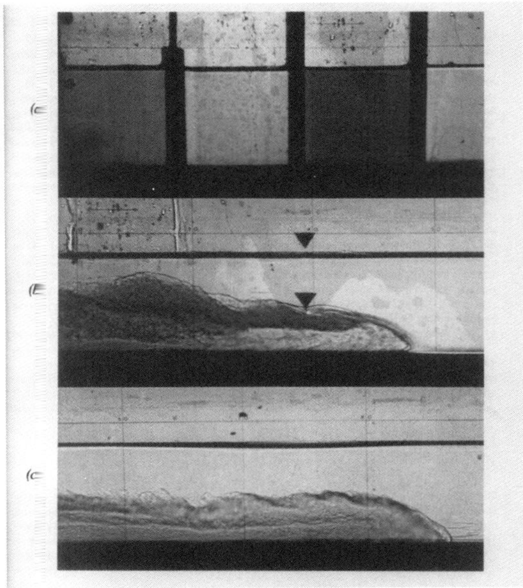

Fig. 18. A laboratory gravity current — the fluid density increases from blue to yellow to red. As the current progresses the densest fluid eventually leads. [*Source*: Linden & Simpson (1986).]

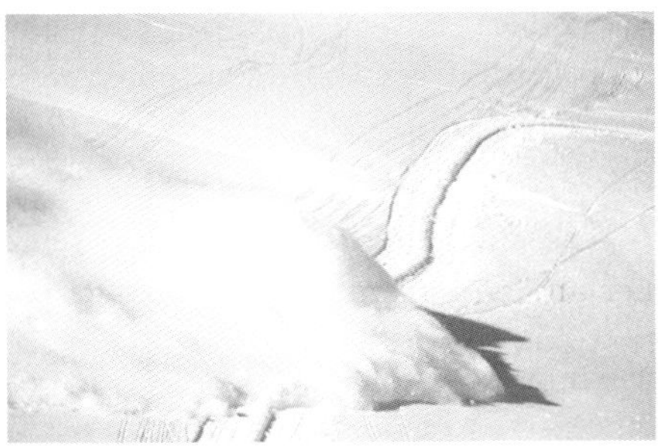

Fig. 19. Avalanche in Tuca, Spain.

dimensions of the buoyancy $[B] = \mathrm{L}^3\mathrm{T}^{-2}$ and so matching dimensions

$$U = cB^{1/3}t^{-1/3}, \tag{5.3}$$

where c is a dimensionless constant. This implies that the speed decreases with time. As a result the Reynolds number of the current decreases until viscous forces become important.

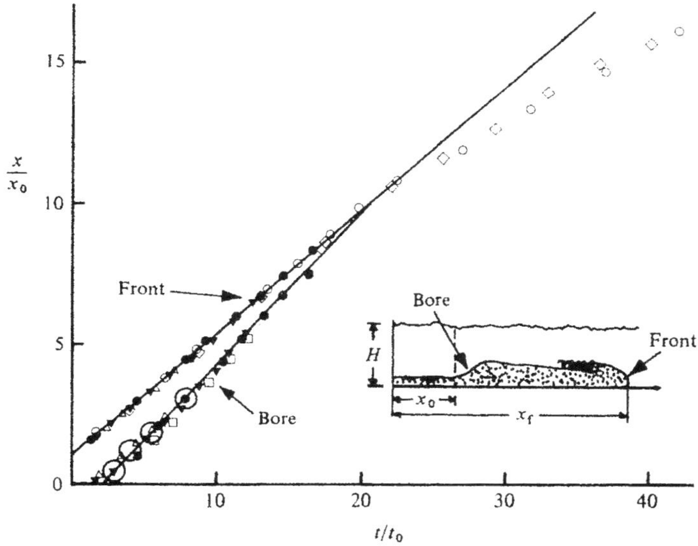

Fig. 20. The current initially travels at constant speed until the bore from the rear wall catches up with the front — this brings the information that the release has a finite volume. [*Source*: Rottman and Simpson (1983).]

5.2.2. *Laboratory verification*

This verification is shown in Fig. 20, as described by Rottman and Simpson (1983).

5.3. *The Front Froude Number*

Both Eqs. (5.2) and (5.3) have unknown constants F and c that need to be determined by further analysis, experiments or numerical simulations. This is a huge topic for gravity currents and one which continues to be an active research question. There are in fact two unknowns associated with the front of a gravity current — the speed and the depth of the current — and so two relations are needed to determine them from an initial release.

We confine our remarks here to noting that these constants F and c are O(1).

6. Rotating Flows

6.1. *Rotating Frame and the Coriolis Force*

The Earth rotates about its spin axis once a day. Therefore, to understand atmospheric and oceanic motion on timescales comparable to or longer

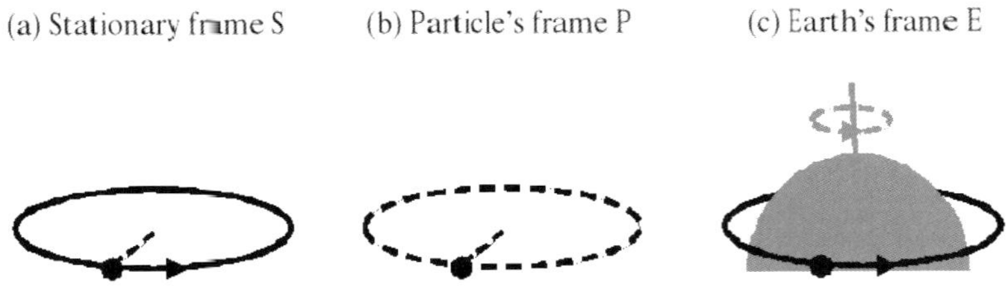

Fig. 21. (a) A particle moves in a circular orbit in a stationary frame S; (b) in the co-moving frame anchored to the particle, there is no motion; (c) in Earth's frame E, which has its own rotation rate, the particle moves at a different velocity than in frame S in a circular orbit.

than a day, the inertial forces that arise from the Earth's rotation must be considered. To clarify the physical origin of these forces, consider a particle moving in a circle of radius r and (absolute) velocity v_a in a stationary frame S, e.g., in space (Fig. 21(a)). The particle undergoes centripetal acceleration due to a net radially inward force F per unit mass

$$F = -\frac{v_a^2}{r}. \tag{6.1}$$

On the other hand, in the co-moving frame P anchored to the particle, there is no motion (Fig. 21(b)) and so the force F must be balanced by an inertial force per unit mass F_a such that

$$F + F_a = 0 \tag{6.2}$$

$$F_a \equiv \frac{v_a^2}{r} = \Omega_a^2 r \tag{6.3}$$

where Ω_a is the rate of rotation of frame P. In this case, F_a is the centrifugal force per unit mass arising from the rotation of frame P and acts radially outwards.

Both the stationary frame S and the particle frame P are rather special. Suppose now the particle is observed in a frame E, e.g., the Earth frame, that has a rotation rate Ω that is not zero and that is also different from the rotation rate Ω_a of frame P (Fig. 21(c)) In frame E, the particle's velocity is $v = v_a - \Omega r$. Since F is a physical quantity, its value must be invariant between the frames of reference. So in frame E, Eq. (6.1) is still valid:

$$\begin{aligned} F &= -\frac{v_a^2}{r} \\ &= -\frac{(v + \Omega r)^2}{r}. \end{aligned}$$

Rearranging,

$$F + \Omega^2 r + 2\Omega v = -\omega^2 r$$

where $\omega = v/r$ is the angular velocity of the particle in frame E. Thus, the centripetal acceleration $-\omega^2 r$ observed in frame E is the resultant of the physical force F, the centrifugal force per unit mass $\Omega^2 r$ arising from frame E's rotation (cf. Eq. (6.3) for the rotation of frame P) and the Coriolis force per unit mass $2\Omega v$ that arises from the motion of the particle in the rotating frame E. If we further transform from frame E to the particle frame P,

$$F + \Omega^2 r + 2\Omega v + \omega^2 r = 0. \tag{6.4}$$

Comparing Eqs. (6.2) and (6.4), it is clear that the centrifugal and Coriolis forces observed in frame E are parts of the inertial force F_a experienced by the particle in its own frame P.

The above derivation could be repeated in vectorial notation to obtain

$$\mathbf{a} = \mathbf{F} + \mathbf{F}_{cf} + \mathbf{F}_{co},$$

$$\mathbf{a} \equiv \omega \times (\omega \times \mathbf{r}), \tag{6.5}$$

$$\mathbf{F}_{cf} \equiv -\boldsymbol{\Omega} \times (\boldsymbol{\Omega} \times \mathbf{r}), \tag{6.6}$$

$$\mathbf{F}_{co} \equiv -2\boldsymbol{\Omega} \times \mathbf{v}, \tag{6.7}$$

where $\mathbf{F}_{cf}$ and $\mathbf{F}_{co}$ are the centrifugal and centripetal force per unit mass and $\mathbf{a}$ is the centripetal acceleration of the particle's circular orbit in frame E. All motions on the Earth experience $\mathbf{F}_{cf}$, $\mathbf{F}_{co}$ and $\mathbf{a}$.

6.2. *Inertial Oscillations*

On Earth, the centrifugal force and gravity can be considered together as one "effective gravitational force" and in this way, the centrifugal force due to the Earth's rotation is not considered separately. Because the ocean and atmosphere are thin layers of fluid compared to the Earth's radius, only the vertical component of the Earth's rotation is dynamically important. At latitude ϕ, this vertical component is $\Omega \sin \phi$. Thus, the Coriolis force per unit mass on the Earth is effectively

$$\mathbf{F}_{co} = f\ \mathbf{v} \times \hat{z} \tag{6.8}$$

$$f \equiv 2\Omega \sin \phi \tag{6.9}$$

where $\hat{z}$ is the unit vector pointing upward and f is the Coriolis parameter. Thus, Coriolis effects are small near the equator and increase toward the poles.

Consider a horizontal ring of fluid at rest in a counterclockwise-rotating frame E. In the stationary frame, the ring would be rotating and possesses angular momentum. If the ring expands, conservation of angular momentum implies that it would rotate more slowly. So an expanding ring will develop clockwise rotation in frame E. Figure 22(a) shows that Coriolis force acting on the ring will tend to resist the expansion. Conversely, if the ring contracts as in Fig. 22(b), conservation of angular momentum causes it to develop counterclockwise rotation in frame E and the Coriolis force will tend to resist the contraction. Therefore, the Coriolis effect can provide a restoring force against perturbations. (The same restoring effect is also achieved if frame E rotates in a clockwise direction.)

For a fluid at rest in a rotating frame, if pressure perturbations are small enough to be neglected, velocity perturbations (u', v') are influenced by the Coriolis force only.

$$\left.\begin{aligned} \frac{Du'}{Dt} &= fv' \\ \frac{Dv'}{Dt} &= -fu' \end{aligned}\right\} \Rightarrow \begin{pmatrix} u' \\ v' \end{pmatrix} \propto \begin{pmatrix} \sin ft \\ \cos ft \end{pmatrix} \tag{6.10}$$

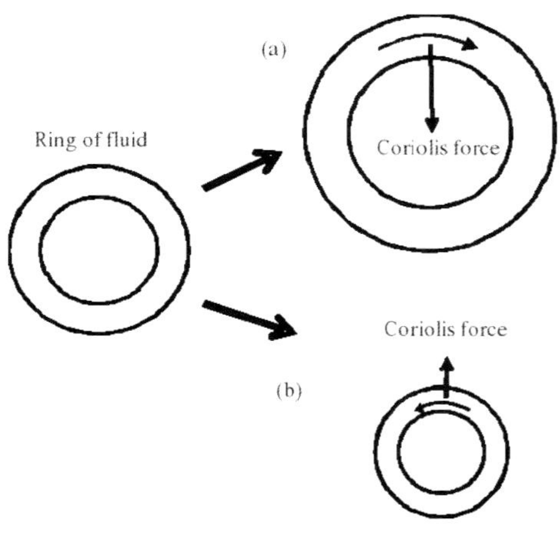

Fig. 22. A ring of fluid is at rest in a frame that rotates counterclockwise. (a) The ring expands and Coriolis force acts radially inward; (b) the ring contracts and Coriolis force acts radially outward. In both cases, Coriolis force has a restoring effect on the perturbation.

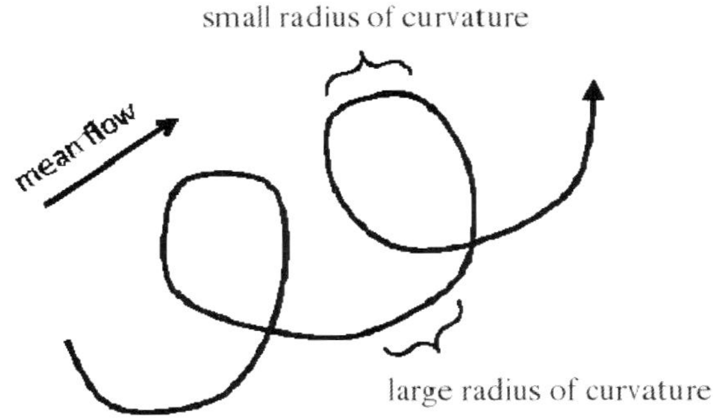

Fig. 23. Schematic diagram of an inertial oscillations superimposed on a mean flow in the northern hemisphere.

Velocity and hence displacement undergo circular oscillations known as inertial oscillations (Fig. 23). In the northern hemisphere where $f > 0$, the oscillation is clockwise; in the southern hemisphere where $f < 0$, the oscillation is counterclockwise. In both cases, the sense of inertial oscillation is opposite to the sense of rotation of the reference frame. Thus, the motion is said to be anticyclonic. (Cyclonic motion would correspond to the same sense as the rotation of the reference frame.) The frequency of inertial oscillations is maximum at f when the displacement is horizontal (compared to that of internal gravity waves which is maximum at buoyancy frequency N when the displacement is vertical). The typical scale of such inertial oscillations observed on the ocean surface is about a few kilometers.

6.3. *Rossby Radius of Deformation and Eddies*

When there is a horizontal density gradient, the resulting pressure gradient will cause the denser fluid to propagate as a gravity current into the less dense fluid. As discussed in Sec. 5, the speed of the gravity current is $\sqrt{g'H}$. When a mass of denser fluid spreads outwards into a surrounding pool of less dense fluid in a rotating frame, the Coriolis force causes the denser fluid to rotate anticyclonically (Figs. 24(a) and (b)). This produces an inward Coriolis force that stops the outward spreading over a timescale of one rotation period (Fig. 24(c)). At equilibrium, the denser fluid will have spread over a characteristic distance called the Rossby radius of deformation:

$$L_R \equiv \frac{\sqrt{g'H}}{f}. \tag{6.11}$$

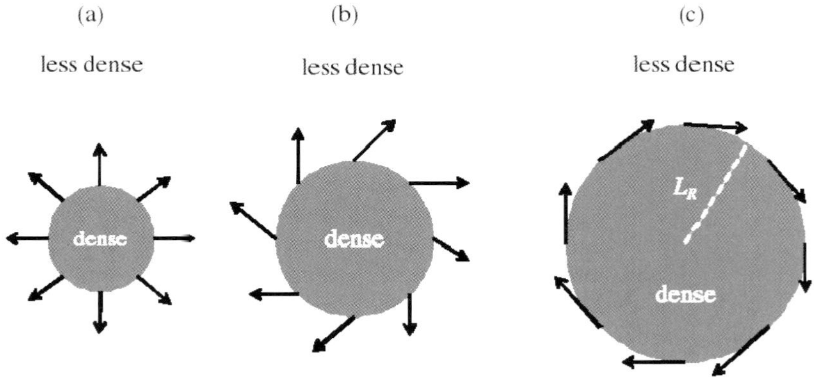

Fig. 24. (a) Dense fluid starts spreading outwards into a mass of less dense fluid in a counterclockwise-rotating frame; (b) Coriolis force causes the outward velocity to deflect right, leading to an anticyclonic flow being set up; (c) eventually, the dense fluid stops spreading when it has spread about one Rossby radius of deformation L_R as the inward Coriolis force acting on the anticyclonic flow is balanced by the outward pressure gradient force.

In a continuously stratified fluid, density or pressure anomalies propagate as an internal gravity wave with speed of the order of NH. Thus, spreading density or pressure anomalies are limited over the Rossby radius of deformation:

$$L_R \equiv \frac{NH}{f}. \tag{6.12}$$

Thus, L_R is a measure of the typical size of density or pressure anomalies in the atmosphere or ocean. These anomalies are associated with anticyclonic or cyclonic flow depending on the core of the anomalies being dense high-pressure fluid or less dense low-pressure fluid, respectively. From the sense of the associated flow, the anomalies are called anticyclonic or cyclonic eddies. For both the atmosphere and ocean, $N \sim 10^{-2}\,\mathrm{s}^{-1}$ and $f \sim 10^{-4}\,\mathrm{s}^{-1}$ are typical and thus $L_R \sim 100H$.

- For the atmosphere, $H \sim 10\,\mathrm{km}$ and so the typical eddy size is $L_R \sim 1000\,\mathrm{km}$.
- For the ocean $H \sim 1\,\mathrm{km}$ and so the typical eddy size is $L_R \sim 100\,\mathrm{km}$.

6.4. *Buoyancy-Driven Coastal Currents*

Consider the same thought experiment as in Fig. 24 except that there is an additional vertical wall aligned in one radial direction as shown in Fig. 25(a).

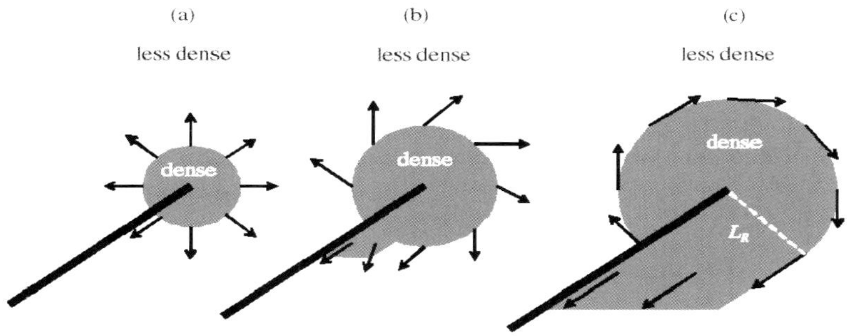

Fig. 25. (a) The dense fluid starts spreading outwards in a counterclockwise-rotating frame, including along the rigid wall aligned along a radial direction; (b) some time later, spreading continues along the rigid wall unaffected by the Coriolis effect. Meanwhile, the protrusion of fluid along the wall also tends to spread in the normal direction to the wall; (c) finally a current flows along the wall with a width of the order of Rossby radius of deformation L_R.

Along the wall, anticyclonic flow cannot be set up due to the physical barrier and the Coriolis force there is balanced by the normal reaction from the wall. The dense fluid continues to spread along the wall in the manner shown in Fig. 25(b). At the same time, the protrusion of denser fluid along the wall would also tend to spread in the normal direction to the wall. But this normal flow is deflected anticyclonically by the Coriolis force. Eventually, the normal flow is stopped by the Coriolis force and a steady current of width L_R flows outwards along the wall as in Fig. 25(c).

In the atmosphere, there are no lateral boundaries. In the ocean, coasts provide lateral boundaries that provide a reaction to the Coriolis forces. One example is the East Greenland Current, where cold, but less-dense arctic water flows southward along the east coast of Greenland. This current is responsible for much of the ice transport from the Arctic Ocean into the warmer Atlantic Ocean and plays an important role in the heat exchange between the two oceans.

References

Baines, W. D. and Turner, J. S. (1969). Turbulent convection from a source in a confined region, *J. Fluid Mech.* **37**, pp. 51–80.

Gill, A. E. (1982). *Atmosphere-Ocean Dynamics* (Academic Press, San Diego).

Gille, J. (1967). Interferometric measurements of temperature gradient reversal in a layer of convecting air, *J. Fluid Mech.* **30**, pp. 371–384.

Herring, J. R. (1963). Investigation of problems in thermal convection. *J. Atoms. Sci.* **97**, pp. 429–439.

Linden, P. F. and Simpson, J. E. (1986). Gravity-driven flows in a turbulent fluid, *J. Fluid Mech.* **172**, pp. 481–497.

Morton, B. R., Taylor, G. I. and Turner, J. (1956). Turbulent gravitational convection from maintained and instantaneous sources, *Proc. Royal Soc. Lond.* **23A**, pp. 1–23.

Rottman, J. W. and Simpson, J. E. (1983). Gravity currents produced by instantaneous release of a heavy fluid in a rectangular channel, *J. Fluid Mech.* **135**, pp. 95–110.

Vallis, G. K. (2006). *Atmospheric and Oceanic Fluid Dynamics* (Cambridge University Press, Cambridge, UK).

Chapter 3

WEATHER AND CLIMATE

Emily Shuckburgh

British Antarctic Survey
High Cross, Madingley Rd,
Cambridge, CB3 0ET, UK
emsh@bas.ac.uk

Fluid dynamics is at the heart of our climate system, governing the motion of the atmosphere, oceans, and ice sheets. Indeed, as will be described here, the equations of motion for a fluid on a rotating sphere form the core of the numerical climate models used to predict future change. The spherical shape of the Earth leads to different approximations being appropriate within and outside the tropical regions. Aspects of weather phenomena such as mid-latitude winter storms and the tropical monsoons can be understood in this way. The Pacific and Atlantic Oceans are bounded in longitude by the continents, whereas the Southern Ocean is not. This leads to the dominance of different physical processes and hence different circulation patterns. Some processes, such as those associated with the El Niño Southern Oscillation, can only be understood in terms of the coupled dynamics of the atmosphere and ocean. In both the atmosphere and the oceans, global-scale overturning circulations transport heat as well as chemical species around the Earth. The physical mechanisms responsible for these aspects of the climate system are reviewed, and their potential for change with increased atmospheric greenhouse gas concentrations is discussed.

1. Introduction

Fluid dynamics is fundamental to our understanding of the atmosphere and oceans and their role in determining the Earth's weather and climate. With ever increasing concern as to how humans are influencing the climate system, it is vitally important to understand the relevant physics. The entire system is immensely complex, but simplifications based on fluid dynamics enable us to make predictions concerning the weather for the next few days and the climate for the next hundred years. The emphasis in this chapter is on the large-scale circulation — that is on scales of several hundred kilometers and more in the atmosphere and a few tens of kilometers and more in the ocean. We will consider the forcing mechanisms that drive these circulations and the dynamics that governs them. Rotating fluids such as the atmosphere and oceans have unusual properties. We will see how fluid-dynamical principles provide strong constraints that organize these fluid flows on Earth and help determine our weather and climate.

The Earth is an almost perfect sphere with a mean radius of $a = 6370\,\mathrm{km}$. It is subject to a gravitational acceleration of $g = 9.81\,\mathrm{ms}^{-1}$ and has a rotational period of $\tau = 24\,\mathrm{h}$, which corresponds to an angular velocity of $\Omega = 2\pi/\tau = 7.27 \times 10^{-5}\,\mathrm{s}^{-1}$. The atmosphere and oceans are thin films of fluids on the spherical Earth under the influence of: (i) Gravity, (ii) The Earth's rotation, and (iii) Heating by solar radiation.

2. Forcing of the Atmosphere and Ocean Circulation

2.1. *Atmospheric Properties*

We first consider the properties of the atmosphere. Global mean surface pressure p_s is 1013 hPa and the global mean density of air at the surface ρ_s is $1.235\,\mathrm{kgm}^{-3}$. The air in the atmosphere is a mixture of "ideal" gases: Nitrogen (N_2) and oxygen (O_2) are the largest by volume, but other gases including carbon dioxide (CO_2), water vapor (H_2O), methane (CH_4) and ozone (O_3) also play significant roles in influencing the atmospheric circulation via their radiative effects. Atmospheric water vapor is present in variable amounts (typically about 0.5% by volume). The amount is strongly dependent on the temperature and it is primarily a consequence of evaporation from the ocean surface. It plays a key role in determining the climate as it strongly absorbs radiation in the infrared, the region of the spectrum at which the Earth radiates energy back out to space.

To a good approximation the atmosphere as a whole behaves as an ideal gas, with each mole of gas obeying the law $pV_a = R_gT$, where p is the pressure, V_a is the volume of one mole, R_g is the universal gas constant, and T is the absolute temperature. If M_a is the mass of one mole, the density is $\rho = M_a/V_a$, and the ideal gas law may be written as

$$p/\rho = RT, \tag{2.1}$$

where $R = R_g/M_a$ is the gas constant per unit mass. The value of R depends on the composition of the sample of air. For dry air it is $R = 287\,\mathrm{J\,kg^{-1}\,K^{-1}}$.

In a mixture of ideal gases, each gas has a "partial pressure," which is the pressure the gas would have if it alone occupied the volume. The total pressure of a gas mixture is the sum of the partial pressures of each individual gas in the mixture (this is Dalton's law). Thus, for example, for moist air with pressure p, $p = p_d + p_w$ where p_d is the partial pressure of dry air and p_w is the partial pressure of the water vapor in the air. If a closed container filled with moist air is brought to temperature T, the amount of the water that is in liquid form and in gaseous form when thermodynamic equilibrium has been reached (i.e., when the rate of evaporation equals the rate of condensation) can be measured. The partial pressure of the water vapor at thermodynamic equilibrium is known as the "saturation vapor pressure." As the temperature increases, this saturation vapor pressure varies according to the Clausius–Clapeyron equation $\frac{dp_w}{dT} = \frac{Lp_w}{R_w T^2}$, where L is the latent heat of vaporization per unit mass and R_w is the specific gas constant for the vapor. If L is a constant (a fairly good approximation at typical atmospheric temperatures), this can be integrated to give $p_w(T) = Ae^{-L/R_w T}$, and hence the saturation vapor pressure increases strongly with temperature in the atmosphere.

Another key property is the "specific heat capacity," c, which is the measure of the heat energy required to increase the temperature of a unit quantity of air by one unit. The specific heat of a substance is typically measured under constant pressure (c_p). However, fluids may instead be measured at constant volume (c_V). Measurements under constant pressure produce greater values than those at constant volume because work must be performed in the former. For dry air, $c_p = 1005\ \mathrm{J\,kg^{-1}\,K^{-1}}$ and $c_V = 718\ \mathrm{J\,kg^{-1}\,K^{-1}}$. Finally, air is compressible, so if p increases at constant T then ρ increases, and it has a large coefficient of thermal expansion, so if T increases at constant p then ρ decreases.

For descriptive purposes, the atmosphere is divided into layers in the vertical direction, according to the variation of temperature with height. The layer from the ground to about 10–15 km, in which temperature decreases with height, is called the "troposphere." It is bounded above by the "tropopause." From the tropopause to about 50 km is a layer in which the temperature increases with height called the "stratosphere." Further layers are found above this, but it is the troposphere and the stratosphere that are the most important for determining weather and climate, and so we will concentrate exclusively on the dynamics of these two regions.

2.2. *Solar Forcing*

The forcing of atmosphere comes from the Sun, however interactions with the land and the oceans are also important. The atmosphere is continually bombarded by solar photons at infrared, visible, and ultraviolet wavelength. Some solar photons are scattered back to space by atmospheric gases or reflected back to space by clouds or the Earth's surface; some are absorbed by atmospheric gases or clouds, leading to heating of parts of the atmosphere; and some reach the Earth's surface and heat it. Atmospheric gases (especially CO_2, H_2O, and O_3), clouds and the Earth's surface also emit and absorb infrared photons, leading to further heat transfer between one region and another, or loss of heat to space.

The amount of energy that the Earth receives from the Sun has varied over history, however at present the incident solar flux, or power/unit area, of solar energy (the so-called "solar constant") is $F = 1370$ Wm^{-2}. Given that the cross-sectional area of the Earth intercepting the solar energy flux is πa^2, where a is the Earth's radius, the total solar energy received per unit times is $F\pi a^2 = 1.74 \times 10^{17}$ W. As noted above, not all this radiation is absorbed by the Earth; a significant fraction is directly reflected. The ratio of the reflected to incident solar energy is called the "albedo" α. Under present conditions of cloudiness, snow and ice cover, on average the albedo is $\alpha = 0.3$; i.e., 30% of the incoming solar radiation is reflected back to space without being absorbed. That means

$$\text{final incoming power} = (1 - \alpha)F\pi a^2. \tag{2.2}$$

In physics, a "black body" is an idealized object that absorbs all radiation that falls on it. No radiation passes through it and none is reflected. Because no light is reflected or transmitted, the object appears black when it is cold. However, a black body emits a temperature-dependent spectrum of light. We have just described how the Earth reflects much of the radiation that is incident upon it, nevertheless as an approximation we can assume that it emits radiation in the same temperature-dependant way as a black body. This emitted radiation is given by the Stefan–Boltzmann law which states that the power emitted per unit area of a black body at absolute temperature T is σT^4, where σ is the Stefan–Boltzmann constant ($\sigma = 5.67 \times 10^{-8}$ Wm^{-2}K^{-4}). This power is emitted in all directions from the surface of the Earth, which has an area of $4\pi a^2$. Thus, in this model, if the Earth has a uniform temperature T_e,

$$\text{final outgoing power} = 4\pi a^2 \sigma T_e^4. \tag{2.3}$$

This gives a definition of the "emission temperature" T_e. It is the temperature one would infer by looking back at the Earth from space if a black-body curve was fitted to the measured spectrum of outgoing radiation. Assuming that the Earth is in thermal equilibrium, the incoming and outgoing power must balance. Therefore, from Eqs. (2.2) and (2.3) we find

$$T_e = \left[\frac{(1-\alpha)F}{4\sigma}\right]^{1/4}. \tag{2.4}$$

On substituting standard values for α, F, and σ, we find $T_e \approx 255\,\mathrm{K}$. This value is in the right ballpark, but is more than 30 K lower than the observed mean surface temperature of about 288 K.

By considering the simplest possible model of the climate system, i.e., that the Earth receives energy from the Sun, directly reflects back about 30%, and emits radiation as a black-body, we have captured some of the key aspects. However, comparison with observations of surface temperature have demonstrated that this very simplified model must have some important missing ingredients which we will now explore.

2.3. *Greenhouse Effect*

As has been mentioned above, atmospheric constituents emit, absorb, and scatter radiation. To develop our simple model of the climate system we need to take this into consideration. Hence we now assume that our system has a layer of atmosphere of uniform temperature T_a. This atmosphere is assumed to transmit a fraction τ_{sw} of incident solar (short-wave) radiation and a fraction τ_{lw} of any incident thermal (long-wave) radiation, and to absorb the remainder. This is to mimic the effects of the atmospheric constituents.

We have seen that taking into account the albedo effects and the difference between the area of the emitting surface $4\pi a^2$ and the intercepted cross-sectional area of the solar beam πa^2, the mean unreflected incoming solar flux at the top of the atmosphere F_S (power per unit area) is

$$F_S = \frac{1}{4}(1-\alpha)F, \tag{2.5}$$

with $F_S \approx 240\,\mathrm{Wm}^{-2}$. Of this, a proportion $\tau_{sw}F_S$ reaches the ground and the remainder is absorbed by the atmosphere. In our revised model, we will assume that the ground has a temperature T_g, and that it emits as a black body. This gives an upward flux of $F_g = \sigma T_g^4$, of which a proportion $\tau_{lw}F_g$

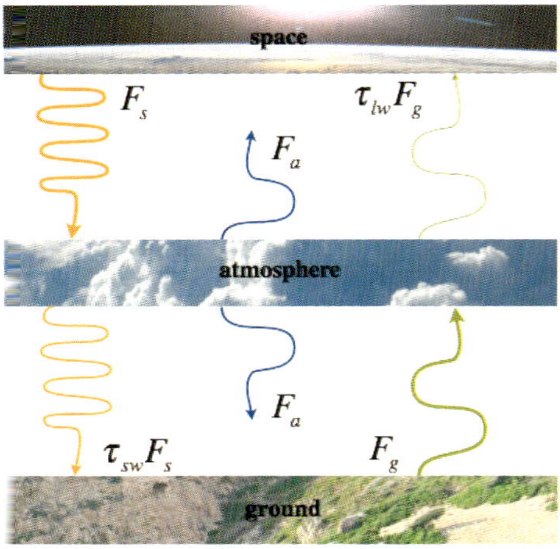

Fig. 1. A simple model of the greenhouse effect. The atmosphere is taken to be a shallow layer at temperature T_a and the ground a black body at temperature T_g. The various solar and thermal fluxes are shown (see text for details).

reaches the top of the atmosphere with the remainder being absorbed by the atmosphere. The atmosphere is not a black body and instead it emits radiation following Kirchhoff's law, such that the emitted flux is $F_a = (1 - \tau_{lw})\sigma T_a^4$ both upwards and downwards, where T_a is the temperature of the atmosphere. This is illustrated in Fig. 1.

Assuming that the system is in equilibrium, these fluxes must balance. At the top of the atmosphere we have $F_S = F_a + \tau_{lw}F_g$ and at the ground we have $F_g = F_a + \tau_{sw}F_S$. By eliminating F_a, and using the Stefan–Boltzmann law for T_g, we find that

$$T_g = \left[\frac{(1-\alpha)F}{4\sigma}\frac{1+\tau_{sw}}{1+\tau_{lw}}\right]^{1/4}, \tag{2.6}$$

or from Eq. (2.4)

$$T_g = T_e\left(\frac{1+\tau_{sw}}{1+\tau_{lw}}\right)^{1/4}. \tag{2.7}$$

This gives us a revised estimate of the temperature at the surface which varies according to the proportion of short-wave and long-wave radiation that the atmosphere transmits. Reasonable values for the Earth's atmosphere are $\tau_{sw} = 0.9$ and $\tau_{lw} = 0.2$, i.e., the atmosphere transmits considerably more short-wave solar radiation than long-wave thermal radiation. When we include these values we find that the surface temperature is

$T_g \approx 286\,\text{K}$, which is much closer to the observed value of $T_E \approx 288\,\text{K}$. Including an atmosphere that allows greater transmission of short-wave radiation than long-wave radiation has had the influence of increasing the surface temperature. This is known as the "greenhouse effect."[a]

2.4. *Radiative Transfer*

We now investigate in a little more detail the transfer of energy within the atmosphere by photons, or "atmospheric radiative transfer." The Sun has a temperature of $T_S \approx 6000\,\text{K}$ and emits short-wave radiation at ultraviolet, visible and infrared wavelengths between about 0.1 and $4\,\mu\text{m}$. The Earth, by contrast, has a much lower temperature, $T_E \approx 288\,\text{K}$, and emits radiation at infrared wavelengths between about 4 and $100\,\mu\text{m}$. Gases in the atmosphere act to absorb radiation of certain wavelengths, notably, O_3 in the ultraviolet, and CO_2 and H_2O in the infrared. The interested reader is referred to Kiehl and Trenberth (1997) for figures showing the absorption of short-wave and long-wave radiation at different wavelengths by the different gases. We noted earlier that temperature and pressure decrease with altitude in the troposphere and this is critical to a full understanding of the greenhouse effect. At the surface, the relatively high temperature and pressure means that gases absorb radiation in broad bands around specific wavelengths (collision-induced broadening), but these bands reduce in width with altitude as the pressure and temperature decrease. As radiation emitted from the Earth's surface moves up layer by layer through the atmosphere, some is stopped in each layer. Each layer then emits radiation back toward the ground and up to higher layers. However, due to the reduction in width of the absorption bands with altitude, a level can be reached at which the radiation is able to escape to space. Adding more greenhouse gas molecules means that the upper layers will absorb more radiation and the altitude of the layer at which the radiation escapes to space increases, and hence the radiation escapes from a layer with lower temperature. Since colder layers do not radiate heat as well, all the layers from this height to the surface must warm to restore the incoming/outgoing radiation balance (see Pierrehumbert, 2010, for more details).

[a]Note that Eq. (2.6) indicates that if the atmospheric constituents were to change in such a way that the transmission of long-wave radiation were reduced further, this simple model would predict that the temperature at the surface would increase. However, an increase in concentration of radiatively active gases will generally also lead to a decrease in the transmission of short-wave radiation and to a change in the albedo, and so detailed considerations are required to determine the exact impact.

It has become standard to assess the importance of a factor (such as an increase in a radiatively active atmospheric constituent) as a potential climate change mechanism in terms of its "radiative forcing." This is a measure of the influence the factor has in altering the balance of incoming and outgoing energy, and it is defined as the change in net irradiance (i.e., the difference between incoming and outgoing radiation) measured at the tropopause. For carbon dioxide the radiative forcing is given to a good approximation by a simple algebraic expression: $\Delta F = 5.35 \times \ln \frac{C}{C_0}$ Wm^{-2}, where C is the concentration of carbon dioxide and C_0 is a reference value taken to be 278 ppm. For a doubling of carbon dioxide values above pre-industrial values, this gives a radiative forcing of approximately 3.7 Wm^{-2}.

Radiative forcing can be used to estimate a subsequent change in equilibrium surface temperature (ΔT_S) arising from that radiative forcing via the equation $\Delta T_S = \lambda \, \Delta F$, where λ is the "climate sensitivity" and ΔF is the radiative forcing. For our very simplest climate model we found that the emitted radiation per unit area was $F = \sigma T_e^4$, from which we inferred a value of $T_e \approx 255$ K. We can estimate the climate sensitivity in the absence of any feedbacks in the system λ^* by considering the change in radiative forcing after a new steady state is reached: $\Delta F = (\frac{\delta F}{\delta T})|_{T=T_e} \Delta T_S^*$. This gives a climate sensitivity in the absence of feedbacks of $\lambda^* = (4\sigma T_e^3)^{-1}$, or 0.26 K/(Wm^{-2}). Using this to estimate the temperature increase due to a doubling of carbon dioxide in the absence of feedbacks gives ~1 K. In reality, feedbacks in the system (for example, changes to the albedo and the water vapor content of the atmosphere) will influence the temperature change. The Fourth Assessment Report (AR4) of the Intergovernmental Panel on Climate Change (IPCC) (Solomon, 2007) stated that the climate sensitivity (taking into account relevant feedbacks "*is likely to be in the range* 2° *C to* 4.5° *C with a best estimate of about* 3° *C, and is very unlikely to be less than* 1.5° *C. Values substantially higher than* 4.5° *C cannot be excluded, but agreement of models with observations is not as good for those values.*" The greatest uncertainty concerns the influence of cloud-related feedbacks.

In addition to the radiatively active gases, aerosols (particles with a size of 1–10 μm diameter) can influence atmospheric radiative transfer. Such aerosols include sulphate, black carbon, organic carbon, mineral dust, and sea salt. They have a direct effect of scattering (leading to a negative radiative forcing) and absorbing both short-wave and long-wave radiation (leading to either positive or negative radiative forcing). They also can have an indirect effect by altering microphysics, amount, and lifetime of clouds (potentially leading to a large negative radiative forcing, although there is considerable uncertainty). Explosive volcanic eruptions can lead to short-lived (a few years) change in the radiative forcing (a negative influence) arising from sulphate aerosol injected into the stratosphere (e.g., the eruption of Mount Pinatubo in 1991).

2.5. *Climate Change*

Global atmospheric concentrations of carbon dioxide, methane and nitrous oxide have increased markedly as a result of human activities since 1750 and now far exceed pre-industrial values determined from ice cores spanning many thousands of years. The global atmospheric concentration of carbon dioxide has increased from a pre-industrial value of about 280 ppm to 379 ppm in 2005; methane has increased from 715 ppb to 1774 ppb, and nitrous oxide from 270 ppb to 319 ppb. The global increases in carbon dioxide concentration are due primarily to fossil fuel use and land use change, while those of methane and nitrous oxide are primarily due to agriculture. Associated with these changes has been a change in the radiative forcing. The total change to the radiative forcing between 1750 and 2005 was estimated to be 1.6 $\text{W}\,\text{m}^{-2}$, with a range of [0.6 to 2.4] $\text{W}\,\text{m}^{-2}$ (see Fig. SPM.2. of IPCC AR4 Working Group I for a breakdown into the different components) (Solomon, 2007). The majority of this increase is due to anthropogenic forcing, although there is also a small estimated increase of approximately 0.12 $\text{W}\,\text{m}^{-2}$ arising from a gradual increase in solar output during the industrial era.

The IPCC AR4 concluded that the "*warming of the climate system is unequivocal, as is now evident from observations of increases in global average air and ocean temperatures, widespread melting of snow and ice, and rising global average sea level.*" The increase in global average surface temperature from 1850–1899 to 2001–2005 was 0.76 [0.57 to 0.95]°C. The average atmospheric water vapor content has increased since at least the 1980s over land and ocean as well as in the upper troposphere, in a manner broadly consistent with the extra water vapor that warmer air can hold. Observations since 1961 show that the average temperature of the global ocean has increased to depths of at least 3000 m and that the ocean has been absorbing more than 80% of the heat added to the climate system. Such warming causes seawater to expand, contributing to a rise in sea level. Mountain glaciers and snow cover have declined on average in both hemispheres. In addition, the global average sea level rose at an average rate of 1.8 [1.3 to 2.3] mm year over 1961 to 2003, with contributions from thermal expansion, melting glaciers and icecaps, and losses from the ice

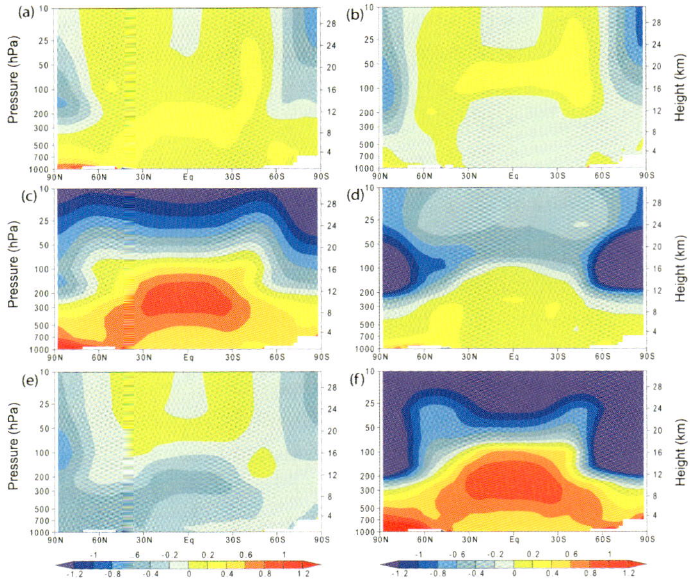

Fig. 2. Zonal mean atmospheric temperature change from 1890 to 1999 (C per century) as simulated by the PCM model from (a) solar forcing; (b) volcanoes; (c) well-mixed greenhouse gases; (d) tropospheric and stratospheric ozone changes; (e) direct sulphate aerosol forcing; and (f) the sum of all forcings. Plot is from 1,000 hPa to 10 hPa (shown on left scale) and from 0 km to 30 km (shown on right). [*Source*: Fig. 9.1 of IPCC AR4 Working Group I.]

sheets of Greenland and Antarctica. The reader is referred to the IPCC report for a full description of the observed changes.

Climate models can be used to predict the expected responses — in terms of patterns of variation in space, time or both — to external forcing. Figure 2 illustrates the zonal average temperature response in one climate model to several different forcing agents over the last 100 years: (a) solar forcing, (b) volcanoes, (c) well-mixed greenhouse gases, (d) tropospheric and stratospheric ozone changes, (e) direct sulphate aerosol forcing, and (f) the sum of all forcings. The simulated response to solar forcing is a general warming everywhere, and to volcanic sulphate aerosol is a surface and tropospheric cooling and a stratospheric warming that peak several months after a volcanic eruption and last for several years. The simulated responses to anthropogenic forcings are different to these natural forcings. Greenhouse gas forcing produces warming in the troposphere, a cooling in the stratosphere, and somewhat more warming near the surface in the northern hemisphere due to its larger land fraction.[b] The combined effect of tropospheric and stratospheric ozone forcing is to warm the troposphere, due to increases in tropospheric ozone, and cool the stratosphere, particularly at high latitudes where stratospheric ozone loss has been greatest, and sulphate aerosol forcing results in cooling throughout most of the globe. The net effect of all forcings combined is a pattern of warming in the northern hemisphere near the surface that is dominated by the greenhouse gas contribution, and cooling in the stratosphere that results predominantly from greenhouse gas and stratospheric ozone forcing.

2.6. *Further Atmospheric Properties*

We now consider some further properties of the atmosphere that are important for determining its dynamics. Each portion of the atmosphere is approximately in what is known as "hydrostatic balance" (usually valid on scales greater than a few kilometers). This means that the weight of the portion of atmosphere is supported by the difference in pressure between the lower and upper surfaces and that the following relationship between density ρ and pressure p holds to a good approximation

$$g\rho = -\frac{\partial p}{\partial z}, \tag{2.8}$$

where g is the gravitational acceleration and z is the height. The ideal gas law (2.1) can be used to replace ρ in this equation by p/RT. In the simplest case of an isothermal temperature profile where $T = T_0 =$ constant, the equation can be integrated to obtain

$$p = p_0 e^{-gz/RT_0}. \tag{2.9}$$

The quantity $H = RT_0/g$, known at the "scale height," is the height over which the pressure falls by a factor of e. If $T_0 = 240\,\text{K}$ (a typical value for the troposphere) then $H \approx 7\,\text{km}$. In general the temperature does not vary greatly in the atmosphere, certainly not by orders of magnitude as the pressure and density do. This means that Eq. (2.9) yields profiles of pressure decreasing exponentially with heights that are a good approximation to reality.

The first law of thermodynamics states that the increase in internal energy of a system δU is equal to the heat supplied plus the work done on

[b] Land regions have a shorter surface response time to the warming than do ocean regions.

the system. This can be written as

$$\delta U = T\delta S - p\delta V, \tag{2.10}$$

where S is the "entropy" of the system, T is the temperature, and V is the volume. For a unit mass of ideal gas, for which $V = 1/\rho$, it can be shown that $U = c_V T$, where c_V is the specific heat capacity at constant volume. Thus

$$\delta S = c_p \frac{\delta T}{T} - R\frac{\delta p}{p}, \tag{2.11}$$

where $c_p = c_V + R$ is the specific heat capacity at constant pressure. (Note that if Q is the amount of heat absorbed by the system, $\delta S = \delta Q/T$.)

An "adiabatic" process is one in which heat is neither lost nor gained so that $\delta S = 0$. In this case, Eq. (2.11) can be integrated. If $T = \theta$ when $p = p_0$, then $\theta = T(p_0/p)^{R/c_p}$. The quantity θ is known as the "potential temperature." It is the temperature a portion of air would have if, starting from temperature T and pressure p, it were compressed until its pressure equaled p_0. Surfaces of constant potential temperature are known as "isentropic" surfaces.

For an incompressible fluid, if the background temperature increases (decreases) with height, then a fluid parcel displaced adiabatically upward will be cooler (warmer) than its surroundings and this will fall back down (continue to rise). Thus the fluid is said to be stable (unstable). For a compressible fluid, the same argument holds, but it is the potential temperature, rather than the temperature that is the relevant quantity. Hence a compressible fluid is said to be stable if the background potential temperature increases with height (this means that it is often useful to use potential temperature as a vertical co-ordinate).

The rate of decrease of temperature with height is known as the "lapse rate" Γ, i.e., $\Gamma = -\frac{dT}{dz}$. From Eq. (2.11) for an adiabatically rising parcel, if the atmosphere is in hydrostatic balance (2.8), then

$$-\left(\frac{dT}{dz}\right)_{\text{parcel}} = \frac{RT}{c_p p}\left(\frac{dp}{dz}\right)_{\text{parcel}} = \frac{g}{c_p} \equiv \Gamma_a. \tag{2.12}$$

Here Γ_a is known as the "dry adiabatic lapse rate" and it is about $10\,\mathrm{K\,km^{-1}}$. Now we can revisit the example of a fluid parcel being displaced adiabatically upward. If the background temperature falls more rapidly with height than the adiabatic lapse rate, i.e., $\Gamma > \Gamma_a$, then the parcel will be warmer than its surroundings and will continue to rise and the atmosphere is unstable. On the other hand, if $\Gamma < \Gamma_a$ then the atmosphere is stable. In general, the atmosphere is stable to this "dry convection," but it can be unstable in hot, arid regions such as deserts. Convection carries heat up and thus reduces the background lapse rate until the dry adiabatic lapse rate is reached.

Water vapor in the atmosphere plays an important role in the dynamics of the troposphere. Latent heating and cooling can transfer heat through the Earth system (e.g., evaporation of droplets of seawater and subsequent condensation into droplets at another location in the atmosphere transfers heat from the ocean to the atmosphere). Water vapor also influences convection. As a moist air parcel rises adiabatically, p falls, so T falls, the water vapor condenses, and latent heat is released. The moist adiabatic lapse rate is less than for dry air (and thus is more easily exceeded).

Convective processes in the atmosphere strongly influence the vertical temperature structure in the troposphere. Simple 1-D radiative equilibrium calculations predict that the temperature would decrease sharply with height at the lower boundary, implying convective instability. Calculations including both radiative and convective effects, adjusting the temperature gradient to neutral stability where necessary, and taking into account the effects of moisture, predict a less sharp decline in temperature through the troposphere in agreement with observations (see (Manabe and Wetherald, 1967)). These calculations indicate two distinct regions of the atmosphere: The troposphere where the temperature structure is strongly influenced by convective processes and the stratosphere where the temperature structure is determined predominantly by radiative processes.

2.7. *Oceanic Properties*

Next we consider the properties of the oceans. Just over 70% of the Earth's surface is covered by water. The average depth of the oceans is about 3.7 km, but in places it exceeds 6 km. The volume of the oceans is $\sim 3.2 \times 10^{17}$ m^3, and the total heat capacity of the oceans is about one thousand times as large as that of the atmosphere. The oceans store 50 times more carbon than the atmosphere and take up roughly one third of the carbon dioxide released by human activities each year. Thus the oceans play a critical role in determining our climate. In addition, changes in sea surface temperature can directly influence the atmosphere and its weather systems (e.g., hurricane formation and the El Niño Southern Oscillation).

The density of fresh water is maximum at 4°C, with a value of 0.999×10^3 kgm^{-3} (fresh water colder than this is less dense). The mean density of seawater is only slightly greater, with a value of 1.035×10^3 kgm^{-3}. The density depends on temperature, salinity, and pressure in a complex and nonlinear way; however, temperature typically influences density more than salinity in the parameter range of the open ocean. In discussions of ocean dynamics, the buoyancy, b is often used with $b = -g\frac{\rho - \rho_o}{\rho}$, where ρ is the density of a parcel of water and ρ_o is the density of the background. Thus if $\rho < \rho_o$ the parcel will be positively buoyant and will rise. Since the density does not vary greatly in the ocean (by only a few %), we can write the buoyancy as $b = -g\frac{\sigma - \sigma_o}{\rho_{\text{ref}}}$, where $\sigma = \rho - \rho_{\text{ref}}$ and $\rho_{\text{ref}} = 1000$ kgm^{-3}. The surface layer of the ocean, known as the "mixed layer," is strongly stirred by the winds. Over this layer (typically 50–100 m) the temperature and salinity, and hence density, vary little with depth. Below this is a layer where the vertical gradients of temperature and density are greatest, called the "thermocline." This is about 600 m deep in mid-latitudes and about 100–200 m deep at low latitudes. The waters of the thermocline are warmer and saltier than the deep ocean below, which is known as the "abyss."

The ocean, like the atmosphere, is a stratified fluid on a rotating Earth, and therefore there are many similarities in terms of their dynamics. However, there are also differences. The oceans are bounded by solid continents; in the ocean the vertical density stratification is weaker than in the atmosphere (and this influences the scales of instabilities) and water is (almost) incompressible. In addition, with the exception of the Southern Ocean, the major ocean basins are laterally bounded by continents, allowing large-scale horizontal pressure gradients to develop in a way that is impossible in the atmosphere. The timescales of variability are typically longer than those in the atmosphere. The surface mixed layer, which is directly influenced by the atmosphere, exhibits variability on diurnal, seasonal, and interannual timescales. However, the ocean interior only varies significantly on decadal to centennial and longer timescales. Finally, water vapor plays an important role in the dynamics of the atmosphere (especially the troposphere) and salinity plays an important role in the dynamics of the oceans.

2.8. *Ocean Forcing*

The forcing of the oceans is rather different to that of the atmosphere. As we have discussed, a significant fraction of solar radiation passes through the atmosphere to heat the Earth's surface and drive convection from below. In the ocean, convection is driven by buoyancy loss from above as the ocean exchanges heat and fresh water at the surface (including through brine rejection in sea ice formation).

The heat flux at the ocean surface has four components: (i) sensible heat flux (which depends on the wind speed and air/sea temperature difference), (ii) latent heat flux (from evaporation/precipitation), (iii) incoming short-wave radiation from the sun, and (iv) long-wave radiation from the atmosphere and ocean. The net freshwater flux is given by evaporation minus precipitation, including the influences of river runoff and ice formation processes.

Winds blowing over the ocean surface exert a stress on it and directly drive ocean circulations, particularly in the upper kilometer or so. This means that at the surface there is a strong similarity between the pattern of ocean currents and the atmospheric winds. The wind stress is typically parametrized by $\tau_{\text{wind}} = \rho_a c_D u_{10}^2$, where ρ_a is the density of air, u_{10} is the wind speed at 10 m and c_D is a drag coefficient (a function of wind speed, atmospheric stability, and sea state). It can be shown that the wind stress gives rise to a force per unit mass on a slab of ocean of

$$\mathcal{F}_{\text{wind}} = \frac{1}{\rho_{\text{ref}}} \frac{\partial \tau_{\text{wind}}}{\partial z}. \tag{2.13}$$

Below the surface, the winds, flow over topography, the tides, and other processes indirectly influence the circulation.

3. Dynamics of the Atmosphere and Oceans

3.1. *Role of Dynamics*

We will now focus on the dynamics of the atmosphere and oceans and the processes that determine the weather and climate. By "weather" we usually mean events associated with atmospheric flows with length scales of 100 m or more and timescales of a few days or less. Different components of the atmosphere vary on different timescales (individual clouds vary on timescales of less than an hour, whereas mid-latitude weather systems vary on timescales of several days). By "climate" we usually mean the state of the atmosphere on longer timescales — years to decades. It can also be described as the probability distribution of the variable weather.

We have seen that the vertical temperature variation in the troposphere and stratosphere can be characterized in terms of the influence of radiative and convective processes. We now consider what determines the pole-to-equator temperature variation. Tropical regions receive more incoming solar

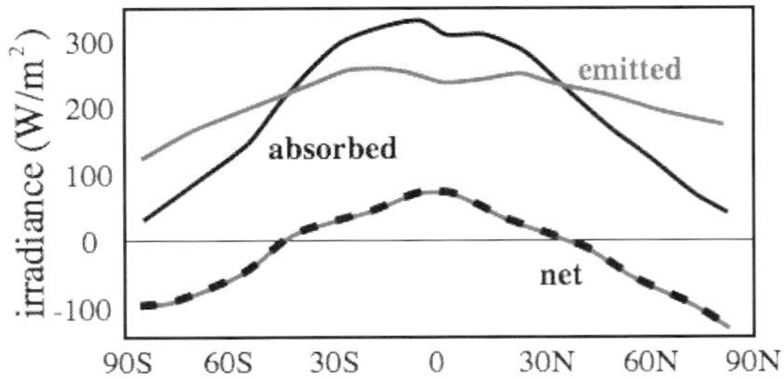

Fig. 3. Annual mean absorbed solar radiation, emitted long-wave radiation, and net radiation. [*Source*: Based on Fig. 5.5 of Marshall and Plumb (2008).]

radiation than polar regions because the solar beam is concentrated over a smaller area due to the spherical curvature of the Earth. However, observations indicate that in the annual mean the tropical regions emit less radiation back to space than they receive; and that the polar regions emit more radiation than they receive (see Fig. 3). This implies that there must be a transport of energy on the Earth from the equator to the pole that takes places in the atmosphere and/or the ocean. By considering the equations of motion for a rotating stratified flow, we will examine the processes driving poleward energy transport. We will see that in the atmosphere, the processes involved also give rise to the jet streams and mid-latitude weather systems.

3.2. *Rotating Fluids*

In considering the flow of a fluid, it is useful to describe the evolution of a parcel of fluid as it follows the fluid. The time variation of properties of such a parcel can be very different to the time variation of the properties at a fixed point in space. A meteorological example of this is given by clouds which form under certain conditions when wind blows over mountains. As the air is forced upward over the mountains, it cools, and this means that if the air is sufficiently saturated in water vapor, water vapor will condense and form a cloud. The location of such a cloud is geographically tied to the mountain and thus the time derivative at a fixed point in space of the cloud amount will be zero. However, following the flow, an air parcel will pass in and out of cloud as it flows over the hill and into its lee. The rate of change of a quantity $C = C(x, y, z, t)$ (where in our example about C would be the cloud amount) following the flow is known as the "Lagrangian derivative" and is given by

$$\frac{DC}{Dt} \equiv \frac{\partial C}{\partial t} + \mathbf{u} \cdot \nabla C. \tag{3.1}$$

When the wind blows it carries properties, such as heat, moisture, and pollutants, with it. This is described by the term $\mathbf{u} \cdot \nabla$ which represents "advection."

We want to write down the equations of motion for the fluid flow. There are five key variables: The velocity (u, v, w), the pressure p, and the temperature T (in the atmosphere humidity is also an important variable and in the ocean salinity is). Correspondingly, there are five independent equations: Newton's second law (three equations), conservation of mass (one equation), and the first law of thermodynamics (one equation).

Newton's second law states that in an inertial frame,

$$\frac{D\mathbf{u}}{Dt} = -\frac{1}{\rho}\nabla p + \mathbf{g}^* + \mathcal{F}, \tag{3.2}$$

where $-\frac{1}{\rho}\nabla p$ is the pressure gradient force of relevance for fluids, $\mathbf{g}^*$ is the gravitational force and $\mathcal{F}$ is the sum of the frictional forces, all per unit mass. To represent the weather we observe, it is natural to describe the flow seen from the perspective of someone on the surface of the Earth, and thus we need to consider the motion in the rotating frame of the Earth. Newton's second law as described above holds in an inertial frame of reference. When it is translated into a rotating frame of reference, additional terms are introduced that are specific to that frame. For example, consider a rotating frame where the angular rotation of the frame is expressed by the vector $\mathbf{\Omega}$ pointing in the direction of the axis of rotation, and with magnitude equal to the angular rate of rotation $\Omega = 7.27 \times 10^{-5}\mathrm{s}^{-1}$ (one revolution per day). The flow velocity in the rotating and inertial frames are related by $\mathbf{u}_{\mathrm{inertial}} = \mathbf{u}_{\mathrm{rotating}} + \mathbf{\Omega} \times \mathbf{r}$ and the Lagrangian derivative is given by

$$\left(\frac{D\mathbf{u}_{\mathrm{in}}}{Dt}\right)_{\mathrm{in}} = \left(\frac{D\mathbf{u}_{\mathrm{rot}}}{Dt}\right)_{\mathrm{rot}} + 2\mathbf{\Omega} \times \mathbf{u}_{\mathrm{rot}} + \mathbf{\Omega} \times \mathbf{\Omega} \times \mathbf{r}. \tag{3.3}$$

Thus the additional terms that are introduced when considering Newton's second law in a rotating frame are $2\mathbf{\Omega} \times \mathbf{u}_{\mathrm{rot}}$, the "Coriolis acceleration" and $\mathbf{\Omega} \times \mathbf{\Omega} \times \mathbf{r}$, the "centrifugal acceleration." It is convenient to combine the centrifugal force with the gravitational force in one term $\mathbf{g} = -g\hat{\mathbf{z}} = \mathbf{g}^* + \mathbf{\Omega} \times \mathbf{\Omega} \times \mathbf{r}$, where $\hat{\mathbf{z}}$ represents a unit vector parallel to the local vertical. The gravity, g, defined in this way is the gravity measured in the rotating frame, $g = 9.81\,\mathrm{ms}^{-2}$.

The thinness of the atmosphere/ocean enables a local Cartesian co-ordinate system, which neglects the Earth's curvature, to be used for many problems. Taking the unit vectors $\hat{\mathbf{x}}$, $\hat{\mathbf{y}}$, and $\hat{\mathbf{z}}$ to be eastward (zonal), northward (meridional) and upward, respectively, the rotation vector can be written in this co-ordinate basis as $\boldsymbol{\Omega} = (0, \Omega\cos\phi, \Omega\sin\phi)$ for latitude ϕ. In the atmosphere $|u| \sim 10\,\mathrm{ms}^{-1}$ (less in the ocean) and so $\Omega u \ll g$. In addition, both in the atmosphere and the ocean, vertical velocities w, typically $\leq 10^{-1}\,\mathrm{ms}^{-1}$, are much smaller than horizontal velocities. Hence $2\boldsymbol{\Omega} \times \mathbf{u} \simeq f\hat{\mathbf{z}} \times \mathbf{u}$, where $f = 2\Omega\sin\phi$. This means we can write Eq. (3.2) as

$$\frac{D\mathbf{u}}{Dt} + \frac{1}{\rho}\nabla p - \mathbf{g} + f\hat{\mathbf{z}} \times \mathbf{u} = \mathcal{F}. \tag{3.4}$$

In the local Cartesian system, considering the vertical direction, if friction $\mathcal{F}_z$ and vertical acceleration Dw/Dt are small (as is generally true for large-scale atmospheric and oceanic systems), then we have

$$\frac{Du}{Dt} + \frac{1}{\rho}\frac{\partial p}{\partial x} - fv = \mathcal{F}_x \tag{3.5}$$

$$\frac{Dv}{Dt} + \frac{1}{\rho}\frac{\partial p}{\partial y} + fu = \mathcal{F}_y \tag{3.6}$$

$$\frac{1}{\rho}\frac{\partial p}{\partial z} + g = 0, \tag{3.7}$$

where the vertical direction gives the equation for hydrostatic balance (2.8) introduced earlier. For large-scale atmospheric motions ($\sim$100 km), the Coriolis force is significant. It deflects moving air to right (left) in northern (southern) hemisphere. More generally, angular momentum from rotation and the constraints it imposes give rotating fluids unusual and sometimes counter-intuitive properties, as we shall see.

The remaining two equations are the conservation of mass (the mass of a fixed volume can only change if ρ changes and this requires a mass flux into the volume)

$$\frac{D\rho}{Dt} + \rho\nabla \cdot \mathbf{u} = 0, \tag{3.8}$$

and the first law of thermodynamics (see Eq. (2.11))

$$\frac{DQ}{Dt} = c_p\frac{dT}{dt} - \frac{1}{\rho}\frac{Dp}{Dt}. \tag{3.9}$$

Here DQ/Dt is "diabatic heating rate" per unit mass. In the atmosphere it is mostly due to latent heating/cooling (condensation/evaporation) and radiative heating/cooling.

A useful quantity for analyzing atmosphere and ocean dynamics is the "potential vorticity" since it is conserved following adiabatic, frictionless flow. There are a number of different formulations of the potential vorticity that are applicable to different problems, but in essence it is a measure of the ratio of the absolute vorticity to the absolute depth of a vortex. The simplest form of relevance to a homogeneous fluid is $PV = (f+\zeta)/H$, where ζ is the relative vorticity and H is the depth of the fluid. For a stratified fluid the simplest form is Ertel's potential vorticity, $PV = \frac{1}{\rho}\zeta \cdot \nabla\theta$.

3.3. *Weather and Climate Models*

The equations of motion we have just derived are a simplified form of the equations that are at the heart of weather and climate models. Such "general circulation models" (GCMs) solve numerically discretized versions of the equations of motion. Computational constraints mean that there is a limit to the scale of motion that these models can directly resolve. In the atmosphere, large-scale motion such as planetary (Rossby) wave disturbances ($\sim 10^4$ km) and synoptic-scale disturbances (weather systems) ($\sim$2000 km) are well captured. However, smaller-scale motion such as deep/shallow convection ($1 \sim 100$ km), gravity waves ($\sim$1–1000 km), and boundary-layer turbulence ($\sim$1 m–2 km) generally need to be parametrized — i.e., represented approximately in terms of the larger-scale resolved variables. Similarly in the ocean, small-scale processes are parametrized (indeed the scales of motion are typically ten times smaller in the ocean than in the atmosphere,[c] making the problem even more challenging). The forcing of an atmospheric general circulation model may include specified solar input, radiatively active gases (including O_3, CO_2, and CH_4), sea surface temperature, and sea ice. What is included or not in the model depends on whether the processes are important over the timescale for which the model is being used to project (hours to weeks for weather models, decades to centuries for climate models). For climate projections, coupled models are usually employed in which separate models of the atmosphere, ocean, cryosphere, biosphere, and some chemical cycles are linked together in such a way that changes in one may influence another.

[c]See later discussion of Rossby radius in Sec. 3.5.

3.4. *Dynamical Processes*

To understand the behavior of different processes, including those related to weather systems, it is useful to consider the relative magnitudes of the different terms in Eq. (3.4). For this purpose we introduce the Rossby number Ro, which is the ratio of acceleration terms to Coriolis terms. For large scales in atmosphere, the typical velocity, length, and timescales are $U \sim 10\ \mathrm{ms}^{-1}$, $L \sim 10^6$ m, and $T \sim 10^5$ s. This means $\frac{\partial \mathbf{u}}{\partial t} \sim \frac{U}{T} \approx 10^{-4}\ \mathrm{ms}^{-2}$, $\mathbf{u} \cdot \nabla \mathbf{u} \sim \frac{U^2}{L} \approx 10^{-4}\ \mathrm{ms}^{-2}$, and $f\hat{\mathbf{z}} \times \mathbf{u} \sim fU \approx 10^{-3}\ \mathrm{ms}^{-2}$ (in mid-latitudes) and hence that $Ro = \frac{U}{fL} \sim 0.1$. In the ocean, in mid-latitude gyres, the typical scales are $U \sim 0.1\ \mathrm{ms}^{-1}$ and $L \sim 10^6$ m, so $Ro \sim 10^{-3}$. Therefore in both cases, because the Ro number is small, we can neglect the acceleration terms in favor of the Coriolis terms. In addition, away from boundaries the friction is negligible, so in the horizontal we have

$$f\hat{\mathbf{z}} \times \mathbf{u} + \frac{1}{\rho}\nabla p = 0. \tag{3.10}$$

This approximation is known as "geostrophic balance." It is a balance between the Coriolis force and the horizontal pressure gradient force and is used to define the "geostrophic wind" $\mathbf{u_g}$ given by

$$(u_g, v_g) = \left(\frac{-1}{f\rho}\frac{\partial p}{\partial y}, \frac{1}{f\rho}\frac{\partial p}{\partial x}\right). \tag{3.11}$$

Geostrophically balanced flow is normal to the pressure gradient, i.e., along contours of constant pressure. In the northern (southern) hemisphere, motion is therefore counterclockwise (clockwise) around the center of low pressure systems (see Fig. 4). Note also that the wind depends on magnitude of pressure gradient: It is stronger when the isobars are closer. When the wind swirls counterclockwise in the northern hemisphere or clockwise in the southern hemisphere, it is called "cyclonic" flow; the opposite direction is called "anticyclonic" flow. A hurricane is a cyclone.

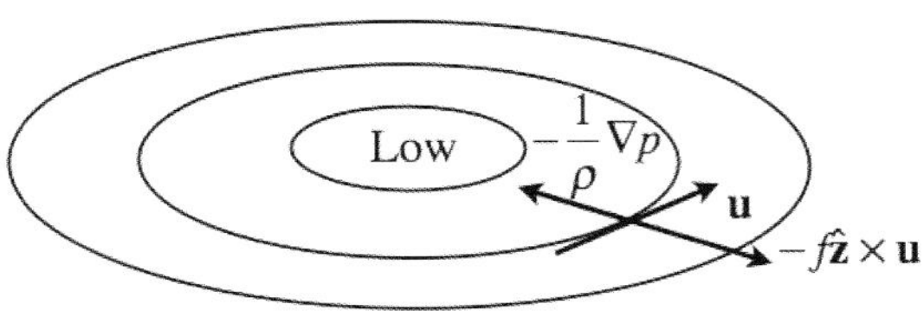

Fig. 4. Cyclonic geostrophic flow around a low pressure center in the northern hemisphere.

Analysis of the equations provides information regarding the vertical variation of the geostrophic velocities. In the case where ρ and f are constant, by taking the horizontal derivatives of the geostrophic wind, it can be shown that it is horizontally non-divergent. Taking vertical derivative of the geostrophic wind and using the hydrostatic balance equation gives that $\left(\frac{\partial u_g}{\partial z}, \frac{\partial v_g}{\partial z}\right) = 0$, and the equation for the conservation of mass (3.8) then gives that $\frac{\partial w_g}{\partial z} = 0$. Under slightly more general conditions, namely for a slow, steady, frictionless, "barotropic" (i.e., density depends only on pressure so $\rho = \rho(p)$) fluid, it can be shown that the horizontal and vertical components of the velocity cannot vary in the direction of the rotation vector Ω. This is known as the "Taylor–Proudman theorem." It means that vertical columns of fluid remain vertical (they cannot be titled or stretched); such columns of fluid are known as "Taylor columns." However, in general in the atmosphere and ocean, density does vary on pressure surfaces as it varies with, e.g., temperature; the fluid is said to be "baroclinic." If the density can be written as $\rho = \rho_{\mathrm{ref}} + \sigma$ where $\sigma \ll \rho_{\mathrm{ref}}$ (this is generally the case in the ocean), then replacing ρ by ρ_{ref} in the denominator of equation for the geostrophic wind (3.11) and taking $\partial/\partial z$, then making use of the hydrostatic balance (3.7), gives

$$\left(\frac{\partial u_g}{\partial z}, \frac{\partial v_g}{\partial z}\right) = \frac{g}{f\rho_{\mathrm{ref}}}\left(\frac{\partial \sigma}{\partial y}, -\frac{\partial \sigma}{\partial x}\right). \tag{3.12}$$

This is a simple form of the "thermal wind equation." Larger density variations in the atmosphere mean that the equivalent expression is most straightforward when written in pressure rather than height co-ordinates in this case. Taking $\partial/\partial p$, using the hydrostatic relation and the ideal gas law, it can be shown that

$$\left(\frac{\partial u_g}{\partial p}, \frac{\partial v_g}{\partial p}\right) = \frac{R}{fp}\left(\frac{\partial T}{\partial y}, -\frac{\partial T}{\partial x}\right). \tag{3.13}$$

As we have discussed before, there is a pole-to-equator temperature gradient in the atmosphere ($f^{-1}\partial T/\partial y < 0$ in both hemispheres) which implies through Eq. (3.13) that $\partial u/\partial p < 0$, i.e., that with increasing height the winds become more eastward (westerly) in both hemispheres. Consistent with this, strong jet streams are observed in mid-latitudes of both hemispheres, with the strongest westerly winds in the upper troposphere (see Fig. 5).

On planetary scales, variations in f become important. It is therefore useful to introduce the "β-plane approximation" for f in which it

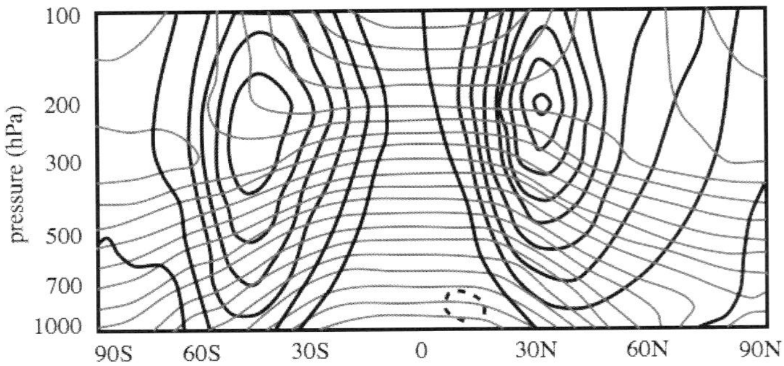

Fig. 5. Zonally averaged temperature (gray contours, interval 5 K, minimum value 200 K at the equator at 100 hPa) and zonally averaged zonal winds (black contours, interval 5 ms^{-1}, maximum value 40 ms^{-1} at 30°N and 200 hPa). Values typical of December–February.

is restricted to vary linearly with the northward direction y such that $f = f_0 + \beta y$, where $\beta = \frac{df}{dy} = \frac{2\Omega}{a}\cos\phi$. Then the divergence of the geostrophic flow (Eq. 3.11) is given by

$$\nabla_h \cdot \mathbf{u}_g = -\frac{\beta}{f} v_g. \tag{3.14}$$

In the incompressible ocean, the horizontal divergence of geostrophic flow is associated with vertical stretching of water columns in the interior (where the flow is geostrophic) because $\nabla_h \cdot \mathbf{u}_g + \frac{\partial w}{\partial z} = 0$. In this case the vertical and meridional (i.e., northward) currents are related by

$$\beta v_g = f\frac{\partial w}{\partial z}. \tag{3.15}$$

Close to boundaries, geostrophic balance no longer holds because frictional effects become important and so

$$f\hat{\mathbf{z}} \times \mathbf{u} + \frac{1}{\rho}\nabla p = \mathcal{F}. \tag{3.16}$$

The result is that there is an "ageostrophic" component of flow (high to low pressure), $\mathbf{u} = \mathbf{u}_g + \mathbf{u}_{ag}$. This effect is important in the lower ~1 km of the atmosphere and the upper ~100 m of the ocean. The geostrophic flow is horizontally nondivergent (except on planetary scales), but the ageostrophic flow is not. Winds that deviates toward low pressure systems near the surface are convergent and through mass continuity there must be an associated vertical velocity away from the surface. This is known as "Ekman pumping." In the atmosphere Ekman pumping produces ascent, cooling, clouds and possibly rain in low pressure systems. In the ocean it is a key component of the circulation in gyres. For an incompressible flow (such as water in the ocean) with $\nabla \cdot \mathbf{u} = 0$, when the geostrophic flow is nondivergent, the vertical velocity w is given by

$$\nabla_h \cdot \mathbf{u}_{ag} + \frac{\partial w}{\partial z} = 0. \tag{3.17}$$

A more accurate approximation to the equations of motion for large-scale low-frequency motions away from the tropics, where the Rossby number is small, $Ro \ll 1$, is given by the "quasi-geostrophic equations." The reader is referred to Andrews (2010) for a derivation and further explanations. Usefully, these equations can be combined into one equation. When friction and diabatic heating are neglected this is given by

$$\frac{D_g q}{Dt} \equiv \frac{\partial q}{\partial t} + u_g \frac{\partial q}{\partial x} + v_g \frac{\partial q}{\partial y} = 0, \tag{3.18}$$

where

$$q = f_0 + \beta y + \frac{\partial^2 \psi}{\partial x^2} + \frac{\partial^2 \psi}{\partial y^2} + \frac{\partial}{\partial z}\left(\frac{f_0^2}{N_B^2}\frac{\partial \psi}{\partial z}\right) \tag{3.19}$$

is the "quasi-geostrophic potential vorticity." Here ψ is the geostrophic streamfunction and N_B is a buoyancy frequency[d], and the β-plane approximation has been used.

Wave-like motions frequently occur in the atmosphere and oceans. One important class of waves are known as "Rossby waves." The Rossby wave is a potential vorticity-conserving motion that owes its existence to an isentropic gradient of potential vorticity. By considering small-amplitude disturbances to a uniform background flow, $(U, 0, 0)$, governed by the quasi-geostrophic equations, a wave-like solution can be found

$$\psi' = \mathbb{R}\, \hat{\psi} \exp[i(kx + ly + mz - \omega t)], \tag{3.20}$$

with a dispersion relation

$$\omega_{\text{Rossby}} = kU - \frac{\beta k}{k^2 + l^2 + f_0^2 m^2 / N_B^2}. \tag{3.21}$$

[d]See previous chapter of this volume, "Geophysical and Environmental Fluid Dynamics," for definition.

The zonal phase speed of the waves $c \equiv \omega/k$ always satisfies $(U-c) > 0$, i.e., the wave crests and troughs move westward with respect to the background flow.

The Coriolis parameter is much smaller in the tropics than in the extra-tropics and consequently the "equatorial β-plane approximation," in which $f \approx \beta y$ (where $\beta \equiv 2\Omega/a$), $\sin\phi \approx y/a$ (where y is the distance from the equator) and $\cos\phi \approx 1$, is used to explore the dynamics. Eastward and westward propagating disturbances that are trapped about the equator (i.e., they decay away from the equatorial region) are possible solutions. Non-dispersive waves that propagate eastward with phase speed $c = \sqrt{gH}$ (where H is an equivalent depth) are known as "equatorial Kelvin waves." Typical phase speeds in the atmosphere are $c_{\text{Kelvin}} = 20$–$80\,\text{ms}^{-1}$ and in the ocean $c_{\text{Kelvin}} = 0.5$–$3\,\text{ms}^{-1}$. Another class of equatorial waves are westward propagating equatorial Rossby waves whose dispersion curves are given approximately by $\omega_{eq\text{Rossby}} = -\beta k/(k^2 + (2n+1)\beta/c_{\text{Kelvin}})$, where n is a positive integer. See Gill (1982) for more detail.

3.5. *General Circulation of the Atmosphere*

The "general circulation" is usually taken to mean the global-scale atmospheric flow that is averaged in time over a period sufficiently long to remove random variations associated with individual weather systems, but short enough to retain monthly and seasonal variations. If the Earth was not rotating, then the circulation would be driven by the pole-to-equator temperature difference, with warm air rising at low latitudes and sinking at high latitudes. On the rotating Earth, as air moves away from the equator in the upper troposphere, it gains an eastward (westerly) velocity component from the Coriolis effect. In the tropics the Coriolis parameter $f = 2\Omega\sin\phi$ is small, but in mid-latitudes it has a significant influence. Thus the overturning circulation is confined to low latitudes — if it extended all the way to the poles, the westerly component arising from the Coriolis effect would become infinite. Moist air rises near the equator in the "inter-tropical convergence zone (ITCZ)," and dry air descends in the subtropical desert regions at about 20–30°. This is known as the "Hadley circulation." In the upper troposphere, at the poleward extent of this circulation, is the strong westerly flow of the jet streams (see Fig. 5). The jet streams are strongest in winter, with average speeds of around $30\,\text{ms}^{-1}$. We have seen that the vertical gradient of this westerly flow is in thermal wind balance with the horizontal temperature gradient.

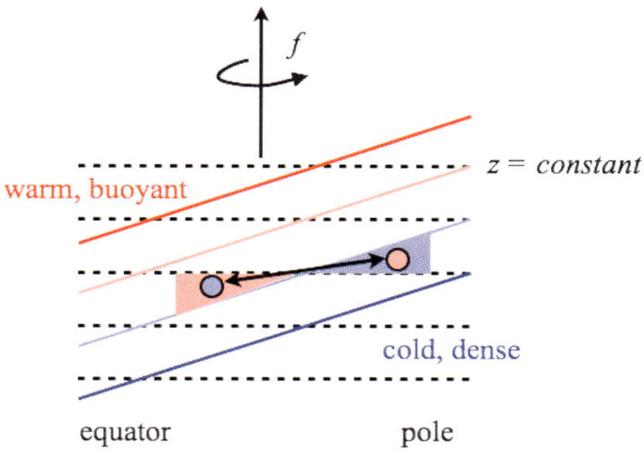

Fig. 6. Slopes of parcel trajectories relative to the zonal mean potential temperature surfaces for a baroclinically unstable disturbance in a rotating frame.

The equatorward return flow at the surface, where friction is important, is weak and ageostrophic, but the Coriolis effect still provides an easterly component to the flow, resulting in the northeasterly (southeasterly) trade winds in the northern (southern) hemisphere. The mean motion of an air parcel in this overturning circulation is given by the time-averaged "residual mean meridional circulation."[e]

The westerly flow in the mid-latitude jet streams is hydrodynamically unstable and can spontaneously break down into "eddies," which manifest themselves as traveling weather systems. Such eddies play a vital role in transporting heat, moisture, and chemical species in the latitude/height plane. So what is the mechanism for the instability? A rotating fluid will adjust to geostrophic equilibrium, rather than to rest. In this configuration it has potential energy which is available for conversion to other forms by a redistribution of mass. However, having available potential energy in the fluid is not sufficient for instability since rotation tends to inhibit the release of this potential energy.

If a parcel of air moves upward in a "wedge of instability" between a sloping surface of constant potential temperature and the horizontal, and is replaced by a similar parcel moving downward, the potential energy will be reduced. The released available potential energy is converted to kinetic energy of eddying motions in a process known as "baroclinic instability."

[e]This is different to the Eulerian mean circulation in which so-called Ferrel cells are observed in mid-latitudes and the reader is referred to, e.g., Vallis (2006) for further detail on these two descriptions.

The westerly flow in the mid-latitude jet streams has both horizontal and vertical mean-flow shears. The vertical shear of the mean flow is in thermal wind balance with a horizontal temperature gradient, providing available potential energy for baroclinic instability.[f] The baroclinic instability process is associated with a poleward and upward transport of heat. This completes the description of the poleward heat transport[g] in the atmosphere that we earlier inferred must exist to explain the observed latitudinal variation of outgoing versus incoming radiation: In the tropics heat is transported poleward by the Hadley circulation and at higher latitudes baroclinic eddies are mainly responsible for the heat transport.

Analysis of idealized flows (the classic example is known as the "Eady problem") can provide an indication of the typical properties of baroclinic instability. The reader is referred to Holton (2004) for more detail. Baroclinic instability is a wave instability. The wavelength at which the instability is greatest in the Eady problem is $L_{\max} \approx 3.9\, L_R$, where $L_R = NH/f_0$ is the "Rossby radius," and the growth rate is $\sigma \approx 0.3U/L_R$. Applying typical values for the atmosphere ($H \sim 10\,\mathrm{km}$, $U \sim 10\,\mathrm{ms}^{-1}$ and $N \sim 10^{-2}\,\mathrm{s}^{-1}$) gives $L_{\max} \approx 4000\,\mathrm{km}$ and $\omega \approx 0.26$ day^{-1}. For the ocean, $H \sim 1\,\mathrm{km}$, $U \sim 0.1\,\mathrm{ms}^{-1}$ and $N \sim 10^{-2}\,\mathrm{s}^{-1}$, giving $L_{\max} \approx 400\,\mathrm{km}$ and $\omega \approx 0.026$ day^{-1}. The atmospheric values are broadly consistent with the observed spatial and growth rates of mid-latitude weather systems. In the ocean, the simple scenario on which these values are based is not quantitatively applicable, but the values give a qualitative sense of the scale and growth rate of instabilities in the ocean relative to the atmosphere. The regions of greatest baroclinic instability (eddy activity) in the oceans are the major currents: The Gulf Stream, the Kuroshio Current, and the Antarctic Circumpolar Current.

3.6. *Ocean Circulation*

In the ocean there is a global-scale "meridional overturning circulation," a system of surface and deep currents that encompasses all basins.[h] This

[f] The horizontal shear in the flow allows for a second instability, known as "barotropic instability" which extracts kinetic energy from the mean-flow field.

[g] In addition to transporting heat, the general circulation also transports angular momentum. The Hadley circulation is associated with a poleward transport of westerly momentum at upper levels and equatorward transport of easterly momentum at lower levels. Mid-latitude eddies also transport angular momentum, mostly transporting westerly momentum poleward.

[h] The meridional overturning circulation used to be called the "thermohaline circulation," due to role of density differences, controlled by temperature and salinity changes,

circulation transports heat, and also salt, carbon nutrients, and other substances around the globe, and connects the surface ocean and atmosphere with the huge reservoirs of the deep ocean. As such, it is of critical importance to the global climate system. We have discussed above the requirement for a poleward heat transport in the atmosphere and/or ocean to explain the observed incoming/outgoing radiation profiles. Detailed calculations indicate that the bulk of the required transport is carried by the atmosphere in the middle and high latitudes, but the ocean makes up a considerable fraction, particularly in the tropics. Heat is transported poleward by the ocean in the overturning circulation if waters moving poleward are compensated by equatorward flow at colder temperatures. In the Atlantic, heat transport is northward everywhere, however, in the Pacific the heat transport is directed poleward in both hemispheres and the Indian Ocean provides a poleward transport in the southern hemisphere. The net heat transport is poleward in each hemisphere.

Ocean surface waters are only dense enough to sink down to the deep abyss at a few key locations, in particular the northern North Atlantic and around Antarctica. Deep ocean convection occurs only in these cold, high-latitude regions, where the internal stratification is small and surface density can increase through direct cooling/evaporation (warm surface water flows northward in the North Atlantic, and in winter, the wind cools and evaporates the water, and North Atlantic Deep Water is formed) or brine rejection in sea ice formation (Antarctic Bottom Water is formed around the Antarctic coast, in particular the Ross and Weddell Seas). North Atlantic Deep Water flows south as a deep western boundary current and eventually enters the Southern Ocean where it mixes with other water masses to become Circumpolar Deep Water. Ultimately the deep water is brought up to the surface by vertical mixing (tides and winds are the primary sources of energy for this) and by the overturning circulation in the Southern Ocean.

The surface ocean currents are dominated by closed circulation patterns known as "gyres." In the northern hemisphere there are anticyclonic gyres in the subtropics of the Pacific and Atlantic with eastward flow at mid-latitudes and westward flow at the equator. The current speed at the interior of these gyres is $\lesssim 10\ \mathrm{cms}^{-1}$, but at the western edge there are strong poleward currents (Kuroshio in the Pacific and Gulf Stream in the Atlantic)

in determining the flow. However, this term is no longer in common oceanographic use since it does not reflect the fact that the winds play a primary role in driving the circulation, and the tides and internal mixing processes are also important for determining the interior density distribution.

with speeds $\gtrsim$100 cms^{-1}. In the subpolar regions of the north Pacific and Atlantic there are cyclonic gyres with southward flowing western boundary currents (Oyashio Current in the Pacific and Labrador Current in the Atlantic). In the tropics the flow is largely zonal (i.e., east–west). Just north of the equator in each ocean basin is an eastward flowing current (counter to the prevailing winds), known as the Equatorial Counter Current, flanked by westward flowing currents to the north and south. In the southern hemisphere, subtropical gyres are also evident, however, the flow is dominated by the Antarctic Circumpolar Current which has typical surface currents of ∼30 cms^{-1}. See Marshall and Plumb (2008) for figures indicating the surface currents.

The sea surface is higher (with respect to the "geoid") to the south of the eastward flowing Gulf Stream than to the north of it, resulting in a pressure gradient force directed northward to balance the southward Coriolis force (the sea surface is about 1 m higher in the subtropics than at the pole). A major contributor to the spatial variation in height of the ocean (known as the "steric effect") is the expansion (contraction) of water columns that are warm (cold) relative to their surroundings. The sea surface is high over the warm subtropical gyres and low over the cool subpolar gyres. Pressure gradients associated with sea-surface tilt are largely compensated by vertical thermocline undulations of about 400 m, ensuring that abyssal pressure gradients and geostrophic flows are much weaker than at the surface.

As noted earlier, winds blowing over the ocean surface exert a stress on it and directly drive ocean circulations close to the surface in the so-called "Ekman layer." At the surface, $z = 0$, the stress is $\tau(0) = \tau_{\text{wind}}$ and this decays over the depth $\delta \simeq 10 - 100\,\text{m}$ of the Ekman layer so $\tau(-\delta) = 0$. The ageostrophic component of motion, $\mathbf{u}_{ag}$, is obtained by substituting the force arising from the wind (2.13) into the equation for motion near a boundary (3.16), giving $f\hat{\mathbf{z}} \times \mathbf{u}_{ag} = \frac{1}{\rho_{\text{ref}}}\frac{\partial \tau}{\partial z}$. By integrating this equation over the depth of the Ekman layer, it can be shown that the lateral mass transport over the layer is given by

$$\mathbf{M}_{Ek} \equiv \int_{-\delta}^{0} \rho_{\text{ref}}\mathbf{u}_{ag}dz = \frac{\tau_{\text{wind}} \times \hat{\mathbf{z}}}{f}. \tag{3.22}$$

Thus the mass transport in the Ekman layer is directed to the right of the wind in the northern hemisphere.[i] Winds at the surface are 45° to the left

[i]Further analysis indicates that the horizontal currents are expected to spiral anticyclonicly with depth from an initial direction of 45° to the right (left) of the wind in the northern (southern) hemisphere, and to decay exponentially in magnitude. Similar

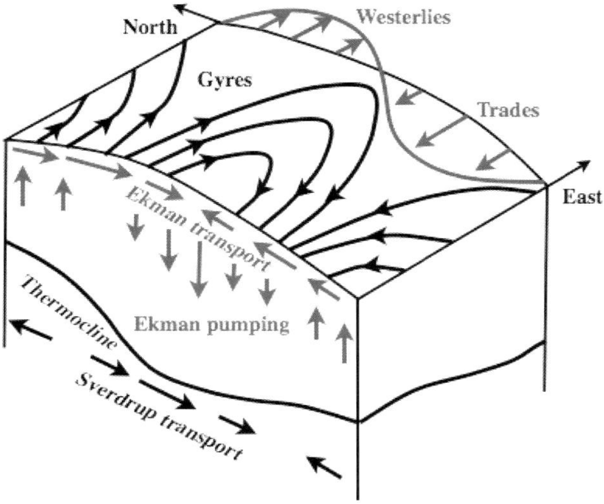

Fig. 7. Schematic (following L. Talley) indicating the Ekman and Sverdrup transports associated with wind-driven ocean gyres.

of the winds aloft, and surface ocean currents are 45° to the right of the wind at the surface. Therefore we expect currents at the sea surface to be nearly in the direction of winds above the planetary boundary layer, which are parallel to lines of constant pressure (see Fig. 7).

In the anticyclonic subtropical gyres, Ekman transport results in convergence. Mass conservation then implies downwelling, or Ekman pumping. In the cyclonic subpolar gyres there is divergence and Ekman suction. The vertical velocity at the surface is zero and so integrating Eq. (3.17) over the Ekman layer gives a vertical velocity at the base of the Ekman layer w_{Ek} of

$$w_{Ek} = \frac{1}{\rho_{\text{ref}}} \nabla_h \cdot \mathbf{M}_{Ek} \tag{3.23}$$

$$= \frac{1}{\rho_{\text{ref}}} \left(\frac{\partial}{\partial x} \frac{\tau_{\text{wind}_y}}{f} - \frac{\partial}{\partial y} \frac{\tau_{\text{wind}_x}}{f} \right). \tag{3.24}$$

In the interior, the flow is in geostrophic balance and the vertical/meridional currents are related by Eq. (3.15). Thus there is expected to

Ekman spirals exist at the bottom of the ocean and the atmosphere. In this case the direction of the flow close to the boundary is 45° to the left (right) of the flow outside the boundary layer in the northern (southern) hemisphere and the direction of rotation is anticyclonic with distance above the bottom.

be a equatorward (poleward) component to the horizontal velocity where $w_{Ek} < 0$ ($w_{Ek} > 0$) with typical values of about 1 cm/s. This means that in the subtropical gyres, the interior flow is equatorward.

The incompressibility condition together with Eq. (3.16) with the forcing coming from the wind can be used to show that the meridional velocity v is given by $\beta v = f\frac{\partial w}{\partial z} + \frac{1}{\rho_{\text{ref}}}\frac{\partial}{\partial z}\left(\frac{\partial \tau_y}{\partial x} - \frac{\partial \tau_x}{\partial y}\right)$. The full depth-integrated flow V can be obtained by integrating this meridional velocity from the bottom of the ocean ($z = -D$, $w = 0$, and $\tau = 0$) to the surface ($z = 0$, $w = 0$, $\tau = \tau_{\text{wind}}$) to give

$$\beta V = \frac{1}{\rho_{\text{ref}}}\hat{\mathbf{z}} \cdot \nabla \times \tau_{\text{wind}}. \tag{3.25}$$

This is known as "Sverdrup balance." The depth-integrated meridional transport is related to the curl of the wind stress and this dictates the sense of motion in the subpolar and subtropical gyres (see Fig. 7). In the Southern Ocean, at levels where no topography exists to support zonal pressure gradients, there can be no mean meridional geostrophic flow and therefore the above Sverdrup approximation does not apply (see Rintoul *et al.*, 2001).

The motion in the interior of the ocean gyres can be understood by considering conservation of potential vorticity, which for a homogeneous fluid is given by $PV = (f + \zeta)/H$. Ekman pumping (suction) squashes (stretches) water columns in the interior, however, by moving equatorward (poleward) these columns are able to conserve their potential vorticity. This is the Sverdrup flow, which is equatorward in the subtropical gyres and poleward in the subpolar gyres and is in geostrophic balance with an east–west pressure gradient. The gyre circulations are closed by strong, narrow boundary currents on the western boundaries where friction means the flow is no longer in geostrophic balance. These currents are on the western, rather than eastern side because, in the subtropical (subpolar) gyres where the Sverdrup transport is equatorward (poleward), the wind puts anticyclonic (cyclonic) vorticity into the ocean which is removed by friction at the boundary as the flow returns poleward (equatorward) on the western side.

3.7. *Tropical Ocean-Atmosphere Coupling*

We now turn our attention to tropical dynamics. The dynamics of the ocean and atmosphere in the tropics are highly coupled. On interannual timescales, the upper ocean responds to the past history of the wind stress and the atmospheric circulation is largely determined by the distribution of sea surface temperatures (SSTs). Latent heat release is the primary energy source for the atmospheric circulation.

The intertropical convergence zone (ITCZ) is a narrow band of deep convective clouds near the equator. Much of the water vapor needed to maintain the convection is supplied by the converging trade winds in the lower troposphere. The convective heating produces large-scale mid-tropospheric temperature perturbations and associated surface and upper level pressure perturbations, which maintain the low-level flow. The zonal mean of the vertical mass flux associated with the ITCZ constitutes the upward mass flux of the mean Hadley circulation. During the course of a year, the pattern of solar forcing migrates, north in northern summer, south in southern summer. Thus the entire Hadley circulation shifts seasonally such that the upwelling branch and associated rainfall are found on the summer side of the equator. The degree to which this happens is strongly controlled by local geography. Seasonal variations over the oceans, whose temperature varies relatively little through the year, are weak, while they are much stronger over land. The migration of the main area of rainfall is most dramatic in the region of the Indian Ocean, where intense rain moves onto the Asian continent during the summer monsoon.

There are strong longitudinal variations associated with variations in the tropical SSTs due mainly to the effects of the wind-driven ocean currents. Several overturning cells exist along the equator associated with diabatic heating over equatorial Africa, Central and South America, and Indonesia. The dominant cell is over the equatorial Pacific and is called the "Walker circulation." The surface pressure is low in the western Pacific and high in the eastern Pacific resulting in a pressure gradient that drives mean surface easterlies (the Coriolis force is negligible in this region). The easterlies provide a moisture source for the convection in the western Pacific in addition to that provided by the high evaporation rates caused by the warm SSTs there (the "warm pool"). The atmospheric circulation is closed by descent over the cooler water to the east.

From Eq. (3.16) with the frictional term being given by the wind stress (2.13), it can be shown that in the tropical region, where $f \simeq \beta y$, the wind stress gives rise to a meridional flow in the ocean:

$$-\beta y v = \frac{1}{\rho_{\rm ref}} \frac{\partial \tau_x}{\partial z}. \tag{3.26}$$

Thus a westward wind stress across the Pacific gives rise to poleward flows either side of the equator in the oceanic Ekman layer, which by continuity drive upwelling near the equator. In addition, since the Pacific is bounded

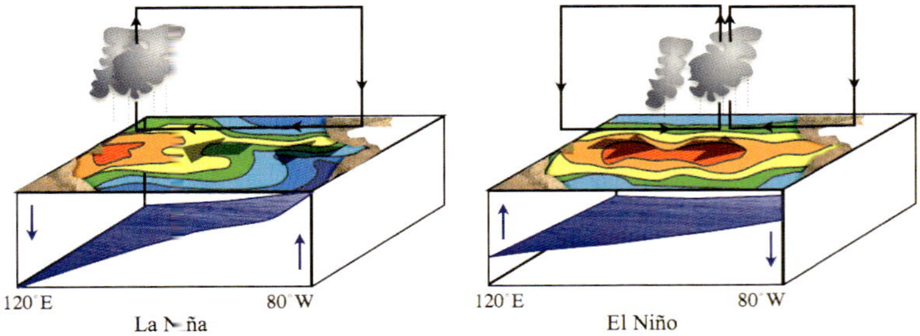

Fig. 8. Schematic (following NOAA/PMEL) of the typical atmosphere/ocean conditions during La Niña and El Niño events.

to the east and west, the westward wind stress results in the thermocline being deeper in the west than the east. Thus the cold (and nutrient-rich) deep water upwells close to the surface in the east, cooling the SSTs there, whereas in the west the cold water does not reach the surface and the SSTs remain warm. The upwelled region is associated with a geostrophic current in the direction of the winds, since in the limit $y \to 0$, Eq. (3.7) gives $\beta u = -\rho_{\text{ref}}^{-1}\partial^2 p/\partial y$. The deepening of the thermocline causes the sea surface to be higher in the west, assuming that flow below the thermocline is weak. Thus there is an eastward pressure gradient along the equator in the surface layers to a depth of a few hundred meters. Away from the equator, below the surface, this is balanced by an equatorward geostrophic flow. At the equator, where $f = 0$, there is a current directly down the pressure gradient, i.e., to the east, the Equatorial Counter Current. At the surface at the equator, the eastward pressure gradient is balanced by the wind stress τ_x.

The east–west pressure gradient across the Pacific undergoes irregular interannual variations with a period in the range is two to seven years. This "see-saw" in pressure, and its associated patterns of wind, temperature and precipitation is called the "Southern Oscillation." An index of the oscillation (the Southern Oscillation Index, SOI) can be obtained by considering the pressure difference between Tahiti in the central Pacific and Darwin, Australia in the western Pacific. The negative phase of the SOI represents below-normal sea level pressure at Tahiti and above-normal sea level pressure at Darwin and vice versa for the positive phase. SSTs in the eastern Pacific are negatively correlated with the SOI, i.e., warm SSTs coincide with anomalously high pressure in the west and low in the east. The phase of the oscillation with anomalously warm SSTs is known as "El Niño," and the phase with anomalously cold SSTs is known as "La Niña." The entire coupled atmosphere-ocean response is known as the El Niño Southern Oscillation (ENSO).

During an El Niño event, the warm pool is shifted eastward from the Indonesian region, and with it the region of greatest convection and the associated atmospheric circulation pattern. The adjustment of the Walker circulation, which corresponds to a negative SOI, leads to a weakening of the easterly trade winds, reinforcing the eastward shift of the warm SSTs. The sea surface slope diminishes, raising sea levels in the east Pacific while lowering those in the west. Ekman-driven upwelling reduces, allowing SSTs to increase (see Fig. 8). The ocean adjusts over the entire basin to a local anomaly in the forcing in the western Pacific through the production of internal waves in the upper ocean. These move both east and west from the anomaly. Eastward propagation of a wave of depression on the thermocline deepens the thermocline in the east Pacific about two months later, relaxing the basin-wide slope of the thermocline. In the coupled system, the ocean forces the atmospheric circulation (through the response to changed boundary conditions associated with the El Niño SST fluctuations) and the atmosphere forces the oceanic behavior (through the response to changed wind stress distribution associated with the Southern Oscillation).

Wave propagation in the ocean is key to the temporal evolution of an El Niño event. The SST anomaly in the western Pacific gives rise to a westerly wind anomaly over the central Pacific which excites oceanic waves. An equatorial Kelvin wave propagates rapidly to the east deepening the thermocline and reinforcing the initial warm SST anomaly in a positive feedback. Equatorial Rossby waves which propagate slowly (with the magnitude of the group velocity being about a third of the Kelvin wave) to the west are also excited. When the Kelvin wave hits the coast on the eastern side (after about two months), its energy feeds westward-propagating Rossby waves and poleward-propagating coastal Kelvin waves. On the western side, when the Rossby waves hit the coast, some energy feeds an eastward propagating Kelvin wave which raises thermocline back towards its original location, reducing the initial SST anomaly and providing a negative feedback. The propagation times of the different waves means that the negative feedback is delayed. A simple model of such a "delayed oscillator" with realistic parameters gives oscillations in the period range of three to four years (see Holton, 2004, for more details).

Recently a new distinct mode of variability has been identified, associated with a particular pattern of central Pacific temperature anomalies. This mode is known as El Niño Modoki and is discussed in detail in the chapter "Dynamics of the Indian and Pacific Ocean" in this volume.

4. Conclusions

In this chapter we have demonstrated how basic fluid dynamical principles can be used to understand the essential elements that determine the Earth's climate and a range of weather and climate processes including mid-latitude storms, ocean gyre circulations, and the El Niño Southern Oscillation. Some of these processes will be examined in more detail in later chapters. Importantly, the equations of fluid dynamics provide the core of numerical models that can be used to predict future weather and climate. We have discussed how changes to the atmospheric concentrations of greenhouse gases lead to changes in the radiative forcing. This in turn leads to changes to the dynamics of the atmosphere and, via changes to the surface forcing, of the oceans. As a consequence, the dynamical processes we have described in this chapter may be subject to change in the future, for example mid-latitude storms, the monsoons, El Niño, and the overturning circulation of the ocean.

References

Andrews, D. G. (2010). *An Introduction to Atmospheric Physics*, 2nd edn. (Cambridge University Press, Cambridge, UK).

Gill, E. A. (1982). *Atmosphere-Ocean Dynamics* (Academic Press, New York).

Holton, J. R. (2004). *An Introduction to Dynamic Meteorology*, 4th edn. (Academic Press, New York).

Kiehl, J. T. and Trenberth, K. E. (1997). Earth's annual global mean energy budget, *Bull. Amer. Meteor. Soc.* **78**, pp. 197–208.

Manabe, S. and Wetherald, R. T. (1967). Thermal equilibrium of the atmosphere with a given distribution of relative humidity, *J. Atmos. Sci.* **24**, 3, pp. 241–25[illegible].

Marshall, J. and Plumb, R. A. (2008). *Atmosphere, Ocean and Climate Dynamics: An Introductory Text* (Elsevier Academic, Singapore), pp. 271–302.

Pierrehumbert, R. T. (2010). *Principles of Planetary Climate* (Cambridge University Press, Cambridge, UK).

Rintoul, S., Hughes, C. and Olbers, D. (2001). Ocean circulation and climate, in G. Siedler, J. Church and J. Gould (eds.), *The Antarctic Circumpolar Current System* (Academic Press, New York).

Solomon, S., Qin, D., Manning, M., *et al.* (2007). *Contribution of Working Group I to the Fourth Assessment Report of the Intergovernmental Panel on Climate Change* (Cambridge University Press, Cambridge, U.K. and New York).

Vallis, G. K. (2006). *Atmospheric and Oceanic Fluid Dynamics* (Cambridge University Press, Cambridge, UK).

Chapter 4

Dynamics of the Indian and Pacific Oceans

Swadhin Behera[1,2,4] and Toshio Yamagata[2,3]

[1] *Climate Variation Predictability and Applicability Research, Research Institute for Global Change/JAMSTEC, Yokohama, Japan*
[2] *Application Laboratory, JAMSTEC, Yokohama, Japan*
[3] *University of Tokyo, Tokyo, Japan*
[4] *behera@janstec.go.jp*

Tropical oceans play a major role in natural variability of the world climate. Anomalous coupled ocean-atmosphere phenomena generated in the tropical oceans produce changes in global atmospheric and oceanic circulation that influence regional climate conditions even in remote regions. On the inter-annual timescale, the El Niño Southern Oscillation (ENSO) of the tropical Pacific Ocean is known as a typical example of such phenomena and has received worldwide attention because of its enormous societal impact. Recently a new mode of variability has been identified with a distinct central Pacific warming pattern. This "El Niño Modoki" mode involves ocean-atmosphere coupled processes, indicating the existence of a unique atmospheric component during the evolution, analogous to the Southern Oscillation in the case of El Niño. The Modoki's impact on world climate is very different from that of ENSO. Interestingly, the Modoki's influences over regions such as the Far East including Japan and the western coast of the USA are almost opposite to those of the conventional ENSO. Modoki events have been more frequent and persistent during recent decades. Inter-annual variability originating in the tropical Indian Ocean includes an ocean-atmosphere coupled phenomenon known as the Indian Ocean Dipole (IOD). The IOD has a unique teleconnection pattern that implies regional climate variability and thus societal impacts in various parts of the globe. These phenomena are described and discussed in detail in this chapter.

1. Introduction

The Earth's climate fluctuates around a normal state, which is generally determined by an average of atmospheric conditions over a 30-year period.

In simple terms this means that the mean climate at a given location is the average weather condition over a long period of time. The significant climate fluctuations that we commonly refer to — while describing unusual characteristics of a season — are on the timescales of years to decades. Any fluctuation beyond these timescales is normally described as climate change, which may arise due to both natural and anthropogenic (human-induced) factors. Thus, climate change is the change in the background state that anchors the climate fluctuations.

The natural elements of climate include atmosphere, hydrosphere, cryosphere (ice), biosphere, and geosphere. Besides these natural elements, the climate system is also influenced by anthropogenic elements arising from increasing use of natural resources since the industrial revolution. Society's awareness of rising global temperature has stimulated intense interest in research on anthropogenic climate change. The global warming related to this anthropogenic climate change remains one of the big challenges for society to manage. However, short-term variations in climate, which are directly related to abnormal weather, extreme phenomena, and associated socio-economic impacts, pose no less of a challenge for seasonal to inter-annual climate predictions. Such short-term climate predictions are required for planning a wide range of weather and climate-sensitive issues and most importantly for the development of adaptation policies.

Our natural climate is made up of several components, each of which is complex and capable of altering the course of the climate system on which civilization is dependent. We must take into account the fact that the state of the atmosphere, often equated with the state of climate, is influenced by numerous processes that are internal to atmosphere as well as arising from interactions with oceans, ice, and ecosystems. On a basic level, the seasonal variation of climate is determined by the 23.4° tilt of the Earth's axis of rotation. This tilt causes the Northern Hemisphere to come closer to the Sun and gain maximum solar energy during boreal summer. Six months later, the Northern Hemisphere is tilted away from the Sun and experiences winter while the Southern Hemisphere experiences the summer conditions.

Leaving apart this north–south variation of seasonal climate, the regional variations of the mean climatic conditions are decided mainly by internal dynamics and physics. For example, the same amount of heat is received over the whole of the tropical oceans, yet ocean upwellings cause cooler sea surface temperature (SST) in the eastern Pacific and along some other coasts (Figs. 1(a), 1(b)). The upwelling is caused by prevailing local winds over those regions, the effect of the Earth's rotation known as the Coriolis effect, and friction in the surface boundary layer of the ocean — all of which cause a net transport of upper ocean water perpendicular to

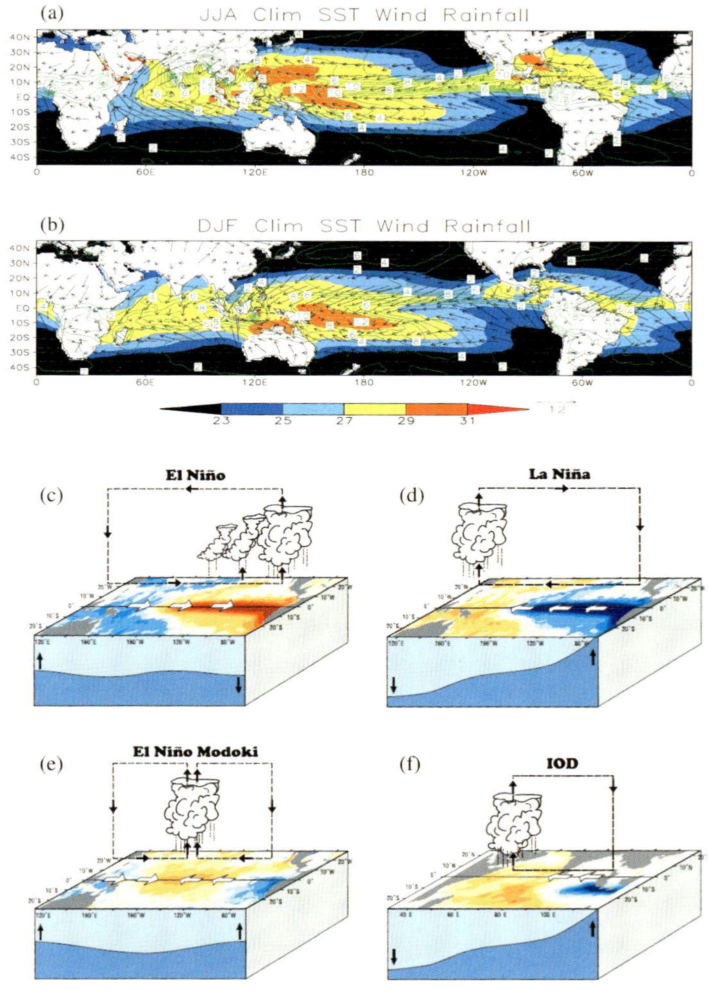

Fig. 1. Seasonal SST, surface wind and rainfall for (a) June–August and (b) December–February. Schematic diagrams of ocean-atmosphere conditions related to (c) El Niño, (d) La Niña, (e) El Niño Modoki, and (f) Positive Indian Ocean Dipole (pIOD).

the winds. This water transport known as Ekman transport is to the right of the wind direction in the Northern Hemisphere, and to the left in the Southern Hemisphere.

The tropical oceans also play a vital role in the global heat budget. Approximately one third of the net solar radiation (about 100 Wm^{-2}) penetrates through the ocean surface, which causes the tropical oceans to be constantly heated by the atmosphere from the surface. The upper 10 m of oceans has the same mass as that of the entire atmosphere and the upper 4 m has a similar capacity to store heat. The tropical oceans circulate the heat through a number of oceanic processes. Upwelling, which brings subsurface waters to the surface, enables the equatorial ocean to absorb atmospheric heat flux. These warm waters are subsequently transported to higher latitudes by the western boundary currents that are largely driven by the Earth's circulation regime. This warm water formation and escape process in the upper tropical oceans is the opposite of the cold water formation process in the polar oceans, both of which are critical for the global heat budget.

In the atmosphere, solar heating in the equatorial region causes air to rise and move poleward. The warm air from the equator begins to cool and sink at about 30° north and south of the equator. The subtropical deserts are associated with this sinking in the atmosphere. Between 30° latitude and the equator, most of the sinking air moves back to the equator and the associated winds are commonly known as the trade winds. The rest of the sinking air moves toward the poles. These air movements are affected by the Coriolis effect, and in the tropics the trade winds appear to curve to the west (Fig. 1(b)) because of this effect.

The trade winds blowing over warm equatorial oceans cause lower atmospheric moisture convergence in what is known as the intertropical convergence zone (ITCZ). The ITCZ often appears disconnected over land and ocean. It is associated with deep atmospheric convection, heavy precipitation (contour lines in Figs. 1(a), 1(b)), and weak mean wind speeds. Together with the atmospheric circulation, the tropical ocean circulation and dynamical conditions create a unique environment for ocean-atmosphere interactions, which are so critically important for global climate variability and change. Because of its large heat capacity, the tropical ocean provides a "long-term memory" for the atmosphere, while in turn the atmosphere helps to drive the slow variations in the ocean through ocean-atmosphere interactions.

2. The Tropical Climate Modes

The regional distributions of seasonal SST and surface winds are determined by a delicate balance among various elements of ocean and atmosphere. For example, the seasonal trade winds and associated Walker circulation in the tropical Pacific cause warm water to pile up on the western side and cold water to upwell on the eastern side (Figs. 1(a), 1(b)). Differences in water masses on either side of the Pacific cause a slope in the thermocline (the depth of water sensitive to climate variations). This slope is maintained by the prevailing winds. The balance between the thermocline slope and the prevailing winds is disturbed by the intermittent development of anomalous ocean-atmosphere coupled modes. In the Pacific, the dominant coupled mode is known as the El Niño Southern Oscillation (ENSO).

2.1. *The ENSO*

El Niño is traditionally known as an abnormal warming of sea surface in the eastern tropical Pacific. This warming is the oceanic component of the ENSO phenomenon. The atmospheric component is the Southern Oscillation defined as the sea level pressure difference between Tahiti and Darwin, and captures the see-saw in the atmospheric sea level pressure between the eastern and western tropical Pacific. Bjerknes (1969) suggested that the El Niño and the Southern Oscillation are in fact just two different aspects of the same phenomenon, and demonstrated a remarkable correlation between Darwin's atmospheric pressure and the water temperature off Peru, two locations separated by the vast span of the Pacific Ocean. He further hypothesized that ocean-atmosphere interaction is at the heart of the ENSO phenomenon (Figs. 1(c), 1(d)) and suggested that an initial change in the ocean could affect the atmospheric conditions, which would in turn induce changes in oceanic conditions to reinforce the initial anomalies. For example, if SSTs in the equatorial eastern Pacific become anomalously warm, it will reduce the east–west gradient in SST. The atmosphere will respond by reducing the east–west gradient in sea level pressure, consequently relaxing the strength of the easterly trade winds that are important to maintain the zonal thermocline slope and the east–west distribution of heat content. The relaxation of the easterly winds in turn will cause an eastward surge of warm water along the equator, positively reinforcing the initial warm SST anomalies. Thus, positive ocean-atmosphere feedback of Bjerknes type amplifies small initial perturbations into large anomalies and eventually evolves as an El Niño event (Fig. 1(c)). The canonical picture of ENSO based on a variety of observations is basically consistent with the Bjerknes hypothesis.

The peak SST warming in the eastern equatorial Pacific associated with an El Niño event is generally observed in December and January. This property has been referred to as the seasonal phase-locking of ENSO to the annual cycle. Furthermore, ENSO events typically last 12–18 months and occur every 2–7 years. While Bjerknes' mechanism explains why the system has two favored states (warm and cold) it does not explain why there is an oscillation between them. That is broadly explained by the equatorial ocean dynamics, involving the depth of the thermocline (or the amount of warm water above the thermocline). The changes in the depth of this warm layer associated with ENSO are a consequence of wind-driven ocean dynamics by which the wind and SST changes in the ENSO cycle are tightly locked together. It is observed that the sluggish thermocline changes are often not in phase with that of SST and wind, and this delay in the response of the thermocline is important for the slow propagation of the ENSO signal.

The equatorial waveguide plays a crucial role in giving rise to the quasi-oscillatory nature of ENSO and many studies have investigated it by simulating the coupled tropical ocean-atmosphere system with models of varying complexity (e.g., Cane and Zebiak, 1985; Philander, 1990; McCreary and Anderson, 1991; Neelin *et al.*, 1998; Chang *et al.*, 2006). The oceanic Kelvin and Rossby waves[a] (Fig. 2) help to propagate energy and momentum received by the ocean from the wind stress. The propagation speeds of similar atmospheric waves are far greater than that of their oceanic counterparts. Therefore, the adjustment timescale of the tropical atmosphere to changes in SST is much shorter (ten days or less) than the adjustment timescale of the equatorial ocean to changes in wind stress (approximately six months). The short adjustment time of the atmosphere supports the assumption that the atmosphere is in a statistical equilibrium with the SST on timescales longer than a few months. Thus, the memory of the state of the climate system primarily resides in the ocean. On the other hand, oceanic Kelvin and Rossby waves can be strongly modified by the

[a]The circulations of the atmosphere and ocean with large spatial scales are influenced by the Earth's rotation or the Coriolis effect and dominantly appear as waves. Most of these waves are known as Rossby waves (or planetary waves). Kelvin waves are found along the equator and along coastlines, where the Coriolis acceleration vanishes. For detailed explanations of these waves, please refer to Gill (1982).

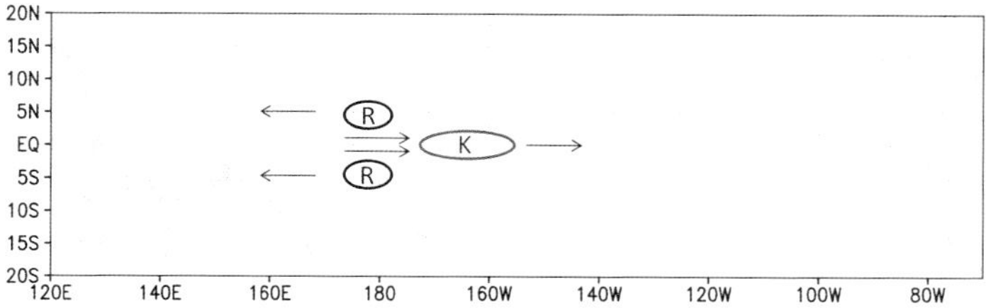

Fig. 2. Schematic diagram showing westward propagating Rossby waves and eastward propagating equatorial Kelvin waves.

air-sea coupling. The Bjerknes feedback can destabilize these waves, giving rise to unstable coupled modes that resemble the slow westward propagating oceanic Rossby mode and the eastward propagating oceanic Kelvin mode. In fact, the coupling between the atmosphere and ocean generates a breed of modes whose characteristics depend on the timescale of dynamical adjustment of the ocean relative to the timescale of the SST anomaly, which is related to the air-sea coupling.

Stability analysis of a simple ENSO model linearized around a given mean state reveals a variety of structures of the coupled modes in a parameter space. The coupled mode most relevant to ENSO appears to reside in a parameter regime where the timescales associated with the local air-sea interaction are comparable to the dynamical adjustment time of the tropical Pacific Ocean. The evolution of the coupled mode in this parameter regime can be described in two phases. During the development phase, the Bjerknes positive feedback dominates and causes the anomalies to grow. During the decay phase, the equatorial wave adjustment process of the ocean delays the termination through a negative feedback. Rossby wave packets carry off equatorial thermocline anomalies of opposite sign to the equatorial anomaly generated by the Bjerknes feedback to the western boundary at which the waves are reflected into equatorial Kelvin waves and the thermocline anomalies propagate eastward (Fig. 2) along the equator. They then counteract the Bjerknes positive feedback and cause the system to turn from warm to cold states and back again. The timescale that is associated with the ocean wave adjustment imparts the "memory" of the coupled system that is essential for the oscillations in this ENSO paradigm.

The above description of ENSO physics is based on a linear, deterministic framework. It offers a basic understanding of the evolution and duration, as well as the oscillatory nature of ENSO events. However, the detailed features of ENSO events can vary considerably from event to event, including when and where the initial warming starts and whether the initial signal propagates eastward or westward. The hypotheses that are used to explain the causes of ENSO irregularity can be broadly grouped into three general categories. The most widely used hypothesis of ENSO variability is often referred to as the delayed oscillator. It relates to the underlying dynamics and emphasizes the oceanic wave propagation. The second category regards the recharge-discharge of the equatorial ocean heat content as the essence of the ENSO oscillation. This hypothesis highlights the importance of nonlinearity (Timmermann *et al.*, 2003) that arises from strong air-sea feedback in an unstable dynamic region. In this regime, not only can ENSO be described as a self-sustained oscillator but it can also interact nonlinearly with either the annual cycle (e.g., (Jin *et al.*, 1994)) or other coupled modes (e.g., (Mantua and Battisti, 1995)) giving rise to deterministic chaos. The loss of predictability in this regime is primarily due to uncertainty in the initial conditions.

The third category of ENSO hypothesis expresses a somewhat opposite view to the second category. The ENSO coupled mode is imagined to be a stable damped regime, and thus the ENSO cycle cannot be self-sustained without external noise forcing (e.g., (Flügel *et al.*, 2004)). Weather noise generated by the internal dynamics of the atmosphere plays a fundamental role in not only giving rise to ENSO irregularity but also in maintaining ENSO variance. In between these two extreme viewpoints lies the hypothesis of the first category, which assumes ENSO to be self-sustained (due to weak nonlinearity) and periodic (Schopf and Suarez, 1988). In this regime, ENSO's behavior is governed by the temporal characteristics of the single most dominant coupled mode together with the influence arising from weather noise (e.g., (Chang *et al.*, 2006)). And the predictability comes from the oscillatory nature of the dominant mode (Chen *et al.*, 2004) while the loss of predictability is primarily due to noise.

2.2. *The ENSO Modoki*

El Niño Modoki has recently been identified as a coupled ocean-atmosphere phenomenon in the tropical Pacific Ocean and has been shown to be quite different from the canonical El Niño in terms of its spatial and temporal characteristics as well as its teleconnection patterns (Ashok *et al.*, 2007; Weng *et al.*, 2007; Ashok and Yamagata, 2009). The definition of "El Niño" has evolved over several decades. Traditionally the term "El Niño" was used for the canonical El Niño associated with warming in the eastern tropical Pacific. This definition is sometimes generalized by considering the warming in the central tropical Pacific as El Niño. However, as we realize now, this broad definition may mix up the canonical El Niño with the El Niño Modoki (Ashok *et al.*, 2007; Weng *et al.*, 2007; Ashok and Yamagata, 2009).

The importance of studying the difference between El Niño Modoki (Fig. 1(e)) and the canonical El Niño (Fig. 1(c)) lies in their unique influences on the surrounding climate. It is recognized that the main characteristics of these two phenomena and their associated climate impacts during recent boreal winters are fundamentally different. Therefore, mixing El Niño Modoki signal with that of the canonical El Niño blurs their characteristic teleconnections and singular impacts on regional climate in the Pacific Rim. Differences in zonal SST gradients in the tropical Pacific associated with those two phenomena (Fig. 3) cause disparities in the Walker circulation. Furthermore, the regional meridional circulations that link these tropical phenomena with the subtropical/extratropical systems will generate different teleconnection patterns associated with the variations in the tropical Walker circulation. Thus, atmospheric circulations arising from variations in zonal SST gradients between the canonical El Niño and the El Niño Modoki in the tropical Pacific cause different types of anomalous climate conditions in the Pacific Rim (Ashok *et al.*, 2007; Weng *et al.*, 2007, 2009a,b).

In Empirical Orthogonal Function/Principal Component (EOF/PC) analysis[b] of the tropical Pacific SST anomalies (derived from Hadley Centre Global Sea Ice and Sea Surface Temperature Analyses), the EOF1 pattern captures the essential features of El Niño. This mode explains about 45% of the tropical Pacific SST variability for the period 1979–2004. The EOF2, which explains 12% of the SST variability for the corresponding period, captures a zonal tripole pattern in the tropical region. In this pattern both eastern and western tropical Pacific SST anomalies have loadings of the same sign, while those of the central tropical Pacific are of opposite sign (cf. Ashok *et al.*, 2007). In higher latitudes, the positive loadings in the central equatorial Pacific spread eastward in both hemispheres, and this boomerang pattern straddles the tongue of negative loadings in the equatorial eastern Pacific (Weng *et al.*, 2007, 2009a). A typical pattern

[b]EOF/PC analysis is a statistical analysis method employed to extract the dominant modes of variability from a dataset. It produces a set of structures in the spatial dimension (EOFs) and a set of corresponding structures in the time dimension (PCs).

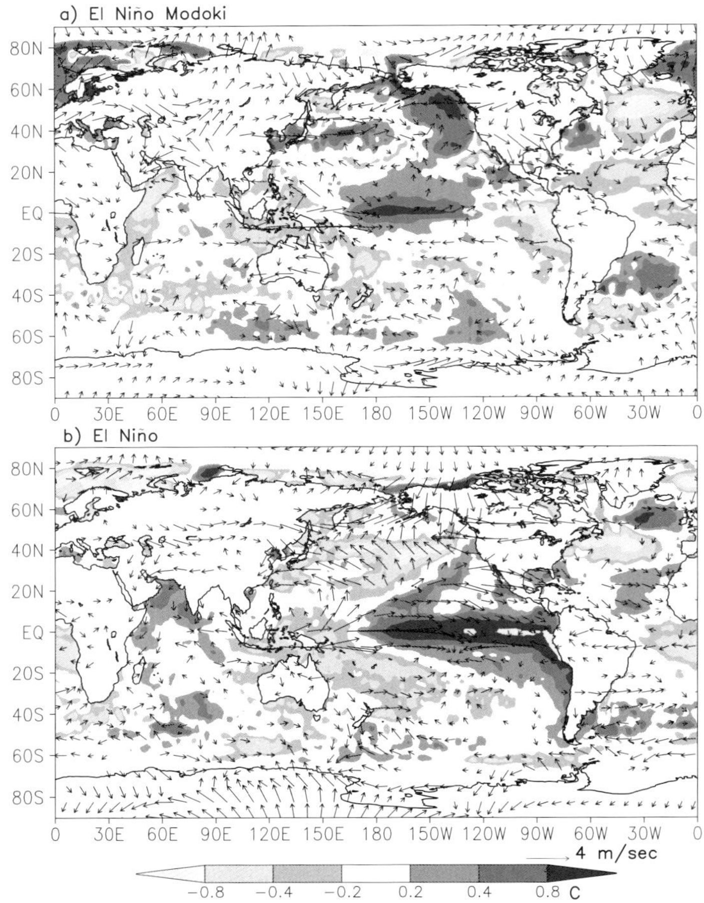

Fig. 3. Composites of summer SSTA of the three strongest events of (a) El Niño Modoki (1994, 2002, and 2004) and (b) El Niño (1982, 1987, and 1997).

of El Niño Modoki is seen in the boreal summer of 2004 and an opposite event, called a La Niña Modoki (Pseudo-La Niña), is observed in 1998; such events are characterized by the anomalously cold SST anomaly on the central Pacific flanked by the warm SST anomalies on either side. Like ENSO, the nomenclature of ENSO Modoki assumes both warm and cold phases of its behavior. Though the correlation between PC1 and Nino3 index (an index of SST anomalies from eastern Pacific used to measure ENSO intensities) is very high (0.98), the correlation between PC2 and Nino3 index is negligible (−0.09) (Ashok *et al.*, 2007). Thus, the El Niño Modoki events are not necessarily related to the conventional El Niño events and an ENSO Modoki index (EMI) is derived based on the unique tripolar nature of the SST anomalies:

$$\mathrm{EMI} = SSTA_{BOX\,A} - 0.5 \cdot SSTA_{BOX\,B} - 0.5 \cdot SSTA_{BOX\,C} \qquad (2.1)$$

The terms in Eq. (2.1) represents the area-averaged SST anomaly (SSTA) over each of the regions A (165°E-140°W, 10°S-10°N), B (110°W-70°W, 15°S-5°N), and C (125°E-145°E, 10°S-20°N), respectively.

2.2.1. *Ocean-atmosphere coupling*

The ocean-atmosphere coupling during ENSO Modoki events is demonstrated by Ashok *et al.* (2007) from lead/lag correlation between the EMI and satellite-derived sea surface height (SSH)[c] anomalies and the regression of the EMI with the wind anomalies. Correlations between EMI and SSH anomalies are seen in the central and western tropical Pacific when the former lags the latter at 12 months' lag (Fig. 4(a)). The signal, apparently excited by westerly wind anomalies in the western Pacific, helps the ENSO Modoki evolution by transporting the warm water from the off-equatorial regions to the equator. This introduces downwelling equatorial Kelvin waves that subsequently deepen the thermocline in the central Pacific. In the following months, positive correlations between EMI and SSH anomalies become larger and the signal propagates westward together with the corresponding correlations of 10 m temperature anomalies. At six months' lag, we observe easterly wind anomalies in the eastern Pacific in addition to the anomalous westerlies in the western Pacific (Fig. 4(c)). Since these winds cause convergence in the central Pacific, the thermocline in the central Pacific further deepens. With increasing easterlies in the eastern Pacific, the equatorial Rossby waves may deepen the thermocline off the equator and thus intensify the warming in the central Pacific. This is indicated by the high correlations around central Pacific at zero lag (Fig. 4(e)). After the peak phase of the event, anomalous easterlies in the eastern Pacific are strengthened, and the equatorial upwelling is also strengthened. The associated downwelling Rossby waves propagate west. Together with

[c]Sea surface height is derived from the altimeter data of several satellites: Geosat, TOPEX/POSEIDON, and Jason (derived from NCEP-NCAR reanalysis data).

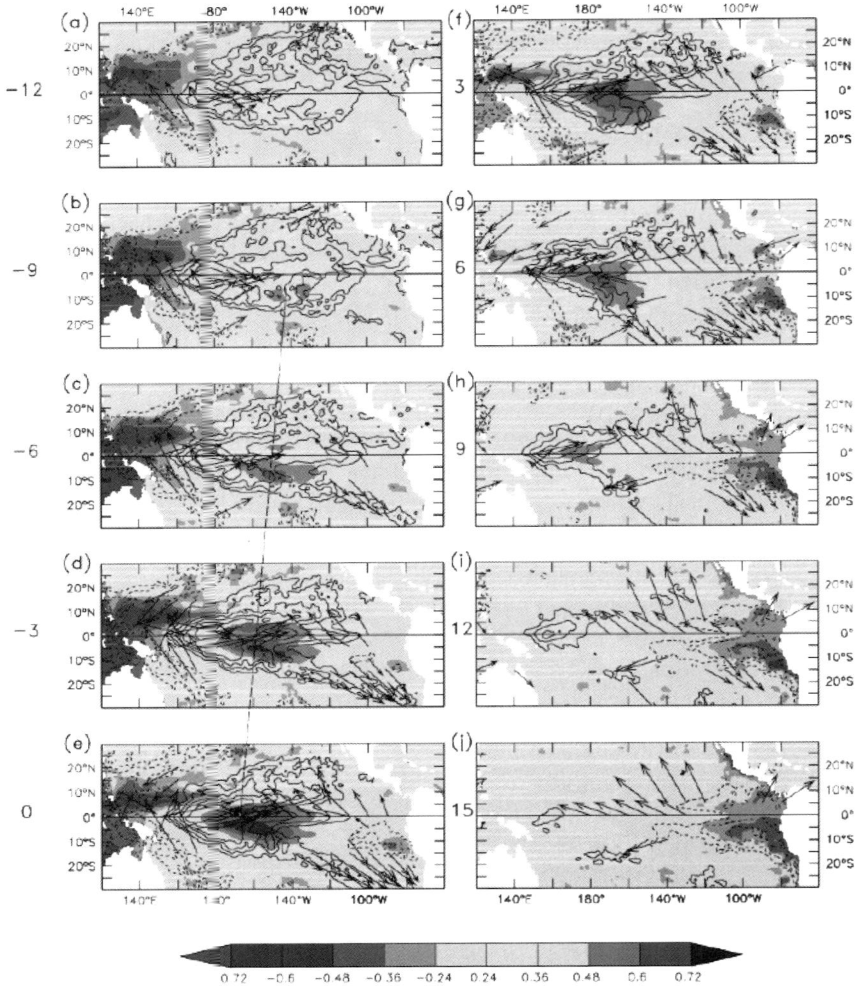

Fig. 4. Lag/lead correlations of monthly EMI with SSH anomalies (shading) and ocean temperature anomalies at 10 m depth (contours). Positive (negative) correlation coefficients correspond to high (low) sea level anomalies. Regressed winds with EMI are shown only if the correlation coefficient between EMI and respective wind components exceeds 0.24. The positive (negative) numbers to the left indicate the months by which the EMI leads (lags) the anomaly distribution fields. [*Source*: Adapted from Ashok *et al.* (2007).]

the weakening of westerlies in the western Pacific, the downwelling Rossby waves smear out the cold anomaly in the western Pacific and eventually terminate the Modoki event (Fig. 4(f–j)).

2.2.2. *ENSO Modoki vs ENSO impacts*

Associated with these two tropical phenomena in the equatorial Pacific, we notice markedly different SSTA patterns in extratropics. For example, a large-scale warm SST anomaly in the extratropical North Pacific is seen during the summer of an El Niño Modoki (Fig. 3(a)) while a large-scale cool anomaly prevails during the summer of an El Niño (Fig. 3(b)). Such a difference may imply that the extratropical low system in the El Niño Modoki case (Fig. 3(a)) could be locally excited by a warm SSTA in that oceanic region, which is synchronized with the warming of the central tropical Pacific (Weng *et al.*, 2007). The extratropical SST anomalies seem to be a forcing for such a cyclonic low, while the cool SST anomalies in the extratropical North Pacific in the El Niño case (Fig. 3(b)) seem to be a response to the anomalous low system when frequent cyclonic activities over the season reduce solar input and cool the SST. Nevertheless, other factors may play important roles when the extratropical regions are weakly associated with weak El Niño Modoki events.

The persistent summer drought in the western United States is caused not only by below-normal rainfall (Fig. 5(a)), but also by above-normal temperature in El Niño Modoki summers (Weng *et al.*, 2007). The surface temperature related to El Niño Modoki is warmer than normal in the western states, while cooler than normal in the central and eastern states. However, the El Niño-related temperature in most areas of the United States, except for the southeastern and northwestern states, is basically cooler than normal (Weng *et al.*, 2007). Combined effects of rainfall and temperature anomalies mean that the El Niño Modoki-related warmer surface temperature anomaly in the western United States exacerbates the drought due to less rainfall there. It is consistent with the relationship between precipitation and surface temperature over land in summer (Trenberth and Shea, 2005). Thus the northwestern USA is easily susceptible to persistent drought if El Niño Modoki events continue from summer to winter seasons. This may have been the situation in the beginning of the 21st century when the drought in the northwestern USA was sustained for several years

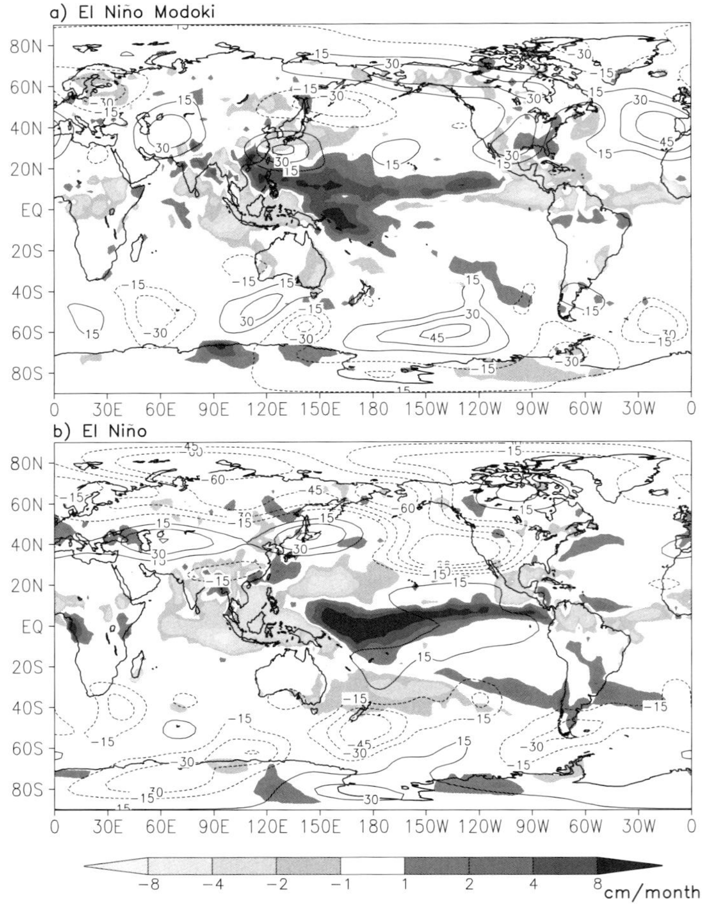

Fig. 5. Composites of June–August rainfall anomalies (shaded) and geopotential height anomalies of the three strongest events of (a) El Niño Modoki (1994, 2002, and 2004) and (b) El Niño (1982, 1987, and 1997).

(Weng *et al.*, 2007). In the eastern North Pacific, El Niño Modoki is associated with a positive modified Pacific-North American (PNA) pattern.[d]

[d]The PNA teleconnection pattern is one of the most prominent modes of low-frequency variability in the northern hemisphere extratropics. The positive phase of the PNA pattern features above-average geopotential heights in the vicinity of Hawaii and over the intermountain region of North America, and below-average heights located south of the Aleutian Islands and over the southeastern United States. The PNA pattern is associated with strong fluctuations in the strength and location of the East Asian jet stream.

The tropical storm activities near Japan and the southeastern United States may be enhanced during El Niño Modoki events. In the tropical region, the SSTA pattern associated with the El Niño Modoki appears as a warming in the central part of the Pacific with cooling in east and west. It seems to have direct and indirect influences on the rainfall anomalies (Fig. 5(a)) in the Pacific Rim countries associated with the anomalous two-cell type Walker circulation. The anomalous SST gradients and the moisture distribution cause anomalous subtropical and extratropical responses (Weng *et al.*, 2007).

The poles of SSTA and associated atmospheric fields in the tropical Pacific are basically the joint regions of multiple "boomerangs" of these fields. The arms of these boomerangs extend eastward and poleward in the Pacific, with the northern arms being stronger in the boreal winter season. In the western North Pacific El Niño Modoki is associated with a positive Pacific-Japan pattern,[e] the enhanced western-north Pacific summer monsoon and weakened East Asian summer monsoon, which causes droughts in much of Japan and central eastern China, and floods in southern China (Weng *et al.*, 2007, 2009b). Different patterns with alternating wet/dry boomerangs between the two phenomena cause different, and even opposite, precipitation and temperature anomalies for a given region. A common feature of the two phenomena is that the outer arms of the boomerangs are discontinuous, suggesting more interactions between tropical and subtropical/extratropical systems there. The boomerang arms appear at lower latitudes in the western Pacific and higher latitudes in the eastern Pacific where the *direct* influence of the two tropical phenomena could reach. This also explains why the East Asian winter monsoon — influenced by the two phenomena — is limited to lower latitudes, including southeastern China, Taiwan, and southern Japan.

The location of a wet boomerang is closely related to the path of moisture transport from the tropics to the subtropics. The wet boomerang arms may cause the so-called "atmospheric river" (Newell *et al.*, 1992; Ralph *et al.*, 2004). During El Niño Modoki (Fig. 6(a)), the anomalous low-level southwesterlies from the central tropical Pacific to the eastern subtropical North Pacific are associated with the northern arm of the wet boomerang, which involves a northward migration of the ITCZ. The low-level flows act like an atmospheric river in the climate sense, carrying the moisture from the tropical Pacific to the southwestern USA through the wet boomerang

[e]The Pacific-Japan (PJ) teleconnection pattern is one of the dominant atmospheric anomaly patterns that influence summertime weather conditions over the Far East. It is characterized by anomalous convective activity over the tropical Northwestern Pacific and a meridional dipole of anomalous circulation in the lower troposphere.

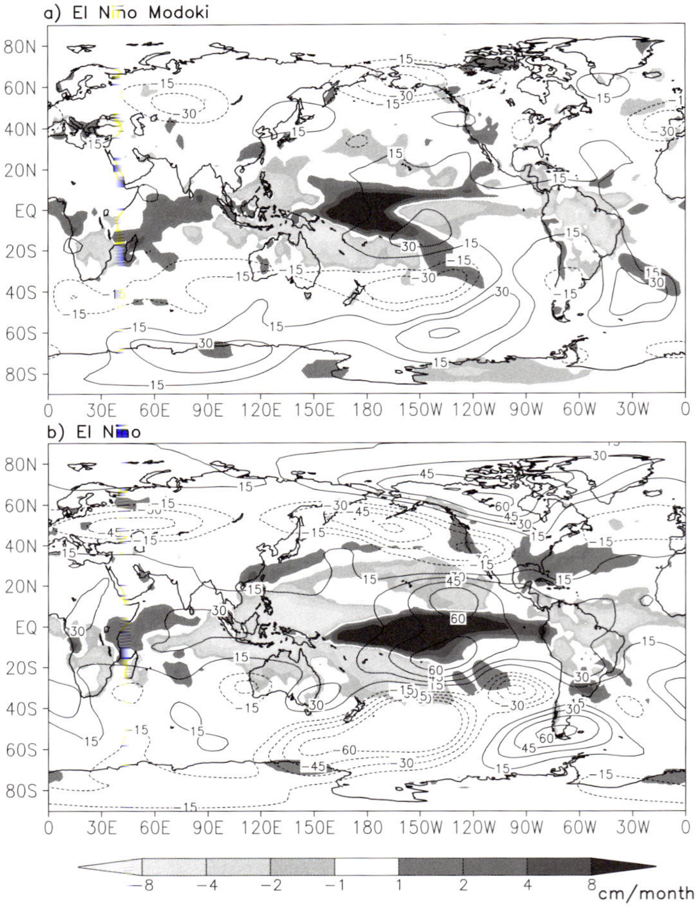

Fig. 6. Same as Fig. 5 but for December–February.

arm. The characteristic atmospheric circulation patterns during El Niño Modoki winter may provide a favorable climate background to anchor the atmospheric river on the weather scale. Such an atmospheric river was seen in southern California in early January, 2005, during an El Niño Modoki winter, which caused a meter of rainfall and massive mudslides (Kerr, 2006). Such a case is less likely to occur during an El Niño winter, because the anomalous southwesterlies carry the moisture from the eastern tropical Pacific to the Caribbean, the Gulf of Mexico, and southern and southeastern

USA (Fig. 6(a)), but unlikely to southwestern USA (Weng *et al.*, 2009a). Although the southwestern USA may also be wet during an El Niño, the moisture is more likely to be transported from the mid-latitudes by the westerlies in the southeastern flank of the anomalous Aleutian Low, which may cause much of the western USA to be wet. Thus, based on this analysis, the northward shift of the ITCZ that brings moisture from the tropical Pacific to the southwestern USA is more likely to occur during an El Niño Modoki winter than during an El Niño winter (Fig. 6(b)).

2.3. *The Indian Ocean Dipole*

Unlike the Pacific Ocean, the inter-annual variability in the tropical Indian Ocean has received less attention. This is mainly because the variability in the basin is more complex due to the changing monsoon winds and the complex geometry. The southwest monsoon winds that dominate the annual cycle produce strong upwelling along the Somali coast of the western Indian Ocean (Fig. 1(a)). However, a weakening of these winds during monsoon transition seasons of spring and fall gives rise to warmer SSTs mainly caused by weaker upwelling and higher solar insolation. During transition seasons, the otherwise weak equatorial winds become stronger and the eastward winds generate the strong equatorial currents known as the Yoshida-Wyrtki jet, which transports the warm waters to the east. Anomalous events evolve, sometimes owing to an imbalance between the equatorial winds and the east–west slope in the equatorial Indian Ocean.

Recent studies show that these anomalous events in the basin are manifestations of an ocean-atmosphere coupled phenomenon known as the Indian Ocean Dipole (IOD) mode (Saji *et al.*, 1999; Webster *et al.*, 1999; Yamagata *et al.*, 2004). During a positive IOD event, the eastern tropical Indian Ocean becomes colder than normal while the western side becomes warmer (Fig. 1(f)). These changes in the SSTs during the IOD events are found to be associated with related changes in the surface wind and rainfall. Equatorial winds reverse direction from westerlies to easterlies during the peak phase of the positive IOD events together with abundant rainfall over the western Indian Ocean/East Africa and scarce rainfall over the eastern Indian Ocean/Indonesia (Fig. 1(f)). This is similar to the Bjerknes type of air-sea interaction in the tropical Pacific (Bjerknes, 1969). However, the dipole pattern is not restricted only to the SST anomalies in an IOD event. The thermocline changes in response to the equatorial winds through the oceanic adjustment process during IOD events (Rao *et al.*, 2002). It rises in the east and deepens in the central and western Indian Ocean. The seasonal southeasterly winds along the Sumatra coast are also strengthened during the positive IOD events and cause SST cooling by coastal upwelling (Vinayachandran *et al.*, 1999, 2002) and evaporation (Behera *et al.*, 1999). The dipole pattern related to IOD is identified in heat content/sea level anomalies (Rao *et al.*, 2002), outgoing long-wave radiation (OLR) anomalies (Behera *et al.*, 1999), and sea level pressure anomalies (Behera and Yamagata, 2003). Therefore, dipole mode indices are derived using several ocean-atmosphere variables: SSTA, wind, sea surface height, satellite derived sea level anomalies, and OLR anomalies (cf. Fig. 1 of Yamagata *et al.*, 2003).

2.3.1. *Ocean-atmosphere coupling*

The dipole mode originally introduced using the SST anomalies, is found to be coupled strongly with subsurface temperature variability. Rao *et al.* (2002) discussed how the evolution of the dominant dipole mode in the subsurface is controlled by equatorial ocean dynamics forced by zonal winds in the equatorial region. The subsurface dipole provides a kind of delayed oscillator mechanism (cf. Schopf and Suarez, 1988) required to reverse the phase of the surface dipole in the following year through propagation of oceanic Rossby/Kelvin waves (Feng and Meyers, 2003), which is also confirmed from coupled model studies (Gualdi *et al.*, 2003; Yamagata *et al.*, 2004). Thus, the turnaround of the subsurface dipole leads to the quasi-biennial oscillation of the tropical Indian Ocean (Rao *et al.*, 2002; Feng and Meyers, 2003). The ocean dynamics may play an important role for the quasi-biennial oscillation in the Indo-Pacific sector through changes in the Asian monsoon.

Xie *et al.* (2003) suggested that Rossby waves in the southern Indian Ocean play a very important role in air-sea coupling and that these coupled Rossby waves are dominantly forced by ENSO. In subsequent studies, Yamagata *et al.* (2004) and Rao and Behera (2005) have distinguished regions that are influenced by IOD and ENSO (Fig. 7). They showed that the wind stress curl associated with the IOD forces the westward propagating downwelling long Rossby waves north of 10°S that increase the heat content of the upper layer in the central and western Indian Ocean during positive IOD events (Fig. 7, left-hand panels). The heat content anomaly maintains the SST anomaly that influences the wind stress anomaly, thereby completing the ocean-atmosphere feedback loop. In contrast, the ENSO influence

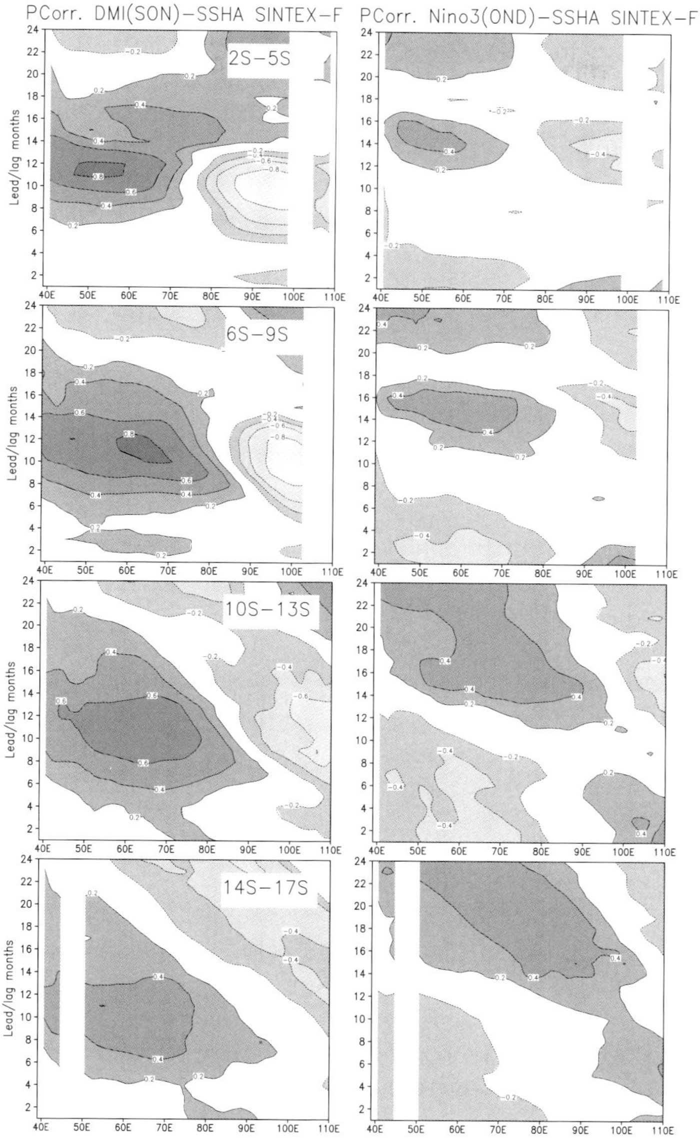

Fig. 7. Partial correlation between the September–November DMI and the south Indian Ocean SSH anomalies (left-hand panels) for different latitude bands from SINTEX-F1 simulation results. The corresponding correlation for the October–December Niño-3 index is shown on the right-hand panels. Values shown are statistically significant at 99% level using a two-tailed t-test.

dominates over the upwelling dome south of 10°S in the southern Indian Ocean (Fig. 7, right-hand panels). A similar north/south displacement in the response of sea level to wind forcing is found in the study by Wijffels and Meyers (2004). The cause of this is not very clear at this stage but the ENSO-related variation of the southern trade winds is one possible candidate. Another possible candidate is the Indonesian throughflow: The oceanic anomaly of the Pacific origin may propagate westward and enhance local air-sea coupling south of 10°S (e.g., Webster *et al.*, 1999).

The Indonesian throughflow apparently plays a role in the decadal variability of ENSO and IOD. Using an output from a 200-year integration of the SINTEX-F1 Coupled Atmosphere-Ocean General Circulation Model (CGCM), Tozuka *et al.* (2007) have found that the first EOF mode of the decadal (9–35 years) SST anomaly represents a basin-wide uniform mode that has a close connection with the Pacific ENSO-like decadal variability. On the other hand, the second EOF mode has shown a clear east–west dipole pattern in the Indian Ocean and has close relations with variations in the Indonesian throughflow and the heat transport in southern Indian Ocean. Since the pattern resembles the inter annual IOD despite the longer timescale, the mode is named as the "decadal IOD." One of the most interesting interpretations found in this study is that the decadal air-sea interaction in the tropics could be a statistical artefact and the decadal IOD may be interpreted as "*decadal modulation of interannual IOD events.*"

2.3.2. *Triggering and termination processes*

The precondition for IOD evolution is another issue that requires more research. Several studies indicate the presence of a favorable mechanism in the eastern Indian Ocean that combines cold SST anomalies, anomalous southeasterlies, and suppression of convection into a feedback loop (Behera *et al.*, 1999; Saji *et al.*, 1999). However, recent studies suggest a few alternatives: Atmospheric pressure variability in the eastern Indian Ocean (Gualdi *et al.*, 2003), favorable changes in winds in relation to the Pacific ENSO and the Indian monsoon (Annamalai *et al.*, 2003), oceanic conditions of the Arabian Sea related to the Indian monsoon (Prasad and McClean, 2004), and influences from the southern extratropical region (Lau and Nath, 2004). It has also been found from observed data that the equatorial winds in the Indian Ocean are related to variabilities in pressure and trade winds of the southern Indian Ocean (Hastenrath and Polzin, 2004). All these studies fall short on more than one occasion to answer the failure (or success) of IOD evolution in spite of favorable (or unfavorable) preconditions (Behera *et al.*, 2006, 2008); e.g., no IOD formation in 1979 (Gualdi *et al.*, 2003) and the aborted IOD event of 2003 (Rao and Yamagata, 2004).

The intraseasonal oscillation (ISO), also known as the Madden–Julian oscillation (MJO), in atmospheric variability of the Indian Ocean shows pronounced seasonality with the strongest activity in boreal winter and spring (Madden and Julian, 1994; Gualdi and Navarra, 1998). Since the ISOs originate in the tropical Indian Ocean they play a significant role in the IOD evolution. In recent studies, Rao and Yamagata (2004) and Rao *et al.* (2007) have examined the possible link between the ISO activity and the IOD termination using multiple datasets. They observed strong 30–60 day oscillations of equatorial zonal winds prior to the termination of all IOD events, except for the event of 1997. This may be a reason why the 1997 IOD event was sustained until early February 1998 instead of usual termination dare around December. Thus the strong westerlies associated with the ISO excite anomalous downwelling Kelvin waves that terminate the coupled processes in the eastern Indian Ocean by deepening the thermocline in the east, as discussed by Fischer *et al.* (2005) for the 1994 IOD event. Gualdi *et al.* (2003) have suggested that the anomalously high ISO activity in the northern summer of 1974 might explain the aborted IOD event in that year.

2.3.3. *IOD impacts*

Like ENSO, the IOD can exert its influence on various parts of the globe via atmospheric teleconnection (Saji and Yamagata, 2003) and by interacting with other modes of climate variability. Through the changes in the atmospheric circulation, IOD influences ENSO (Izumo *et al.*, 2010), rainfall variability during the Indian summer monsoon (Behera *et al.*, 1999; Cherchi *et al.*, 2007), the summer climate condition in East Asia (Guan *et al.*, 2003), the African rainfall (Behera *et al.*, 2005), the Sri Lankan Maha rainfall (Zubair *et al.*, 2003), the Australian rainfall (Ashok *et al.*, 2003), and the Brazil rainfall (Chan *et al.*, 2008).

The precipitation over the northern part of India, the Bay of Bengal, Indochina, and the southern part of China was enhanced during the 1994 positive IOD event (Behera *et al.*, 1999; Saji and Yamagata, 2003). The positive IOD and El Niño have opposite influences in the Far East, including Japan and Korea (Saji and Yamagata, 2003). Positive IOD events give

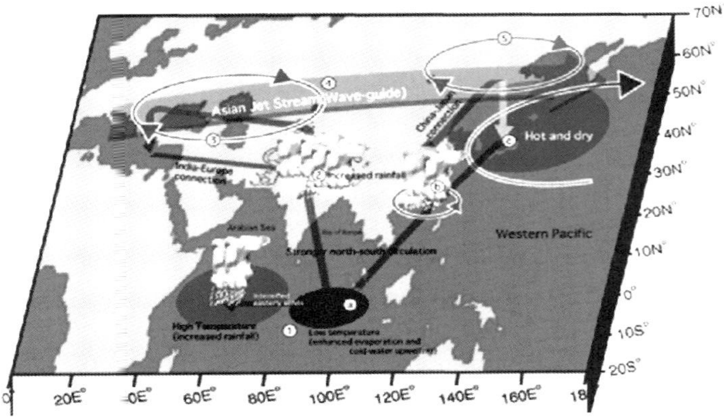

Fig. 8. Schematic diagram showing the IOD influence on the East Asian summer conditions.

rise to warm and dry summers in East Asia as is seen during 1961 and 1994 (Guan *et al.*, 2003; Yamagata *et al.*, 2004).

Several studies have found that the equivalent barotropic high known as the Bonin High was strengthened during positive IOD events over East Asia (e.g., (Yamagata *et al.*, 2004)). The anomalous pressure pattern that is often linked to the unusually hot summer is recognized as a whale tail pressure pattern by the Japanese weather forecasters. The tail part, which is corresponding to the Bonin High, is the equivalent barotropic in contrast to the larger head part, which corresponds to the baroclinic Pacific High. The IOD-induced divergent flow over the Tibetan Plateau (Sardeshmukh and Hoskins, 1988) excites a Rossby wavetrain, which propagates northeastward from the southern part of China to Japan (Fig. 8). This is quite similar to the Pacific-Japan pattern (Nitta, 1987), although the whole system is shifted a little westward to give rise to the Indian Ocean-Japan pattern. In the latter case, the convective anomalies in the eastern Indian, associated with the IOD, give rise to anomalies in the Philippines region and those anomalies subsequently influence Japan. For example, in a positive IOD event, subsidence over the eastern Indian Ocean will give rise to higher convective activity over Philippines and that in turn will cause subsidence over Japan like the Pacific-Japan pattern.

In another process, the IOD-induced diabatic heating around India excites a long atmospheric Rossby wave to the west of the heating. The latter is similar to the monsoon-desert mechanism that connects the circulation changes over the Mediterranean Sea/Sahara region with the heating over India (Rodwell and Hoskins, 1996). The westerly Asian jet acts as a waveguide for the eastward propagating tropospheric disturbances to connect the circulation change around the Mediterranean Sea with the anomalous circulation changes over East Asia (Fig. 8). This mechanism is called the "Silk Road process" and may contribute to strengthening the equivalent barotropic Bonin High in East Asia.

The IOD has a paramount impact on the October–December short rains variability of East Africa. Behera *et al.* (2003, 2005), using observed data and SINTEX-F simulations, found that positive IOD (El Niño) events are related to enhanced (reduced) rainfall in East Africa. The anomalous westward low-level winds in response to the anomalous zonal gradient of SST enhance the moisture transport to the western Indian Ocean and augment seasonal atmospheric convection in East Africa. Simulated correlation patterns are consistent with the observed variation of rainfall anomalies in the East African region (Behera *et al.*, 2005). The relationship is so robust that the raw Dipole Mode Index (DMI)[f] values of July and August could predict 92% of anomalous years of short rains. This is possible because the slow propagation of the air-sea coupled mode in the western Indian Ocean gives rise to predictability of the IOD-induced short rains at least a season ahead (Yamagata *et al.*, 2004; Behera *et al.*, 2005; Rao and Behera, 2005). On the eastern side, over the Indonesian region, the model rain anomaly suggests a stronger influence of the IOD as compared to the ENSO.

In the Southern Hemisphere, the impact of the IOD is notable in several parts of Australia (Ashok *et al.*, 2003) and Brazil (Chan *et al.*, 2008); positive IOD events cause warm and dry conditions over northern Brazil. Rossby wavetrains that are prominent in the Southern Hemisphere are shown to be responsible for the IOD teleconnection.

2.3.4. *IOD predictions*

As in the case of ENSO, CGCMs are proving to be useful in predictability experiments of the IOD. Using the NASA Seasonal-to-Interannual Prediction Project (NSIPP) coupled model system, Wajsowicz (2005) has shown a remarkable predictability of the IOD at three months' lead-time for the

[f]The DMI describes the variability of the Indian Ocean Dipole (IOD) and is defined as the SST anomaly difference between the eastern and the western tropical Indian Ocean (Saji *et al.*, 1999).

decade 1993–2002. In her results, the forecast skill of the eastern pole deteriorates at six months' lead-time. In another study using the SINTEX-F prediction system, Luo *et al.* (2007) showed higher skills for the IOD predictions at a lead of four months. They also found that the seasonal predictability suffers in the Indian Ocean due to an intrinsic winter prediction barrier. In the western Indian Ocean, several warm SST anomalies (during 1983, 1987, 1991, 1997–98, and 2003) and cold anomalies (1985, 1989, 1996, and 1999) can be predicted reasonably well up to nine to twelve months ahead. This is mostly associated with the large influence of ENSO on the western Indian Ocean SST anomalies, which their model predicts very well at longer lead times. In contrast, they reported a rather challenging task to predict the SST anomaly in the eastern Indian Ocean. Nevertheless, the model can skilfully predict the signal there up to about two seasons ahead.

Considering the complicated and delicate physical processes governing the IOD and the sparse subsurface ocean observations available in the Indian Ocean, it is encouraging to find that current state-of-the-art ocean-atmosphere coupled models are capable of predicting the extreme IOD episodes at a lead of 23 seasons. In the presence of chaotic and energetic intraseasonal oscillations in the Indian Ocean, it is understood that a large number of ensemble members of model predictions could improve the long-range forecasts of IOD (Luo *et al.*, 2007). Nevertheless, substantial amounts of effort are required to improve the performance of both atmospheric and oceanic GCMs in simulating the tropical Indian Ocean climate. The flat zonal thermocline, a bias found in several prediction models, in the equatorial Indian Ocean associated with too weak westerly winds in the model predictions may affect the probability density function of IOD predictions, favoring the occurrence of strong events.

The Indonesian throughflow that carries the water from the western Pacific to the eastern Indian Ocean must also be resolved in a more precise way in next-generation coupled models. Errors in the initial subsurface conditions in the tropical Indian Ocean may largely affect the IOD predictions in some circumstances. This situation is expected to improve with the improvements of ocean observations in this area. Current international efforts by the World Climate Research Program/Climate Variability and Predictability (WCRP/CLIVAR) and the Earth Observing System/Global Earth Observation System of Systems/Global Ocean Observing System (EOS/GEOSS/GOOS) to establish a long-term monitoring system in the tropical Indian Ocean (similar to its counterpart the Tropical Atmosphere Ocean/Triangle Trans-Ocean Buoy Network (TAO/TRITON) in the

Pacific) will increase the forecast skills of IOD by providing better initial conditions.

3. IOD, ENSO and ENSO Modoki Interactions

The importance of the remote influence of the Pacific ENSO on the Indian Ocean has long been recognized. In fact, traditionally it was assumed that the variability in the Indian Ocean sector is completely dominated by the remote influence of ENSO with very little variability arising from the local air-sea feedback. Indeed, a basin-wide SST anomaly of almost uniform polarity, which is highly correlated with ENSO in the Pacific, is present as the most dominant interannual mode in the Indian Ocean (Cadet, 1985). The basin-wide anomaly is often first established in the west and spreads eastward as the ENSO event matures.

While the ENSO-related basin-wide warming in the Indian Ocean is easier to understand and predict, the ENSO and the IOD interaction cannot be explained easily. The mechanism through which ENSO exerts its influence on the IOD is not clearly understood. One possible candidate is the zonal Walker circulation. Yamagata *et al.* (2003) have demonstrated that an anomalous Walker cell exists only in the Indian Ocean during pure IOD events. In another study based on data analyses, Meyers *et al.* (2007) also reported independent evolution of IODs. Those linear analyses do not exclude the possibility of nonlinear interaction between the anomalous Walker cells of the Indian and Pacific Oceans associated with IOD and ENSO when they co-occur. From a case study of the 1997–98 El Niño event, Ueda and Matsumto (2000) suggested that the changes in the Walker circulation related to the El Niño could influence the evolution of IOD through changes in the monsoon circulation. Conversely, Behera and Yamagata (2003) showed that IOD modulates the Darwin pressure variability, i.e., one pole of the Southern Oscillation.

The other mechanism is related to oceanic processes and the passage of ENSO signals through the Indonesian throughflow. It is understood that the mature ENSO signal in the western Pacific intrudes into the eastern Indian Ocean through the coastal waveguide around the Australian continent (Clarke and Liu, 1994; Meyers, 1996). The associated changes in SST due to the propagation of coastal Kelvin waves along the west coast of Australia, which is known as the Clarke–Meyers effect (Yamagata *et al.*, 2004), apparently cause some local air-sea interaction in boreal fall in this region just like the annual coupled mode in the eastern Pacific. Note that the SST anomalies around the Australian coast related to the Indonesian through-flow are not necessarily linked to the evolution of IOD, though some models suggest this possibility due to apparent model biases. The exact nature of the impact of the ENSO and IOD interactions through the oceanic route is not fully understood at this stage.

The IOD and ENSO interaction has been extensively studied using the SINTX-F coupled model. Behera *et al.* (2006) reported results from a Pacific Ocean/atmosphere decoupled (noENSO) experiment in addition to a globally coupled control experiment. In the former, the ENSO variability is suppressed. The ocean-atmosphere conditions related to the IOD are realistically simulated by both experiments, including the characteristic east–west dipole in SST anomalies. In the EOF analysis of SST anomalies from the noENSO experiment, the IOD takes the dominant seat instead of the basin-wide monopole mode. Moreover, the coupled feedback among anomalies of the upper-ocean heat content, SST, wind, and the Walker circulation over the Indian Ocean are reproduced. This demonstrates that the dipole mode in the Indian Ocean is mainly determined by intrinsic processes within the basin.

The amplitudes of SST anomalies in the western IOD pole of co-occurring IODs are aided by dynamical and thermodynamical modifications related to the ENSO-induced wind variability. Anomalous latent heat flux and vertical heat convergence associated with the modified Walker circulation contribute to the alteration of western pole anomalies (Behera *et al.*, 2006). Ocean dynamics also play a role in deciding the strength of western warming. Though the Rossby wave phase speed remains unchanged in both model experiments, the amplitude of the downwelling Rossby waves in the western part is stronger in the positive IODs that co-occur with El Niños.

In the absence of ENSO variability in the noENSO experiment, the inter-annual IOD variability is dominantly biennial (Behera *et al.*, 2006). In another sensitivity experiment, where the ocean and atmosphere are decoupled in the tropical Indian Ocean (noIOD experiment), the ENSO periodicity is protracted to a five to six year spectral peak (cf. Fig. 6 in Behera *et al.*, 2006). These model experiments show that the frequency modulation of IOD and ENSO to a great extent is determined by their interaction. In the absence of such an interaction, the basin size and land-sea distribution perhaps decide their intrinsic periodicity. It is noted that the Walker cell in the Indian Ocean intensifies during the peak season of the IOD with upper (lower) tropospheric divergence (convergence) in the west and opposite conditions in the east during IOD and ENSO co-occurrences (Behera *et al.*, 2006). However, lower tropospheric convergence is stronger for the noENSO IODs as compared to that for the co-occurring IODs. This suggests that stronger winds related to the lower tropospheric convergence in the west lead to higher evaporative loss and colder SST in the noENSO IODs.

These results are supported by several other model studies. In one of the early CGCM studies, Iizuka *et al.* (2000) found a remarkable similarity between observed and model IODs from their moderately high resolution CGCM. In another CGCM study, Yu *et al.* (2002) decoupled the Pacific ENSO from IOD and have demonstrated that the model IOD evolves without the ENSO forcing. Most of the other model studies (Gualdi *et al.*, 2003; Yamagata *et al.*, 2004; Behera *et al.*, 2006) report an independent IOD mode except for the model study of Baquero-Bernal *et al.* (2002). The origin of this discrepancy apparently lies in the interpretation of model results in the later study, rather than the model resolutions, as the IOD is well-simulated in moderate resolution CGCMs developed for long-term climate studies (Lau and Nath, 2004). From a 900-year GFDL CGCM experiment, Lau and Nath (2004) found recurrent evolution of IOD patterns. As in the observation, some strong IOD episodes are found in their model results in the absence of ENSO influences.

Besides ENSO and IOD interactions, the ENSO Modoki also interacts with the IOD. The condition in April–May 2007, following the IOD and El Niño of 2006, had an El Niño Modoki, with warm SST anomalies just west of the dateline flanked by cold anomalies in the eastern Pacific and in the seas surrounding the Maritime Continent. At this time, the eastern Indian Ocean was warmer than normal in the phase of turning to the negative IOD. The associated SST gradient and atmospheric conditions with subsidence over the Maritime Continent then favored easterlies to develop in the eastern equatorial Indian Ocean (Behera *et al.*, 2008). The unusual incidence of the 2007 IOD in relation to the interaction between the Indian and Pacific Oceans has been verified using 200-year SINTEX-F1 model results from which four incidences of successive positive IOD events were identified. From the observation and model results, it has been found that the atmospheric conditions related to warm anomalies in the central Pacific caused the easterly anomalies in the equatorial Indian Ocean in April–May. This then led to the formation of a successive positive IOD event during boreal fall.

4. Discussions

Variability of the atmospheric and oceanic conditions in Indian-Pacific Oceans gives rise to an array of naturally occurring ocean-atmosphere coupled modes. Regional climate variations in different parts of the world are influenced by one or a combination of these climate modes. In the past, major attention has been paid to the El Niño/Southern Oscillation (ENSO), which is the dominant coupled mode of the tropical Pacific Ocean, and its impact on climate variability. We now realize that the tropical Indian Ocean has a unique mode of climate variability known as the Indian Ocean Dipole, and that there is a second mode in the Pacific called the El Niño Modoki. Both these modes have significant impacts on regional climate variations worldwide. Research initiatives are needed to understand the real impacts of these modes on society by understanding their roles in the generation of extreme weather events and by improving their predictability at longer lead times.

Several studies have already shown the ENSO influence on the Indian Ocean, but how the IOD influences El Niño and its predictability remained until recently an important issue to be understood (Izumo *et al.*, 2010). On the basis of various forecast experiments, by activating and suppressing air-sea coupling in the individual tropical ocean basins using SINTEX-F prediction results, Luo *et al.* (2009) have shown that the extreme IOD played a key role in driving the 1994 El Niño Modoki, in contrast to the traditional El Niño theory. The El Niño Modoki has been occurring frequently in recent decades (Tozuka *et al.*, 2008), coincident with a weakened atmospheric Walker circulation in response to anthropogenic forcing. Luo *et al.* (2009) suggested that the extreme IOD may significantly enhance El Niño and its onset forecast. Usually, most of the strong El Niño events are accompanied by positive IOD events. The co-occurrence of positive IODs helps to strengthen the Walker circulation in the Pacific associated with an El Niño. However, this relationship is dependent on the phase of the IOD evolution as was found from the 2006 El Niño, which remained weak in spite of its co-occurrence with a positive IOD event. In this case the late development of the 2006 IOD event could not strengthen the El Niño amplitude. Therefore, future changes of the seasonality in the IOD evolution might be important for understanding and predicting future El Niño amplitudes.

Extreme positive IODs have significant contributions to El Niño onset and its long-lead predictability, and hence may have large indirect worldwide climate impacts. The Indo-Pacific inter-basin coupling is crucial to the evolution of both El Niño and extreme IOD and their predictions at long-lead times. After their onsets, however, contributions of the inter-basin coupling to their subsequent growth become limited owing to the dominant role of the local Bjerknes feedback in the individual ocean basins. It is surprising that an El Niño-like signal can be fully generated by extreme IOD as in 1994, in contrast with classical ENSO theory (see Neelin *et al.* (1998) for a review). A better understanding of how El Niño and IOD might evolve and influence each other under global warming may have important implications for the future projection of the climate on Earth. Noticing the more frequent occurrences of extreme IOD and El Niño Modoki in recent decades (Ashok and Yamagata, 2009), perhaps in association with the weakened Walker circulation in response to anthropogenic forcing (Vecchi *et al.*, 2006), it is conceivable that the intensified IOD activity (Abram *et al.*, 2008; Behera *et al.*, 2008) will play a more important role in El Niño evolution under the present global warming trend. This may have implications for our future projection of ENSO in a warmer world.

References

Abram, N. J., Gagan, M. K., Cole, J. E., *et al.* (2008). Recent intensification of tropical climate variability in the Indian Ocean, *Nature Geoscience*, pp. 849–853.

Annamalai, H., Murtugudde, R., Potemra, J., *et al.* (2003). Coupled dynamics over the Indian Ocean: spring initiation of the zonal mode, *Deep-Sea Res. II* **50**, pp. 2305–2330.

Ashok, K., Behera, S. K., Rao, S. A., *et al.* (2007). El Niño Modoki and its possible teleconnection, *J. Geophys. Res.* **112**, C11007, doi:doi:10.1029/2006JC003798.

Ashok, K., Guan, Z. and Yamagata, T. (2003). Influence of the Indian Ocean Dipole on the Australian winter rainfall, *Geophys. Res. Lett.* **30**, doi:10.1029/2003GL017926.

Ashok, K. and Yamagata, T. (2009). Climate change: the El Niño with a difference, *Nature* **461**, pp. 481–484.

Baquero-Bernal, A., Latif, M. and Legutke, S. (2002). On dipole-like variability in the tropical Indian Ocean, *J. Climate* **15**, pp. 1358–1368.

Behera, S. K., Krishnan, R. and Yamagata, T. (1999). Unusual ocean-atmosphere conditions in the tropical Indian Ocean during 1994, *Geophys. Res. Lett.* **26**, pp. 3001–3004.

Behera, S. K., Luo, J.-J., Masson, S., *et al.* (2003). Impact of the Indian Ocean Dipole on the East African short rains: a CGCM study, *CLIVAR Exchanges* **27**, pp. 43–45.

Behera, S. K., Luo, J.-J., Masson, S., *et al.* (2005). Paramount impact of the Indian Ocean Dipole on the East African short rain: a CGCM study, *J. Climate* **18**, pp. 4514–4530.

Behera, S. K., Luo, J.-J., Masson, S., *et al.* (2006). A CGCM study on the interaction between IOD and ENSO, *J. Climate* **19**, pp. 1688–1705.

Behera, S., Luo, J.-J. and Yamagata, T. (2008). The unusual IOD Event of 2007, *Geophys. Res. Lett.* **35**, L14S11, doi:10.1029/2008GL034122.

Behera, S. K. and Yamagata, T. (2003). Influence of the Indian Ocean Dipole on the Southern Oscillation, *J. Meteor. Soc. Japan* **81**, pp. 169–177.

Bjerknes, J. (1969). Atmospheric teleconnections from the equatorial Pacific, *Mon. Wea. Rev.* **97**, pp. 163–172.

Cadet, D. L. (1985). The Southern Oscillation over the Indian Ocean, *J. Climatol.* **5**, pp. 189–212.

Cane, M. A. and Zebiak, S. E. (1985). A theory for El Niño and the Southern Oscillation, *Science* **228**, pp. 1085–1087.

Chan, S. C., Behera, S. K. and Yamagata, T. (2008). Indian Ocean Dipole influence on South American rainfall, *Geophys. Res. Lett.* **35**, L14S12.

Chang, P., Yamagata, T., Schopf, P., *et al.* (2006). Climate fluctuations of tropical coupled system — the role of ocean dynamics, *J. Climate* **19**, pp. 5122–5174.

Chen, D., Cane, M. A., Kaplan, A., *et al.* (2004). Predictability of El Niño for the past 148 years, *Nature* **428**, pp. 733–735.

Cherchi, A., Gualdi, S., Behera, S., *et al.* (2007). The influence of tropical Indian Ocean SST on the Indian summer monsoon, *J. Climate* **20**, pp. 3083–3105.

Clarke, A. J. and Liu, X. (1994). Interannual sea level in the northern and eastern Indian Ocean, *J. Phys. Oceanog.* **24**, pp. 1224–1235.

Feng, M. and Meyers, G. (2003). Interannual variability in the tropical Indian Ocean: a two-year time scale of IOD, *Deep-Sea Res. II* **50**, 12/13, pp. 2263–2284.

Fischer, A., Terray, P., Delecluse, P., *et al.* (2005). Triggers for tropical Indian Ocean variability and links to ENSO in a constrained coupled climate model, *J. Climate* **18**, pp. 3428–3449.

Flügel, M., Chang, P. and Penland, C. (2004). The role of stochastic forcing in modulating ENSO predictability, *J. Climate* **17**, pp. 3125–3140.

Gill, E. A. (1982). *Atmosphere-Ocean Dynamics* (Academic Press, New York).

Gualdi, S., Guilyardi, E., Navarra, A., *et al.* (2003). The interannual variability in the tropical Indian Ocean as simulated by a CGCM, *Climate Dyn.* **20**, pp. 567–582.

Gualdi, S. and Navarra, A. (1998). A study of the seasonal variability of the tropical intraseasonal oscillation, *Global Atmos. Ocean Syst.* **6**, pp. 337–372.

Guan, Z., Ashok, K. and Yamagata, T. (2003). Summer-time response of the tropical atmosphere to the Indian Ocean dipole sea surface temperature anomalies, *J. Meteor. Soc. Japan* **81**, pp. 531–561.

Hastenrath, S. and Polzin, D. (2004). Dynamics of the surface wind field over the equatorial Indian Ocean, *Quart. J. Roy. Meteor. Soc.* **130**, pp. 503–517.

Iizuka, S., Matsuura, T. and Yamagata, T. (2000). The Indian Ocean SST dipole simulated in a coupled general circulation model, *Geophys. Res. Lett.* **27**, pp. 3369–3372.

Izumo, T., Vialard, J., Lengaigne, M., *et al.* (2010). Role of Indian Ocean in extending El Niño predictability, *Nature Geoscience* (in press).

Jin, F.-F., Neelin, D. and Ghil, M. (1994). ENSO on the devil's staircase, *Science* **264**, pp. 70–72.

Kerr, R. (2006). Rivers in the sky are flooding the world with tropical waters, *Science* **313** pp. 313–435.

Lau, N.-C. and Nath, M. (2004). Coupled GCM simulation of atmosphere-ocean variability associated with zonally asymmetric SST changes in the tropical Indian Ocean, *J. Climate* **17**, pp. 245–265.

Luo, J.-J., Masson, S., Behera, S., *et al.* (2007). Experimental forecasts of Indian Ocean dipole using a coupled OAGCM, *J. Climate* **20**, 10, pp. 2178–2190.

Luo, J.-J., Zhang, R., Behera, S., *et al.* (2009). Interaction between El Niño and extreme Indian Ocean dipole, *J. Climate* (in press).

Madden, R. and Julian, P. (1994). Observations of the 40–50 day tropical oscillation: A review, *Mon. Wea. Rev.* **122**, p. 5.

Mantua, N. J. and Battisti, D. S. (1995). Aperiodic variability in the Zebiak-Cane coupled ocean-atmosphere model: air-sea interaction in the western equatorial Pacific, *J. Climate* **8**, pp. 2897–2927.

McCreary, J. and Anderson, D. (1991). An overview of coupled models of El Niño and the Southern Oscillation, *J. Geophys. Res.* **96**, pp. 3125–3150.

Meyers, G. (1996). Variation of Indonesian throughflow and El Niño-Southern Oscillation, *J. Geophys. Res.* **101**, pp. 12255–12263.

Meyers, G., McIntosh, P., Pigot, L., *et al.* (2007). The years of El Niño, Laña and interactions with the tropical Indian Ocean, *J. Climate* **20**, pp. 2872–2880.

Neelin, J. D., Battisti, D. S., Hirst, A. C., Jin, F., Wakata, Y., Yamagata, T. and Zebiak, S. E. (1998). ENSO theory, *J. Geophys. Res.* **102**, pp. 14261–14290.

Newell, R., Newell, N., Zhu, Y., *et al.* (1992). Tropospheric rivers? A pilot study, *Geophys. Res. Lett.* **19**, pp. 2401–2404.

Nitta, T. (1987). Convective activities in the tropical western Pacific and their impact on the Northern Hemisphere summer circulation, *J. Meteor. Soc. Japan* **65**, pp. 373–390.

Philander, S. (1990). *El Niño, La Niña, and the Southern Oscillation* (Academic Press, San Diego, CA).

Prasad, T. G. and McClean, J. L. (2004). Mechanisms for anomalous warming in the western Indian Ocean during dipole mode events, *J. Geophys. Res.* **109**, doi:10.1029/2003JC001872.

Ralph, F., Neiman, P. and Wick, G. (2004). Satellite and CALJET aircraft observations of atmospheric rivers over the eastern North Pacific Ocean during the winter of 1997/98, *Mon. Wea. Rev.* **132**, pp. 1721–1745.

Rao, S. A. and Behera, S. K. (2005). Subsurface influence on SST in the tropical Indian Ocean structure and interannual variabilities, *Dyn. Atmos. Oceans* **39**, pp. 103–135.

Rao, S. A., Behera S. K., Masumoto, Y., *et al.* (2002). Interannual variability in the subsurface tropical Indian Ocean with a special emphasis on the Indian Ocean dipole, *Deep-Sea Res. II* **49**, pp. 1549–1572.

Rao, S. A., Masson, S., Luo, J.-J., *et al.* (2007). Termination of Indian Ocean dipole events in a general circulation model, *J. Climate* **20**, pp. 3018–3035, doi:10.1175/JCLI4164.1.

Rao, S. A. and Yamagata, T. (2004). Abrupt termination of Indian Ocean dipole events in response to intra-seasonal oscillations, *Geophys. Res. Lett.* **31**, doi:10.1029/2004GL020842.

Rodwell, M. J. and Hoskins, B. J. (1996). Monsoons and the dynamics of deserts, *Quart. J. Roy. Meteor. Soc.* **122**, pp. 1385–1404.

Saji, N. H., Goswami, B. N., Vinayachandran, P. N., *et al.* (1999). A dipole mode in the tropical Indian Ocean, *Nature* **401**, pp. 360–363.

Saji, N. H. and Yamagata, T. (2003). Possible impacts of Indian Ocean dipole mode events on global climate, *Climate Res.* **25**, pp. 151–169.

Sardeshmukh, P. and Hoskins, B. (1988). The generation of global rotational flow by steady idealized tropical divergence, *J. Atmos. Sci.* **45**, pp. 1228–1251.

Schopf, P. S. and Suarez, M. J. (1988). Vacillations in a coupled ocean-atmosphere model, *J. Atmos. Sci.* **45**, pp. 549–566.

Timmermann, A., Jin, F.-F. and Abshagen, J. (2003). A nonlinear theory for El Niño bursting, *J. Atmos. Sci.* **60**, pp. 152–165.

Tozuka, T., Luo, J.-J., Masson, S., *et al.* (2007). Decadal Indian Ocean dipole simulated in an ocean-atmosphere coupled model, *J. Climate* **20**, pp. 2881–2894.

Tozuka, T., Luo, J.-J., Masson, S., *et al.* (2008). Tropical Indian Ocean variability revealed by self-organizing maps, *Climate Dyn.* **31**, pp. 333–343.

Trenberth, K. and Shea, D. (2005). Relationships between precipitation and surface temperature, *Geophys. Res. Lett.* **32**, L14703, doi:10.1029/2005GL022760.

Ueda, H. and Matsumoto, J. (2000). A possible triggering process of east–west asymmetric anomalies over the Indian Ocean in relation to 1997/98 El Niño, *J. Meteor. Soc. Japan* **78**, pp. 803–818.

Vecchi, G., Soden, B., Wittenberg, A., *et al.* (2006). Weakening of tropical Pacific atmospheric circulation due to anthropogenic forcing, *Nature* **441**, pp. 73–76.

Vinayachandran, P., Saji, N. and Yamagata, T. (1999). Response of the equatorial Indian Ocean to an anomalous wind event during 1994, *Geophys. Res. Lett.* **26**, pp. 1613–1616.

Vinayachandran, P. N., Iizuka, S. and Yamagata, T. (2002). Indian Ocean Dipole mode events in an ocean general circulation model, *Deep-Sea Res. II* **49**, pp. 1573–1596.

Wajsowicz, R. C. (2005). Potential predictability of tropical Indian Ocean SST anomalies, *Geophys. Res. Lett.* **32**, L24702, doi:10.1029/2005GL024169.

Webster, P. J., Moore, A., Loschnigg, J., *et al.* (1999). Coupled ocean-atmosphere dynamics in the Indian Ocean during 1997–1998, *Nature* **40**, pp. 356–360.

Weng, H., Ashok, K., Behera, S. K., *et al.* (2007). Impacts of recent El Niño Modoki on dry/wet conditions in the Pacific Rim during boreal summer, *Climate Dyn.* **29**, pp. 113–129.

Weng, H., Behera, S. K. and Yamagata, T. (2009a). Anomalous winter climate conditions in the Pacific rim during recent El Niño Modoki and El Niño events, *Climate Dyn.* **32**, pp. 663–674.

Weng, H., Wu, G., Liu, Y., *et al.* (2009b). Anomalous summer climate in China influenced by the tropical Indo-Pacific Oceans, *Climate Dyn.* doi:10.1007/s00382-009-0658-9.

Wijffels, S. and Meyers, G. (2004). An intersection of oceanic waveguides: Variability in the Indonesian throughflow region, *J. Phys. Oceanog.* **34**, pp. 1232–1253.

Xie, S.-P., Annamalai, H., Schott, F., *et al.* (2003). Structure and Mechanisms of South Indian Ocean climate variability, *J. Climate* **15**, pp. 864–878.

Yamagata, T., Behera, S. K., Luo, J.-J., *et al.* (2004). Coupled ocean-atmosphere variability in the tropical Indian Ocean, in C. Wang, S.-P. Xie and J. Carton (eds), *Earth Climate: The Ocean-Atmosphere Interaction*, Geophys. Monogr., 147 (AGU, Washington DC), pp. 189–212.

Yamagata, T., Behera, S. K., Rao, S. A., *et al.* (2003). Comments on "Dipoles, Temperature Gradient, and Tropical Climate Anomalies", *Bull. Amer. Meteor. Soc.* **84**, pp. 1418–1422.

Yu, J.-Y., Mechoso, C., McWilliams, J., *et al.* (2002). Impacts of the Indian Ocean on the ENSO cycle, *Geophys. Res. Lett.* **29**, p. 1204, doi:10.1029/2001GL014098.

Zubair, L., Rao, S. A. and Yamagata, T. (2003). Modulation of Sri Lankan Maha rainfall by the Indian Ocean Dipole, *Geophys. Res. Lett.* **30**(2), 1063, doi:10.1029/2002GL015639.

Chapter 5

The Hurricane-Climate Connection

Kerry Emanuel[a,b]

Program in Atmospheres, Oceans, and Climate
Massachusetts Institute of Technology
Cambridge, MA, USA
emanuel@mit.edu

Tropical cyclone activity has long been understood to respond to changing properties of the large-scale atmospheric and oceanic environment. In this essay, I review evidence for changing tropical cyclone activity, and the controversy surrounding the quality of the data itself and the attribution of these environmental changes to various natural and anthropogenic causes. At the same time, there is growing evidence that global tropical cyclone activity may itself affect climate in such a way as to mitigate tropical climate change but amplify climate change at higher latitudes. I will review this evidence and suggest possible routes forward in exploring these effects.

1. Introduction

It has been understood for some time (Palmén, 1984) that tropical cyclones respond to climate change on a variety of timescales. Empirical studies (Gray, 1968) have established that tropical cyclone activity is sensitive to a variety of environmental conditions, including the magnitude of the shear of the horizontal wind through the depth of the troposphere, sea surface temperature (SST), low level vorticity, and the humidity of the lower and middle troposphere. Theory has so far established only a bound on the intensity

[a]This chapter first appeared in the *Bulletin of the American Meteorological Society* ES10ES20. http://journals.ametsoc.org/doc/pdf/10.117s/BAMS-ES20-89-J-Emanuel May 2008; ©American Meteorological Society. Reprinted with permission.
[b]Corresponding author contact information: Rm 54-1620 MIT, 77 Mass Ave., Cambridge, MA 02139. Phone: (617) 253-2462. Email: emanuel@mit.edu.

of tropical cyclones (Emanuel, 1987), though empirically, this bound has been shown to provide the relevant scaling for the intensity of real storms (Emanuel, 2000). This bound, referred to as the "potential intensity," has the units of velocity and is a function of the SST and the profile of temperature through the troposphere and lower stratosphere (Bister and Emanuel, 2002); it is a far more physically based quantity than SST.

While there has been some advance in the theory of tropical cyclone intensity, the question of frequency is more vexing. About 90 tropical cyclones develop each year around the globe, with a standard deviation of 10; at present, we lack a theory that predicts even the order of magnitude of this number. Although there has been little progress in developing a theory governing the rates of occurrence of tropical cyclones, a number of empirical indices have been developed, beginning with that of Gray (1979). Recently, the author and David Nolan (Emanuel and Nolan, 2004) incorporated potential intensity in an empirical index of the frequency of tropical cyclone genesis, called the Genesis Potential Index (GPI):

$$GPI \equiv |10^5\eta|^{3/2}\left(\frac{H}{50}\right)^3\left(\frac{V_{\mathrm{pot}}}{70}\right)^3(1+0.1V_{\mathrm{shear}})^{-2}, \tag{1.1}$$

where η is the absolute vorticity in s^{-1}, H is the relative humidity at 600 hPa in percent, V_{pot} is the potential intensity in ms^{-1}, and V_{shear} is the magnitude of the vector shear from 850 to 250 hPa, in ms^{-1}.[c] This index was fitted to the annual cycle of genesis in each hemisphere, and to the spatial distributions of storms each month of the year, as described in some detail in Camargo *et al.* (2007), who also showed that the GPI captures some of the dependence of genesis rates on El Niño Southern Oscillation (ENSO). The high power with which the potential intensity enters this empirical index suggests that it plays an important role in the frequency as well as intensity of tropical cyclones, but it must be stressed that a good theoretical understanding of the environmental control of storm frequency is lacking.

While theory is still deficient, there has been some progress in using climate models to simulate the effects of climate change on tropical cyclone activity, as reviewed in Sec. 4. At present, global models are too coarse

[c]The numerical factors in Eq. (1.1) are designed to yield values of the GPI of order unity, but the absolute magnitude of the GPI is regarded as arbitrary.

to resolve the inner cores of intense tropical cyclones, and their ability to simulate the full intensity of such storms is therefore seriously compromised. Yet this approach is beginning to yield interesting and possibly useful insights into the effect of climate change on storm activity.

In this essay, I will review evidence from the instrumental record of changing tropical cyclone activity, including a discussion of various problems with the tropical cyclone data itself, and also briefly review the budding new field of paleotempestology. Section 4 describes the debate over attribution. The fifth section reviews the use of global models to deduce the effects of climate change on tropical cyclones, and presents some results of a new method of deriving tropical cyclone climatology from global gridded data, such as contained in the output of global climate simulations, and in the final section I argue that global tropical cyclone activity is responsible for some or perhaps most of the observed poleward heat transport by the oceans, thereby constituting an essential element of the global climate system. A summary is provided in Sec. 7.

2. Tropical Cyclone Variability in the Instrumental Record

Beginning shortly after World War II, aircraft have surveyed tropical cyclones in the North Atlantic and western North Pacific, though aircraft reconnaissance in the latter basin ended in 1987. During the 1960s, earth-orbiting satellites began to image some tropical cyclones, and by about 1970 it can be safely assumed that hardly any events were missed. Before the aircraft reconnaissance era, tropical cyclone counts depended on observations from ships, islands, and coastal locations. Detection rates were reasonably high only in the North Atlantic, owing to dense shipping, but even here, the precise rate of detection remains controversial (Holland, 2007; Landsea, 2007). Estimates of the intensity of storms as measured, for example, by their maximum surface wind speeds, are dubious prior to about 1958, and some would say, prior to 1970 in the Atlantic and western North Pacific. Elsewhere, there are only very spotty estimates prior to the satellite era. Satellite-based estimates of intensity commenced in the 1970s and have improved along with the spatial resolution of satellite imagery, but the accuracy of such estimates is still debated. Intensity estimates based on aircraft measurements are prone to a variety of biases owing to changing instrumentation and means of inferring wind from central pressure, as

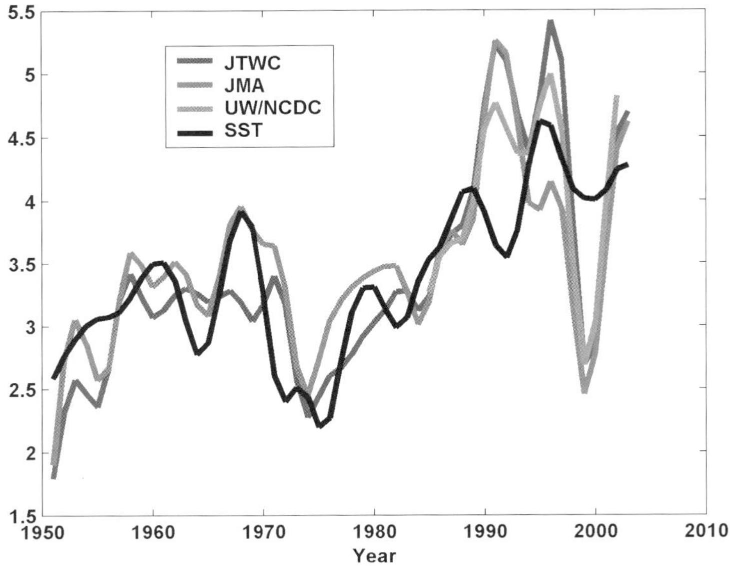

Fig. 1. Power dissipation (colored curves) in the western North Pacific according to data from the U.S. Navy Joint Typhoon Warning Center as adjusted by Emanuel (2005a) (blue), unadjusted data from the Japanese Meteorological Agency (green), and re-analyzed satellite data from Kossin and Vimont (2007) (red). The black curve represents a scaled July–October sea surface temperature in the tropical western North Pacific region. All quantities have been smoothed using a 1-3-4-3-1 filter.

described in the online supplement to Emanuel (2005a).[d] Some indication of the nature of these problems is evident in Fig. 1, which shows a variety of estimates of tropical cyclone power dissipation in the western North Pacific since 1949. (The power dissipation is defined as the integral over the life of each storm of its maximum surface wind speed cubed, also accumulated over each year; see (Emanuel, 2005a)).

Note that the adjusted estimate from the Joint Typhoon Warning Center agrees well with the unadjusted estimate from the Japanese Meteorological Agency and that both are well correlated with the SST prior to the cessation of aircraft reconnaissance in 1987; after that time, there is much more divergence in the estimates and less correlation with SST. There is a

[d]The online supplement is available at: ftp://texmex.mit.edu/pub/emanuel/PAPERS/NATURE03906_suppl.pdf.

general upward trend in SST and tropical cyclone power dissipation, but there are also prominent decadal fluctuations in both. The general upward trend in power dissipation was pointed out by the author (Emanuel, 2005a) and is consistent with the finding by Webster *et al.* (2005) that the global incidence of intense tropical cyclones is generally trending upward.

In the North Atlantic, tropical cyclone records extend back to 1851, but are considered less reliable early in the period, and intensity estimates are increasingly dubious as one proceeds back in time from 1970. (A discussion of the sources or error may be found in the online supplement to Emanuel (2005a[d]) and in (Emanuel, 2007)). A vigorous debate has ensued over the quality of the wind data (Emanuel, 2005b; Landsea, 2005; Landsea *et al.*, 2006), and even the annual frequency of storms is open to question prior to 1970 (Holland, 2007; Holland and Webster, 2007; Landsea, 2007). Similar questions have been raised about the veracity and interpretation of the record of storms in the western North Pacific (Chan, 2006).

Here, on the premise that storms were more likely to be detected near the time of their maximum intensity, we define a "storm maximum power dissipation" as the product of the storm lifetime maximum wind speed cubed and its duration, summed over all the storms in a given year. Figure 2 compares this quantity to the SST of the tropical Atlantic in July through October, going back to 1870. Except for the period 1939–1945, the correspondence between power dissipation and SST is remarkable, even early in the period. Since 1970, the r^2 between the two series is 0.86.

The very low power dissipation during World War II may reflect a dearth of observations owing to enforced radio silence on ships during the war. In the Atlantic, variations in the power dissipation reflect variations in numbers of storms to a large degree (Emanuel, 2007). While some have argued that the number of Atlantic storms may have been grossly underestimated prior to the aircraft and/or satellite eras (Landsea, 2007), statistical analyses of the likelihood of ships encountering storms suggest that the counts are good to one or two storms per year back to 1900 (Tom Knutson, personal communication), and it is also possible to overestimate storm counts owing to multiple counting of the same event encountered infrequently. In addition, Holland and Webster (2007) point out that the large increases during the 1930s and 1990s both occurred during periods when measurement techniques were relatively stable; the advent of aircraft reconnaissance in the 1940s and the introduction of satellites during the 1960s were not accompanied by obvious increases in reported activity. Even with fairly liberal estimates of storm undercounts in the early part of the Atlantic record,

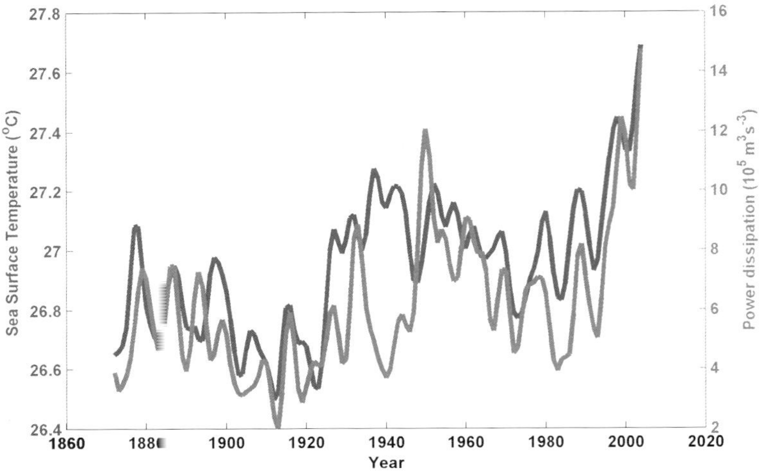

Fig. 2. Storm lifetime maximum power dissipation in the North Atlantic according to data from the NOAA National Hurricane Center as adjusted by Emanuel (2005a) (green). The blue curve represents August–October sea surface temperature in the tropical North Atlantic, from 20 to 60 W and from 6 to 18 N. Both quantities have been smoothed using a 1-3-4-3-1 filter. The sea surface temperature is the HADSST1 data from the United Kingdom Meteorological Office Hadley Center.

the correlation with tropical Atlantic SST remains remarkably high (Mann *et al.*, 2007).

3. Paleotempestology

A number of remarkable efforts are under way to extend tropical cyclone climatology into the geological past by analyzing paleo proxies for strong wind storms. One technique looks at storm surge-generated overwash deposits in near-shore marshes and ponds; this was pioneered by Liu and Fearn (1993) and has been followed up with analyses of such deposits in various places around the western rim of the North Atlantic (Liu and Fern, 2000; Donnelly *et al.*, 2001a,2001b; Donnelly, 2005; Donnelly and Woodruff, 2007). Another technique makes use of dunes of sand, shells, and other debris produced along beaches by storm surges (Nott and Hayne, 2001; Nott, 2003). Very recently, new techniques have been perfected that make use of the anomalous oxygen isotope content of hurricane rainfall (Lawrence and Gedzelman, 1996) as recorded in tree rings (Miller *et al.*, 2006) and speleothems (Frappier *et al.*, 2007a). Collectively, these methods are beginning to reveal variability of tropical cyclone activity on centennial to millennial timescales. For example, the recent work of Donnelly and Woodruff (2007) analyzing overwash deposits near Puerto Rico reveals centennial variability of Atlantic tropical cyclones that is highly correlated with proxies recording long-term variability of ENSO; the same record shows a pronounced upswing over the last century that may reflect a global warming signal. The interested reader is directed to reviews by Nott (2004), Liu (2007), and Frappier *et al.* (2007b).

4. Attribution

The North Atlantic is the only basin with a reasonably long time series of tropical cyclone records, and it is clear from Fig. 2 that there is variability on a broad spectrum of timescales. A Fourier decomposition of the de-trended, unfiltered time series of storm maximum power dissipation shows prominent spectral peaks at around 3, 5, 9, and 80 years. Similar spectral peaks are evident in the de-trended SST data. The first two of these are likely associated with ENSO and known to have a strong effect on Atlantic hurricanes (Gray, 1984). The longest period spectral peak at 80 years is of dubious significance, given that the time series is only ~130 years long, but it is clear from inspection of Fig. 2 that both SST and tropical cyclone power have see-sawed up and down on a multi-decadal timescale over the past century or so.

Mestas-Nuñez and Enfield (1999) examined rotated empirical orthogonal functions (EOFs) of the de-trended global SST and identified the first six of these with modes[e] of the ocean-atmosphere system. The first EOF had timescales of many decades and maximum amplitude in the North Atlantic; this was later identified as a prominent cause of both SST and Atlantic tropical cyclone variability on multi-decadal timescales (Goldenberg *et al.*, 2001) and christened the "Atlantic Multi-Decadal Oscillation," or "AMO" (Kerr, 2000). What began as an EOF ended up as a mode, even though there are only two troughs and one peak in the time series. It is important to recognize that this EOF is global, and while it has large amplitude in the North Atlantic, its amplitude is almost as large in the North Pacific. Furthermore, it turns out that the time series of the amplitude of this first EOF is barely distinguishable from the de-trended time series of August–October

[e]This is technically an incorrect term, as modes are not mathematically equivalent to EOFs.

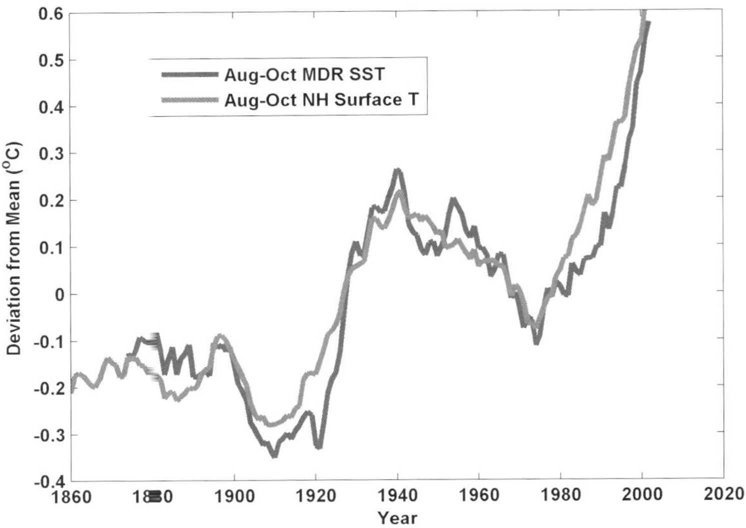

Fig. 3. Ten-year running averages of the Atlantic Main Development Region (MDR) SST (blue) and the northern hemispheric surface temperature (green), both averaged over August–October. The long-term mean has been subtracted in both cases. The United Kingdom Meteorological Office Hadley Center supplies the SST data (HADSST1) and the northern hemispheric surface temperature (HADCRU).

tropical North Atlantic SST, so that there is little advantage in referring to this EOF versus the raw SST. We can ask the somewhat more direct question: What caused the tropical North Atlantic SST (and tropical cyclone power) to see-saw as it did during the 20th century, as evident in Fig. 2?

Figure 3 provides one clue. This compares the ten-year running averages of the August–October SST of the so-called "Main Development Region" (MDR) of the tropical North Atlantic (between Africa and the eastern Caribbean) with the northern hemisphere mean surface temperature (including land). The excellent correspondence between the two time series would seem to imply that on decadal timescales, over the last 100 years or so, the tropical North Atlantic is simply co-varying with the rest of the northern hemisphere. Occam's Razor would lead one to suspect that variations of the two series have a common cause, though it has been suggested that the North Atlantic might be forcing the rest of the hemisphere (Zhang *et al.*, 2007).

The decadal variability in the northern hemispheric surface temperature has been addressed in a number of studies, as summarized in the most recent report of the Intergovernmental Panel on Climate Change (IPCC, 2007). In contrast to Mestas-Nuñez and Enfield (1999), Goldenberg *et al.* (2001) and others, the IPCC report attributes most of the decadal variability to time-varying radiative forcing associated principally with varying solar radiation, major volcanic eruptions, and anthropogenic sulfate aerosols and greenhouse gases. This also helps explain the overall trend, which was disregarded in the EOF analyses. In particular, the warming of the last 30 years or so is attributed mostly to increasing greenhouse gas concentrations, while the cooling from around 1950 to around 1980 is ascribed, in part, to increasing concentrations of anthropogenic sulfate aerosols. Mann and Emanuel (2006) pointed out that the cooling of the northern hemisphere relative to the globe from about 1955 to 1980, evident in Fig. 3, might very well be explained by the concentration of sulfate aerosols in the northern hemisphere. While there is still a great deal of uncertainty about the magnitude of the radiative forcing due to sulfate aerosols, the time series of sulfate concentration is strongly correlated with the difference between global and northern hemisphere surface temperature (Mann and Emanuel, 2006). The important influence of anthropogenic effects in the time history of SST is also emphasized in the work of Hoyos *et al.* (2006); Trenberth and Shea (2006); Santer *et al.* (2006); Elsner (2006); Elsner *et al.* (2006). The author (Emanuel, 2007) emphasizes that the thermodynamic control on tropical cyclone activity is exercised not through SST but through potential intensity, which in the North Atlantic has increased by 10% over the past 30 years. This increase, which is greater than predicted by single-column models for the observed increase in SST, can be traced to increasing greenhouse gases, decreasing surface wind speed in the Tropics, and also to decreasing lower stratospheric temperature (Emanuel, 2007).

Thus there are two schools of thought about the decadal variability of tropical North Atlantic SST and tropical cyclone activity. The first holds that the multi-decadal variability is mostly attributable to natural oscillations of the ocean-atmosphere system (Goldenberg *et al.*, 2001; Kossin and Vimont, 2007), while the second attributes it to time-varying radiative forcing, some of which is natural. These two schools are not mutually exclusive, as the response to time-varying radiative forcing can be greatly modified by natural modes of the system.

Those who attribute Atlantic SST and tropical cyclone variability to a putative AMO often refer to paleo proxy evidence for its existence. The most prominently cited among this evidence is the work of Gray *et al.* (2004), who looked at the first five principal components of variability in tree rings around the rim of the North Atlantic. They fitted these components to the instrumental record of North Atlantic SST, capturing the prominent variability apparent in Fig. 2, and then, using the same curve fit, inferred North Atlantic SST from tree rings back to the middle of the 16th century. Although the record obviously shows the observed multi-decadal variability of the 20th century, close inspection of the reconstructed SST prior to this shows that most of the variability was on somewhat longer timescales, casting doubt on the existence of a quasi-periodic mode. In fact, the only real evidence for the existence of such a mode comes from coupled climate models (Delworth and Mann, 2000), and although many of them exhibit prominent quasi-periodic variability on timescales greater than a decade, the period of such oscillations varies greatly from model to model. It is left to the reader to judge whether the existence of such modes in freely run coupled climate models constitutes strong or weak evidence for such modes in nature.

5. Simulating Global Warming Effects on Tropical Cyclones

Another approach to understanding how climate change might affect tropical cyclone activity is to simulate changing tropical cyclone activity using global climate models. Unfortunately, the horizontal resolution of today's generation of global models is nowhere near sufficient to resolve the intense inner cores of tropical cyclones, and numerical resolution experiments. Chen *et al.* (2007) suggest that grid spacing of no more than a few kilometers is necessary for convergence. Nevertheless, there are quite a few studies of the response of tropical cyclone activity to global warming using global models (Bengtsson *et al.*, 1996; Sugi *et al.*, 2002; Oouchi *et al.*, 2006; Yoshimura and Noda, 2006; Bengtsson *et al.*, 2007). A related approach involves embedding finer resolution regional models within global climate models, so as to better simulate tropical cyclones (Knutson *et al.*, 1998; Knutson and Tuleya, 2004; Knutson *et al.*, 2007). Although results can differ greatly from model to model, there is a general tendency for global warming to reduce the overall frequency of events, to increase the incidence of the most intense storms, and to increase tropical cyclone rainfall rates.

Another approach to downscaling global models to derive tropical cyclone climatologies was presented by the author and his colleagues (Emanuel, 2006; Emanuel *et al.*, 2006); this has recently been extended to account for varying genesis rates (Emanuel *et al.*, 2008). Using certain key statistics from the output of climate models, this technique synthesizes very large numbers (∼104) of tropical cyclones using a three-step process. In the first step, the climate state is "seeded" with a large number of candidate tropical cyclones, consisting of warm-core vortices whose maximum wind speed is only $12\,\mathrm{ms}^{-1}$. These candidate storms then move according to a "beta-and-advection" model (Marks, 1992), which postulates that tropical cyclones move with a weighted tropospheric mean large-scale flow in which they are embedded, plus a correction owing to gyres generated by the storm's advection of planetary vorticity; here the large-scale flow is taken as the climate model-simulated flow. Finally, in the third step, the storm's intensity evolution is simulated using a deterministic, coupled ocean-atmosphere tropical cyclone model phrased in angular momentum coordinates, which achieve very high spatial resolution in the critical central core region. In practice, most of the seeds die a natural death owing to small potential intensity, large wind shear, and/or low humidity in the middle troposphere. We show that the climatology of the survivors is in good accord with observed tropical cyclone climatology.

While details of this technique and the results of applying it to a suite of global climate models are presented in Emanuel *et al.* (2008), we here present one critical result of comparing tropical cyclone activity in the late 20th century to that of the late 22nd century as simulated by global climate models under IPCC scenario A1b, in which atmospheric CO_2 concentrations continue to increase to 720 ppm by 2100, after which they are held constant. Figure 4 shows the percentage increase in "coastal power dissipation" in five ocean basins using seven climate models, deduced using 2000 synthetic events in each model, in each basin, and for each of the 20th century and A1b simulations. Coastal power dissipation is just the sum over a given year of the cube of the maximum wind speed at the time that a storm makes landfall, and is a rough measure of potential destructiveness of tropical cyclones. Results vary greatly from model to model, reflecting the general uncertainties remaining in the field of climate modeling. There is a general tendency for the frequency of events (not shown) to decline, as is the case with direct simulations using global models; this is partially offset by a tendency for increased intensity. In addition, changes in the general circulation result in changing storm tracks, which also influence coastal

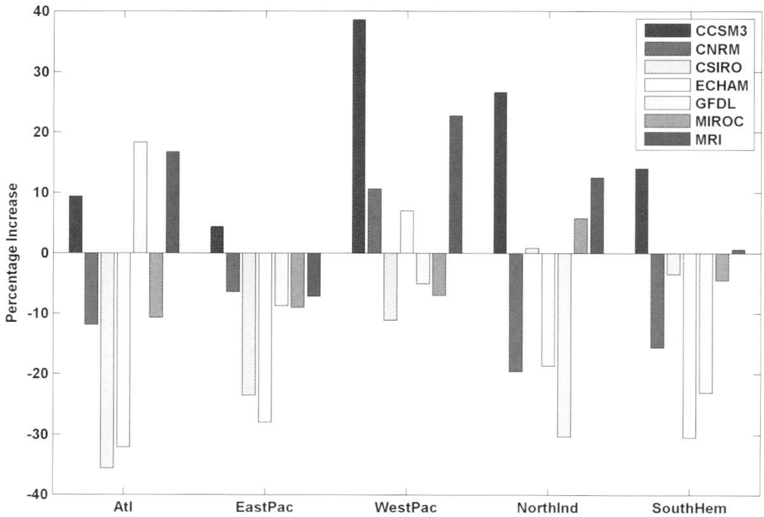

Fig. 4. Percentage increase in coastal tropical cyclone power dissipation between the last 20 years of the 20th century and the last 20 years of the 22nd century, based on 2000 synthetic storms in each of five ocean basins for each of seven global climate models. The 22nd century statistics are taken from models forced according to IPCC scenario A1b. [*Source*: Emanuel *et al.* (2008).]

power dissipation. The decline in frequency of events in these simulations is owing to the increase in the magnitude of an important non-dimensional parameter in the intensity model. This parameter, χ_m, is defined

$$\chi_m \equiv \frac{s_m - s_m^*}{s_0^* - s_b}, \tag{5.1}$$

where s_m is the moist entropy of the middle troposphere (near the level where it attains a minimum value), s_m^* is its saturation value, s_0^* is the moist entropy of air saturated at sea surface temperature and pressure, and s_b is the moist entropy of the boundary layer. This quantity is non-positive, and its magnitude measures the degree of thermodynamic inhibition to tropical cyclone formation. It is easy to show that at constant relative humidity, the numerator of (5.1) scales with the saturation specific humidity, as dictated by Clausius–Clapeyron. On the other hand, the denominator measures the air-sea thermodynamic disequilibrium which, at constant surface wind speed, is proportional to the surface turbulent energy flux into the atmosphere. This, in turn, rises only slowly with global warming, since surface evaporation is constrained to balance the net surface radiative flux, which changes only slowly, once the surface temperature becomes fairly large. Thus global warming has the effect of decreasing tropical cyclone frequency. At the same time, potential intensity generally increases with warming, so that some increase in the frequency of the most intense events is to be expected.

6. Effect of Tropical Cyclones on Climate

Discussions of tropical cyclones and climate almost always assume that any changes in tropical cyclone activity are passive; i.e., there is little or no feedback of tropical cyclones on the climate system. Globally, tropical cyclones contribute only a few percent of the total precipitation (and thus latent heat release) in the tropics; on the other hand, their precipitation efficiency is anomalously high, so that they may serve to dehydrate the tropical atmosphere to some degree. This might serve to cool the tropics, owing to the decline of the greenhouse effect of water vapor. Because of the very high specific entropy content of the tropical cyclone eyewall, they can extend further into the lower stratosphere than most convection; so it is possible that they play a role in the regulation of stratospheric water vapor. But perhaps their greatest influence on climate is exerted through the oceans.

Tropical cyclones are observed to vigorously mix the upper ocean (Leipper, 1967). The mechanism for doing this is somewhat indirect. Because of their horizontal scale and translation speeds, tropical cyclones are particularly efficient in exciting near-inertial oscillations in the upper ocean (Price, 1981). Vertical shear of ocean currents across the base of the mixed layer is almost invariably unstable, resulting in small-scale turbulence that mixes colder thermocline waters across into the mixed layer, thereby cooling it and warming the upper thermocline (Price, 1981). This mixing occurs on timescales of 6–24 hours associated with the passage of storms and the near-inertial response to the time-varying wind stress they produce. The mixing itself does not change the column-integrated enthalpy; enthalpy is merely redistributed in the vertical. However, the cold anomaly produced at the surface is observed to recover over a period of about ten days (Nelson, 1996), owing to a reduction in the turbulent enthalpy flux to the atmosphere. This wake recovery is associated with a net, column-integrated enthalpy increase in the ocean. Assuming that all of the cold anomaly recovers, the author (Emanuel, 2001) estimated that global tropical cyclone

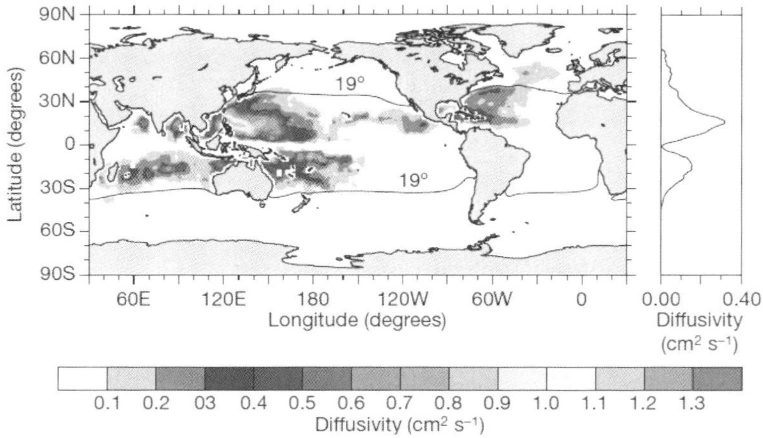

Fig. 5. Vertical diffusivity induced by tropical cyclones, estimated from European Center for Medium Range Weather Forecasts re-analyses. [*Source*: Sriver and Huber (2007). The panel at right shows the zonal average.]

activity results in an average net heat input rate of 1.4 ± 0.7 PW to the tropical oceans, a number comparable to the total poleward heat transport by the oceans. Recently, a more conservative estimate of around was made by Sriver and Huber (2007), who used European Center re-analyses to estimate cold wake recovery.[f] Figure 5, from that paper, shows the estimated vertical diffusivity induced by global tropical cyclone activity.

Experiments with ocean models show that spatially and temporally isolated mixing events are as effective as broadly distributed mixing in inducing a poleward heat transport in the ocean (Scott and Marotzke, 2002; Boos *et al.*, 2004), so that much of the upper ocean heat uptake induced by tropical cyclone mixing is exported toward higher latitudes, though some may return to the atmosphere locally in the tropics in the subsequent cool season.

It is possible that the cold wake may recover only through a very shallow depth, leaving a dipole temperature anomaly (or "heton") in the

[f]Cold wakes were assumed to penetrate only to 50 m depth, and SSTs are updated as infrequently as seven days in the re-analyses, leading to underestimation of cold wake magnitude, so that this estimate is conservative.

upper ocean, with very little change in the column-integrated enthalpy. In principle, the total heat uptake during wake recovery should be reflected in an elevation of the sea surface, which is detectable using satellite-based sea surface altimetry. Using the hydrostatic equation, the change in column enthalpy content, Δk, should be related to the change in sea surface elevation, Δk, by

$$\Delta k = \frac{\rho c_l}{\alpha} \Delta z, \tag{6.1}$$

where ρ is the density of seawater, c_l its heat capacity, and α its coefficient of thermal expansion. (Fig. 5 of Emanuel, 2001), reproduced here as Fig. 6, shows sea surface elevation as a function of time and cross-track distance during the wake recovery of Atlantic Hurricane Edouard in 1996.

One observes that the sea surface rises mostly to the right of the storm track, where the largest near-inertial response and cooling occurs, and that the surface rises by about 5 cm. According to Eq. (6.1), this gives a heat uptake of about 8×10^8 Jm^{-2}, which, when integrated over the approximately 800 km width and 3000 km length of the wake, yields a total heat uptake of around 2×10^{21} J. If there were 15 such events globally each year, the average rate of induced heat uptake would be about 1×10^{15} W, consistent with the earlier estimate by Emanuel (2001). In particular, the magnitude of the sea surface height response evident in Fig. 6 suggests that wake recovery was deep in this case.

The implications of this for climate dynamics should not be understated. As pointed out by the author (Emanuel, 2001), increased tropical cyclone activity in a warmer climate would result in increased tropical heat export by the oceans, mitigating tropical warming but amplifying the warming of higher latitudes. This inference is supported by recent numerical simulations using a coupled climate model in which upper ocean mixing is related to a proxy for tropical cyclone activity (Korty *et al.*, 2008). This effect offers a potential explanation for the equable nature of very warm climates such as that of the early Eocene; high levels of tropical cyclone activity in such warm climates could drive a strong poleward heat flux in the ocean, even in the face of relatively weak pole-to-equator temperature gradients, thus helping to keep such gradients weak. (Today's coupled climate models are notoriously bad at reproducing such weak temperature gradients, perhaps because they have no representation of tropical cyclone-induced ocean mixing.) It may also help explain why most of the observed heat

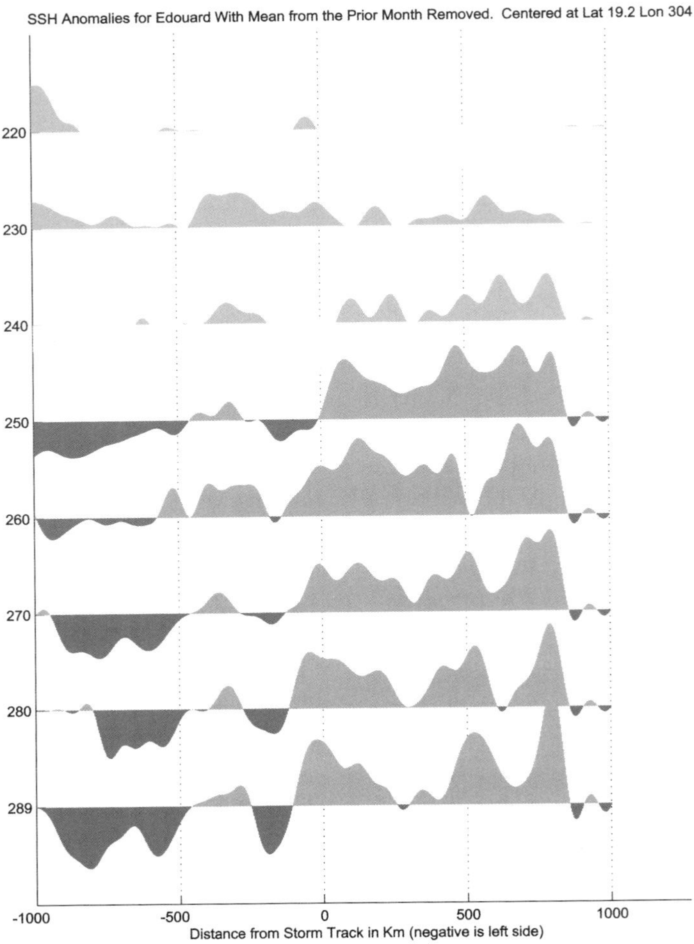

Fig. 6. Cross-track sections of the sea surface height anomaly from TOPEX/POSIDEN at ten day intervals in August–September, 1996. Hurricane Edouard passed this transect on Julian Day 239. The height anomaly corresponding to the vertical separation between the transects is 20 cm. The transect is centered at 19.2 N, 56 W and the time of the transect is indicated at left by Julian Day. The anomalies represent differences from the sea surface height averaged over the month preceding Julian Day 220. (Analysis and figure courtesy of Peter Huybers (private communication).)

uptake by the oceans over the past 50 years has been in the subtropics and middle latitudes (Levitus *et al.*, 2005), whereas coupled models typically show most of the heat uptake occurring in subpolar regions (Manabe *et al.*, 1991).

Clearly, tropical cyclone-induced heat uptake may be an important element of climate dynamics and should remain an active research topic for the next few years at least.

7. Summary

Tropical cyclones respond to climate change in a number of ways. Their level of activity appears to be controlled primarily by four factors: Potential intensity, vertical shear of the horizontal environmental wind, low-level vorticity, and the parameter defined by (5.1) and measuring the specific humidity deficit of the middle troposphere. Records of tropical cyclones are best and longest in the North Atlantic, are somewhat less reliable in the western North Pacific, and are dubious elsewhere, particularly before the satellite era. In the North Atlantic region, tropical cyclone power dissipation is highly correlated with tropical SST during hurricane season, on timescales of a few years and longer. The tropical North Atlantic SST is in turn highly correlated with northern hemisphere surface temperature, at least during hurricane season, on timescales of a decade and longer. The weight of available evidence suggests that multi-decadal variability of hurricane season tropical Atlantic SST and northern hemispheric surface temperature, evident in Fig. 3, is controlled mostly by time-varying radiative forcing owing to solar variability, major volcanic eruptions, and anthropogenic sulfate aerosols and greenhouse gases, though the response to this forcing may be modulated by natural modes of variability. The increase in potential intensity of about 10% in the North Atlantic over the last 30 years was driven by increasing greenhouse gas forcing, declining lower stratospheric temperature, and decreasing surface wind speed (Emanuel, 2007); this increase is consistent with the ~60% increase in tropical cyclone power dissipation during this time.

Explicit simulations of tropical cyclones using global climate models as well as a variety of downscaling techniques all show a general tendency toward decreasing tropical cyclone frequency and increasing intensity and rainfall rates, although there is much variability from model to model and from ocean basin to ocean basin. The increased intensity is related to increasing potential intensity as the climate warms, while the increased rainfall rate is a straightforward consequence of increased atmospheric humidity, according to Clausius–Clapeyron. The decreasing frequency of tropical cyclones appears to be owing to an increase in the magnitude of the thermodynamic inhibition to genesis, as given by the parameter χ_m defined by (5.1); this is a predictable consequence of global (as opposed to local) warming.

Tropical cyclones may affect climate through drying of the troposphere and especially by mixing the upper tropical oceans. Available evidence suggests that global tropical cyclone activity may be an important or even dominant mechanism in maintaining poleward heat flux by the oceans. Since tropical cyclones both respond to and affect climate change, their existence modifies climate dynamics in a way that may help explain both the pattern of recent heat uptake by the oceans, and the peculiar features of very warm climates such as that of the early Eocene. Further research needs to be undertaken to explore these ideas.

Acknowledgments

I thank the modeling groups for making their simulations available for analysis, the Program for Climate Model Diagnosis and Intercomparison (PCMDI) for collecting and archiving the CMIP3 model output, and the WCRP's Working Group on Coupled Modeling (WGCM) for organizing the model data analysis activity. The WCRP CMIP3 multi-model dataset is supported by the Office of Science, U.S. Department of Energy. I was supported by grant ATM-0432090 from the National Science Foundation.

References

Bengtsson, L., Botzet, M. and Esch, M. (1996). Will greenhouse-induced warming over the next 50 years lead to higher frequency and greater intensity of hurricanes? *Tellus* **48A**, pp. 57–73.

Bengtsson, L., Hodges, K. I., Esch, M., *et al.* (2007). How may tropical cyclones change in a warmer climate? *Tellus* **59**, pp. 539–561.

Bister, M. and Emanuel, K. A. (2002). Low frequency variability of tropical cyclone potential intensity, 1: Interannual to interdecadal variability, *J. Geophys. Res.* **107**, p. 4801, doi:10.1029/2001JD000776.

Boos, W., Scott, J. and Emanuel, K. (2004). Transient diapycnal mixing and the meridional overturning circulation, *J. Phys. Oceanogr.* **34**, pp. 334–341.

Camargo, S. L., Emanuel, K. A. and Sobel, A. H. (2007). Use of genesis potential index to diagnose ENSO effects on tropical cyclone genesis, *J. Climate* **20**, pp. 4819–4834.

Chan, J. C. L. (2006). Comments on "Changes in tropical cyclone number, duration, and intensity in a warming environment", *Science* **311**, p. 1713.

Chen, S. S., Price, J. F., Zhao, W., *et al.* (2007). The CBLAST-hurricane program and the next-generation fully coupled atmosphere-wave-ocean models for hurricane research and prediction, *Bull. Amer. Meteor. Soc.* **88**, pp. 311–317.

Delworth, T. L. and Mann, M. E. (2000). Observed and simulated multidecadal variability in the Northern Hemisphere, *Climate Dyn.* **16**, pp. 661–676.

Donnelly, J. P. (2005). Evidence of past intense tropical cyclones from backbarrier salt pond sediments: a case study from Isla de Culebrita, Puerto Rico, USA, *J. Coastal Res.* **I42**, pp. 201–210.

Donnelly, J. P., Bryaut, S. S. and Butler, J. (2001a). 700 yr sedimentary record of intense hurricane landfalls in southern New England, *Geo. Soc. Am. Bull.* **114**, pp. 714–727.

Donnelly, J. P., Roll, S. and Butter, J. (2001b). Sedimentary evidence of intense hurricane strikes from New Jersey, *Geology* **29**, pp. 615–618.

Donnelly, J. P. and Woodruff, J. D. (2007). Intense hurricane activity over the past 5,000 years controlled by El Niño and the West African monsoon, *Nature* **447**, pp. 465–468.

Elsner, J. B. (2006). Evidence in support of the climate change-Atlantic hurricane hypothesis, *Geophys. Res. Lett.* **33**, L16705, doi:10.1029/2006GL026869.

Elsner, J. B., Tsonis, A. A. and Jagger, T. H. (2006). High-frequency variability in hurricane power dissipation and its relationship to global temperature, *Bull. Amer. Meteor. Soc.* **87**, pp. 763–768.

Emanuel, K. (2001). The contribution of tropical cyclones to the oceans' meridional heat transport, *J. Geophys. Res.* **106**, D14, pp. 14771–14782.

Emanuel, K. (2006). Climate and tropical cyclone activity: a new model downscaling approach, *J. Climate* **19**, pp. 4797–4802.

Emanuel, K. (2007). Environmental factors affecting tropical cyclone power dissipation, *J. Climate* **20**, pp. 5497–5509.

Emanuel, K. and Nolan, D. (2004). Tropical cyclone activity and the global climate system, *26th Conference on Hurricanes and Tropical Meteorology, Miami, Amer. Meteor. Soc.*

Emanuel, K., Sundararajan, R. and Williams, J. (2008). Hurricanes and global warming, *Bull. Amer. Meteorol. Soc.* **89**, 3, pp. 347–367.

Emanuel, K. A. (1987). The dependence of hurricane intensity on climate, *Nature* **326**, pp. 483–485.

Emanuel, K. A. (2000). A statistical analysis of tropical cyclone intensity, *Mon. Wea. Rev.* **128**, pp. 1139–1152.

Emanuel, K. A. (2005a). Increasing destructiveness of tropical cyclones over the past 30 years, *Nature* **436**, pp. 686–688. (online suppl<ftp://texmex.mit.edu/puk/emanuel/PAPERS/NATORE03906.suppl.pdf.)

Emanuel, K. A. (2005b). Meteorology: Emanuel replies, *Nature* **438**, p. 7071.

Emanuel, K. A., Ravela, S., Vivant, E., *et al.* (2006). A statistical-deterministic approach to hurricane risk assessment, *Bull. Amer. Meteor. Soc.* **19**, pp. 299–314.

Frappier, A. B., Knutson, T. R., Liu, K.-B., *et al.* (2007a). Perspective: coordinating paleoclimate research on tropical cyclones with hurricane-climate theory and modelling, *Tellus A* **59**, pp. 527–529.

Frappier, A. B., Sahagian, D., Carpenter, S. J., *et al.* (2007b). Stalagmite stable isotope record of recent tropical cyclone events, *Geology* **35**, pp. 111–114.

Goldenberg, S. B., Landsea, C. W., Nez, A. M. M.-N., *et al.* (2001). The recent increase in Atlantic hurricane activity: causes and implications, *Science* **293**, pp. 474–479.

Gray, S. T., Graumlich, L. J., Betancourt, J. L., *et al.* (2004). A tree-ring based reconstruction of the Atlantic Multidecadal Oscillation since 1567 A.D., *Geophys. Res. Lett.* **31**, doi:10.1029/2004GL019932.

Gray, W. M. (1968). Global view of the origin of tropical disturbances and storms, *Mon. Wea. Rev.* **96**, pp. 669–700.

Gray, W. M. (1979). Hurricanes: their formation, structure, and likely role in the tropical circulation, in D. B. Shaw (ed.), *Meteorology Over the Tropical Oceans* (Roy. Meteor. Soc. Bracknell, Enqlan) pp. 155–199.

Gray, W. M. (1984). Atlantic seasonal hurricane frequency. Part I: El Niño and 30 mb quasi-biennial oscillation influences, *Mon. Wea. Rev.* **112**, pp. 1649–1668.

Holland, G. J. (2007). Misuse of landfall as a proxy for Atlantic tropical cyclone activity, *EOS* **88**, 36, pp. 349–350.

Holland, G. J. and Webster, P. J. (2007). Heightened tropical cyclone activity in the North Atlantic: natural variability or climate trend? *Phil. Trans. Roy. Soc.* doi:10.1098/rsta.2007.2083.

Hoyos, C. D., Agudelo, P. A., Webster, P. J., *et al.* (2006). Deconvolution of the factors contributing to the increase in global hurricane intensity, *Science* **312**, pp. 94–97.

IPCC (2007). WG1 AR4 Report, Tech. rep., World Meteorological Organization, Geneva.

Kerr, R. A. (2000). A North Atlantic climate pacemaker for the centuries, *Science* **288**, pp. 1984–1985.

Knutson, T. R., Sirutis, J. J., Garner, S. T., *et al.* (2007). Simulation of the recent multi-decadal increase of Atlantic hurricane activity using an 18 km grid regional model, *Bull. Amer. Meteor. Soc.* **88**, p. 1549.

Knutson, T. R., Tuleya, R. E. and Kurihara, Y. (1998). Simulated increase of hurricane intensities in a CO2-warmed climate, *Science* **279**, pp. 1018–1020.

Knutson, T. R. and Tuleya, R. E. (2004). Impact of CO2-induced warming on simulated hurricane intensity and precipitation: sensitivity to the choice of climate model and convective parameterization, *J. Climate* **17**, pp. 3477–3495.

Korty, R., Emanuel, K. A. and Scott, J. R. (2008). Tropical cyclone-induced upper ocean mixing and climate: application to equable climates, *J. Climate* **21**, pp. 638–654.

Kossin, J. P. and Vimont, D. J. (2007). A more general framework for understanding Atlantic hurricane variability and trends, *Bull. Amer. Meteor. Soc.* **88**, 11, pp. 1767–1781.

Landsea, C. W. (2005). Meteorology: hurricanes and global warming, *Nature* **438**, 7071, pp. E11–E12.

Landsea, C. W., Harper, B. A., Hoarau, K., *et al.* (2006). Can we detect trends in extreme tropical cyclones? *Science* **313**, pp. 452–454.

Landsea, C. W. (2007). Counting Atlantic tropical cyclones back to 1900, *EOS* **88**, pp. 197–200.

Lawrence, J. R. and Gedzelman, S. D. (1996). Low stable isotope ratios of tropical cyclone rains, *Geophys. Res. Lett.* **23**, pp. 527–530.

Leipper, D. F. (1967). Observed ocean conditions and Hurricane Hilda, 1964, *J. Atmos. Sci.* **24**, pp. 182–196.

Levitus, S., Antonov, J. I. and Boyer, T. P. (2005). Warming of the world ocean, 1955–2003, *Geophys. Res. Lett.* **32**, doi:10.1029/2004GL021592.

Liu, K.-B. (2007). Paleotempestology, in S. Elias (ed.), *Encyclopedia of Quaternary Science* (Elsevier, Amsterdam), pp. 1978–1986.

Liu, K.-B. and Fearn, M. L. (1993). Lake-sediment record of late Holocene hurricane activities from coastal Alabama, *Geology* **21**, pp. 793–796.

Liu, K.-B. and Fern, M. L. (2000). Reconstruction of prehistoric landfall frequencies of catastrophic hurricanes in northwestern Florida from lake sediment records, *Quat. Res.* **54**, pp. 238–245.

Manabe, S., Stouffer, P. J., Spelman, M. J., *et al.* (1991). Transient responses of a coupled ocean-atmosphere model to gradual changes of atmospheric CO2. Part I. Annual mean response, *J. Climate* **4**, pp. 785–818.

Mann, M. E. and Emanuel, K. A. (2006). Atlantic hurricane trends linked to climate change, *EOS* **87**, pp. 233–244.

Mann, M. E., Emanuel, K. A., Holland, G. J., *et al.* (2007). Atlantic tropical cyclones revisited, *EOS* **88**, p. 349.

Marks, D. G. (1992). *The Beta and Advection Model for Hurricane Track Forecasting* (Natl. Meteor. Center, Camp Springs, Maryland).

Mestas-Nuñez, A. M. and Enfield, D. B. (1999). Rotated global modes of non-ENSO sea surface temperature variability, *J. Climate* **12**, pp. 2734–2746.

Miller, D. L., Mora, C. I., Grissino-Mayer, H. D., *et al.* (2006). Tree-ring isotope records of tropical cyclone activity, *Proc. Natl. Acad. Sci.* **103**, pp. 14294–14297.

Nelson, N. B. (1996). The wake of Hurricane Felix, *Int. J. Rem. Sens.* **17**, pp. 2893–2895.

Nott, J. F. and Hayne, M. (2001). High frequency of "super-cyclones" along the Great Barrier Reef over the past 5,000 years, *Nature* **413**, pp. 508–512.

Nott, J. F. (2003). Intensity of prehistoric tropical cyclones, *J. Geophys. Res.* **108**, doi:10.1029/2002JD002726.

Nott, J. J. (2004). Palaeotempestology: the study of prehistoric tropical cyclones — a review and implications for hazard assessment, *Environ Int.* **30**, pp. 433–447.

Oouchi, K., Yoshimura, J., Yoshimura, H., *et al.* (2006). Tropical cyclone climatology in a global-warming climate as simukated in a 20 km-mesh global atmospheric model: frequency and wind intensity analyses, *J. Meteor. Soc. Japan* **84**, pp. 259–276.

Palmén, E. (1984). On the formation and structure of tropical hurricanes, *Geophysica* **3**, pp. 26–39.

Price, J. F. (1981). Upper ocean response to a hurricane, *J. Phys. Oceanog.* **11**, pp. 153–175.

Santer, B. D., Wigley, T. M. L. and Gleckler, P. J. (2006). Forced and unforced ocean temperature changes in Atlantic and Pacific tropical cyclogenesis regions, *Proc. Natl. Acad. Sci.* **103**, pp. 13905–13910.

Scott, J. R. and Marotzke, J. (2002). The location of diapycnal mixing and the meridional overturning circulation, *J. Phys. Oceanog.* **32**, pp. 3578–3595.

Sriver, R. L. and Huber, M. (2007). Observational evidence for an ocean heat pump induced by tropical cyclones, *Nature* **31**, pp. 577–580.

Sugi, M., Noda, A. and Sato, N. (2002). Influence of the global warming on tropical cyclone climatology: an experiment with the JMA global climate model, *J. Meteor. Soc. Japan* **80**, pp. 249–272.

Trenberth, K. E. and Shea, D. J. (2006). Atlantic hurricanes and natural variability in 2005, *Geophys. Res. Lett.* **33**, L12704, doi:10.1029/2006GL026894.

Webster, P. J., Holland, G. J., Curry, J. A., *et al.* (2005). Changes in tropical cyclone number, duration and intensity in a warming environment, *Science* **309**, pp. 1844–1846.

Yoshimura, S. M. J. and Noda, A. (2006). Influence of greenhouse warming on tropical cyclone frequency, *J. Meteor. Soc. Japan* **84**, pp. 405–428.

Zhang, R., Delworth, T. L. and Held, I. M. (2007). Can the Atlantic Ocean drive the observed multidecadal variability in northern hemisphere mean temperature? *Geophys. Res. Lett.* **34**, L02709, doi:10.1029/2006GL028683.

Chapter 6

Transport and Mixing of Atmospheric Pollutants

Peter H. Haynes

Department of Applied Mathematics and Theoretical Physics
University of Cambridge
Wilberforce Road, Cambridge, CB3 0WA, UK
phh@damtp.cam.ac.uk

It is now realized that air quality is determined not only by local emissions and local meteorology, but also by long-range atmospheric transport of chemical species from emission regions that may be thousands of kilometers from the region of interest.

Predicting and understanding air quality requires consideration of many different processes, including emissions, boundary layer physics, chemical reactions, and interactions with clouds and particles. These lecture notes focus on the role of atmospheric transport and mixing, emphasizing the fundamental ideas and describing relevant mathematical models. In many parts of the atmosphere large-scale quasi-horizontal flow appears to play the dominant role in transport and in the stirring process that leads ultimately to true (molecular) mixing at very small scales. This means that calculations based on large-scale meteorological datasets can give valuable quantitative information on transport. To make local predictions requires more detailed information on transport, e.g., from regional-scale models.

1. Motivation

The effect of pollution on air quality has been a concern for at least the last 150 years or so and in some cases for longer. Poor air quality originally resulted primarily from coal burning, both domestic and industrial. By the 1950s a distinct problem of photochemical smog resulting primarily from car exhausts had been identified. This was particularly serious in large cities such as Los Angeles where topography favored the trapping (and subsequent photochemical evolution) of polluted air. In such cases the problem is essentially one of local emissions in a given urban area (perhaps a very large urban area such as the Los Angeles basin) leading to an adverse effect on air quality in that same area.

However, more recently the non-local effects of air pollution have been recognized. One example is that of acid rain where the effect of emission of sulphur compounds as part of coal burning is felt hundreds of kilometers away through rainfall that is significantly more acidic than normal, with adverse effects on vegetation and on soil and freshwater ecosystems. Another example is that of low-level ozone, which results from emissions of nitrogen oxides and hydrocarbons (known as ozone precursors). Low-level ozone is potentially harmful to human health and to agriculture (Amann *et al.*, 2008; Royal Society, 2008). Ozone concentrations in emission regions are sometimes relatively low, e.g., in the center of cities, because high concentrations of nitrogen oxides limit ozone concentrations. However, away from these regions, e.g., in suburbs and surrounding rural areas, nitrogen oxide concentrations decrease and ozone concentrations therefore increase. (A corollary is that reduction of nitrogen oxide emissions in city centers, e.g., through installation of catalytic convertors on car exhausts, has actually increased local ozone concentrations.) It has been recognized for some time that since, away from the Earth's surface, the lifetime of tropospheric ozone is relatively long (perhaps 20 days or more) ozone concentrations in Europe, for example, are determined not locally, but by precursor emissions over a broad continental region and efforts to limit or even reduce ozone concentrations have had to focus on continent-wide emissions of the ozone precursors. Indeed it is now clear that local concentrations of ozone are affected by intercontinental transport, (e.g., Akimoto, 2003) and the hemispheric, or indeed global, aspects of air quality are being now being recognized. Ensuring that air quality standards are met therefore requires not just regional but global policies on emissions (Derwent *et al.*, 2006).

It has also been recognized that emissions from non-industrial sources such as agricultural waste burning and forest fires make a significant contribution to gases such as carbon monoxide. In Southeast Asia such biomass burning has had a large-scale effect on air quality, particularly during 1997, but also in several subsequent years, with the effects being felt well over 1000 km away from the primary burning regions. The problems experienced in 1997 and subsequently have prompted the formulation of an ASEAN agreement on Transboundary Haze Pollution.

Quantitative prediction of air quality is therefore now seen to require not only modeling of local emission and transport, but also transport on a regional and indeed even global scale. Of course incorporation of processes on this range of scales may not be possible in a single numerical model, and specific problems will require a specific focus. But an effective overall scientific perspective does need to take account of the global as well as local aspects of the problem.

The recognition that local, regional, and hemispheric-scale processes are relevant to air quality broadens the range of physical processes that are relevant and that must be incorporated in predictive numerical models. Such models must represent emissions, chemical evolution, and transport and mixing by the atmospheric flow. Numerical modeling is a well-developed field and many sophisticated approaches have been devised to represent the effects of different processes. The purpose of these notes is not to describe state-of-the-art numerical modeling, but to set out some of the basis physical processes of atmospheric transport and mixing that must be represented in models. A much broader review of atmospheric composition change, its implications for global and regional air quality, and modeling approaches is provided in Monks *et al.* (2009).

2. Transport and Mixing in the Atmosphere

The dynamics of the atmospheric flow, which determine the transport and mixing properties, are discussed in earlier chapters of this volume and only a brief summary of some key points is given below. Detailed treatments may be found in textbooks such as those by (in increasing levels of sophistication and detail) Houghton (2002), Holton (2004) and Vallis (2006).

A major role in determining the nature of the atmospheric flow is played by stable density stratification, which tends to inhibit vertical motion. The primary physical quantity determining density variations is temperature; however, density variations associated with pressure also need to be taken into account and it turns out that the most appropriate density variable is potential temperature, $\theta = T(p/p_*)^{-\kappa}$ where T is temperature, p is pressure, p_* a constant reference pressure, and κ is a constant, equal to $2/7$. θ is conserved by an air parcel in adiabatic motion — it therefore allows assessment of the effects of vertical displacements. If θ increases upward then an air parcel that is displaced upward will find itself denser than its surroundings and will therefore tend to return to its original level — i.e., the density stratification is stable. On the other hand if θ decreases upward then an air parcel displaced upward will find itself lighter than its surroundings and will tend to move further upward — i.e., the density stratification is unstable. More precisely an appropriate measure of stability is the square of the buoyancy frequency $(g/\theta)(d\theta/dz) = (g/T)(dT/dz + \kappa g/R)$ where g is the gravitational acceleration, R is the gas constant and the second term in the sum follows from the hydrostatic equation and the gas law. If $dT/dz + \kappa g/R$ is positive then the stratification is stable and if it is negative the stratification is unstable.

This dependence of stability on vertical temperature gradient motivates the conventional division of the atmosphere into layers according to the vertical temperature gradient. In the *troposphere* (the lowest 10 km or so of the atmosphere) the temperature decreases with height and whilst the associated density stratification is stable (since $-dT/dz < \kappa g/R$), the stability is relatively weak. In the *stratosphere* (roughly 10–50 km) the temperature is constant with height or increases with height and the stability is much stronger than in the troposphere. The dynamical differences between troposphere and stratosphere, due to the differing stability, are mirrored in chemical differences with, for example, water vapor concentrations in the troposphere being much higher than in the stratosphere and, conversely, ozone concentrations being much lower. (The explanation lies in transport and mixing together with the different sources and sinks of different chemical species.) The transition from troposphere to stratosphere has conventionally been viewed as sharp and the location of the transition is called the *tropopause*. However, for many purposes it is better to consider the transition as taking place over a *tropopause layer* of finite thickness.

The stabilization due to density stratification (and to some extent rotation) mean that three-dimensional turbulence (i.e., the sort of turbulence that would be observed in a wind tunnel, or in a strongly stirred laboratory tank) is confined to relatively localized regions of the atmosphere. In the troposphere these regions include the atmospheric boundary layer (the lowest kilometer or so of the atmosphere where dynamical effects of direct contact with the Earth's surface overcome the stabilization) and to convective clouds (where the stabilization is overcome by dynamical effects of moist processes such as condensation). However, even in the tropics, where moist dynamics are most important, convective clouds fill a relatively small fraction of the total area. Elsewhere in the troposphere and stratosphere there are localized regions of turbulence resulting from dynamical instabilities such as the breaking of inertia-gravity waves. Even in the troposphere, the timescale on which air masses encounter these turbulent regions might be relatively long — several days or more. Evidence for this comes, for example, from the observations of thin layers with a distinct chemical signature, which are likely to have been transported thousands of kilometers from their formation regions (Newell *et al.*, 1999).

In considering flow outside of regions of three-dimensional turbulence, potential temperature θ (which increases upward) is a useful vertical coordinate. Processes which change the potential temperature of an air parcel are relatively weak (molecular dissipation in three-dimensional turbulence, radiative transfer) and therefore to a reasonable approximation — on a timescale of a few days in the troposphere and longer in the stratosphere — air parcels move along surfaces of constant θ. The implication is that air parcels can move rapidly along θ surfaces, but only slowly across them.

Figure 1 shows longitudinally averaged temperature and potential temperature (θ) fields for the atmosphere, which gives a good impression of the typical configuration of the θ-surfaces in a latitude-height cross section. Note that in the weakly stable troposphere the θ-surfaces are relatively widely separated in the vertical, whilst in the strongly stable stratosphere

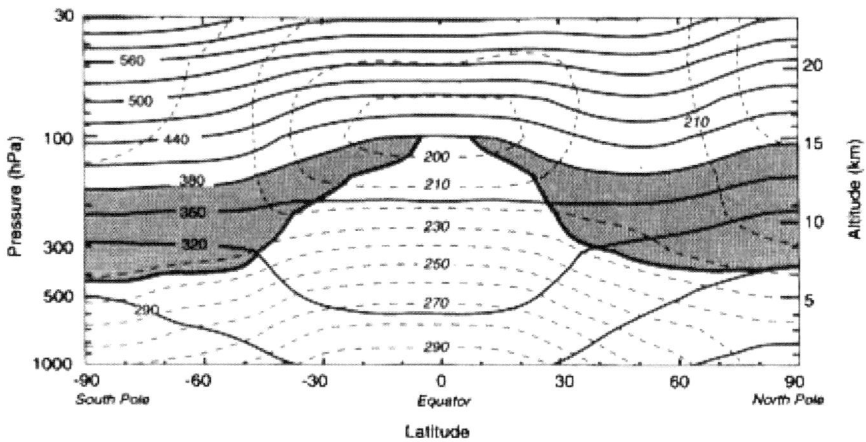

Fig. 1. Latitude-altitude cross section for January 1993 showing longitudinally averaged potential temperature (θ) (solid contour) and temperature (dashed contours). Contours, e.g., θ-contours, in this cross section correspond to surfaces in the three-dimensional atmosphere. The heavy solid contour (cut off at the 380 K θ contour) denotes the tropopause defined as the 2-PVU potential vorticity contour. (See Holton *et al.* (1995) for more detail.) Shaded areas denote the "lowermost stratosphere," being the part of the stratosphere which contains θ-surfaces which enter the troposphere. [*Source*: Holton *et al.* (1995). Data are from United Kingdom Meteorological Office analyses. Copyright 2003 American Geophysical Union. Reproduced by permission of American Geophysical Union.]

they are closer together. In the extratropical troposphere the θ-surfaces slope strongly, indicating a rapid route for transport from the surface (or the boundary layer) to the upper troposphere, or vice versa. This is an important aspect of the intercontinental transport of pollution mentioned in Sec. 1. The tropopause is marked in Fig. 1 by the thick curve. Note that outside of the tropics the θ-surfaces, e.g., the 320 K surface, cut across the tropopause implying the possibility of rapid transport from the stratosphere to the troposphere and even to the surface and, again, vice versa. The part of the stratosphere, marked as shaded in Fig. 1, that is accessible from the troposphere via θ surfaces is sometimes called the *lowermost stratosphere* (Holton *et al.*, 1995). If there were unrestricted rapid motion along θ-surfaces then one might expect significant differences between this part of the stratosphere and that above. The differences are not so great, otherwise the shaded region might have been historically identified as troposphere rather than stratosphere. This illustrates the important point that whilst rapid transport along θ-surfaces is possible, it is not guaranteed. It turns out, for example, that transport along the 320 K θ-surface is relatively inhibited in the region of the tropopause (and if it was not then the dynamical and chemical contrast between troposphere and stratosphere on that θ-surface would disappear). This inhibition of transport is consistent with the presence of the subtropical jet in this region in each hemisphere, i.e., at 30°–40° of latitude and between 8–12 km in altitude. (See further comments in Sec. 3.6.)

The processes that are involved in quasi-horizontal transport along θ-surfaces include synoptic-scale weather systems, larger-scale "planetary" waves that modulate the circulation on scales of thousands of kilometers and, in the tropics, large-scale circulations associated with features such as monsoons, driven by spatial variations in sea surface temperatures (SSTs) and by heating contrasts between land and ocean. All of these flows vary strongly in longitude. Local variations are, of course, important in determining local chemical distributions, but it is also the case that the averaged effect of the longitudinally varying flows needs to be taken into account in explaining the height-latitude variation of the distributions of different chemical species. The longitudinally varying flow has a dual character, with some aspects of its behavior appearing organized and wave-like and other aspects exhibiting considerable nonlinearity and randomness. In the latter respect the flow might therefore be regarded as a kind of turbulence, closely related to the two-dimensional turbulence studied in idealized numerical simulations and in laboratory experiments where there is rapid rotation or strong density stratification (and very different from three-dimensional turbulence). (See Chapters 8 and 9 of Vallis, 2006.)

3. Fundamentals of Transport and Mixing

3.1. *Definitions*

The atmospheric flow (or any other flow) affects the distribution of chemical species through at least three distinct processes. Firstly, it moves chemical species away from their source regions, where they might be emitted by natural processes or by human activity or produced *in-situ* by suitable chemical reactions, e.g., photochemical production, to other regions where they might be detected by suitable measurement, or indeed they might be destroyed by chemical reaction or absorbed at the land or sea surface or onto cloud particles. This process by which chemical species are carried away from source regions to some other part of the flow is called *transport*. Unless the flow is uniform in space, it not only carries chemical species from one location to another, but it also distorts the spatial structure of chemical concentration fields, typically making the spatial structure more complicated by drawing it out into thin filaments or sheets. This process of distortion is called *stirring*. Ultimately, molecular diffusion acts to homogenize chemical concentration fields. This latter process is called *mixing*. Note that if two chemical species A and B, which potentially react together, are emitted in different regions then the final state of mixing is essential for the reaction to proceed. Stirring may lead to thin interleaved filaments or sheets containing either A or B, but the molecules of A or B are separate. It is only when mixing occurs at the molecular level, through the action of molecular diffusion, that the reaction may proceed. (Note that in some descriptions the term "mixing" is used without the requirement for molecular diffusion — e.g., a flow may be described as strongly mixing if it is strongly stirring, since only a small molecular diffusion is needed to change from a "stirred" state to a "mixed" state. But in the chemical context the distinction is very important.)

3.2. *Evolution Equations*

In considering transport, stirring and mixing it can be useful to consider the evolution in time t of either the position $\mathbf{X}(t)$ of a marked particle or the concentration field $\chi(\mathbf{x}, t)$ of a chemical species, with $\mathbf{x}$ being position. Given a velocity field $\mathbf{u}(\mathbf{x}, t)$ the position of a marked particle evolves according to

$$\frac{d\mathbf{X}}{dt} = \mathbf{u}(\mathbf{X}, t), \tag{3.1}$$

and the concentration field evolves according to

$$\frac{D\chi}{Dt} = \frac{\partial \chi}{\partial t} + \mathbf{u} \,.\, \nabla\chi = \kappa\nabla^2\chi \tag{3.2}$$

$$\text{(Estimate)} \qquad \frac{\chi U}{L} \qquad \frac{\kappa\chi}{L^2}$$

where κ is the molecular diffusivity. The operator $D/Dt = \partial/\partial t + \mathbf{u} \,.\, \nabla$ is called the *advective derivative* and represents the rate of change following a fluid particle. Note that if the molecular diffusivity were zero then Eq. (3.2) would simply imply that concentration following a fluid particle is constant. Equation (3.1) would then provide all the information needed to predict the evolution of the chemical concentration.

A rough estimate of the magnitude of each of the terms on the right-hand side of the concentration Eq. (3.2) is given below the equation, assuming that U is a velocity scale and L is a length scale. Note that the ratio of the first (advective) term to the second (molecular diffusive) term is given by dimensionless number, conventionally named the *Peclet number*.

$$Pe = \frac{UL}{\kappa}. \tag{3.3}$$

$Pe \gg 1$ means that diffusion is weak relative to advection. Note that Pe is the ratio between the time for diffusion over distance L, equal to L^2/κ divided by the time for advection L/U.

For the near-surface atmosphere the molecular diffusivity $\kappa \sim 10^{-5}$ $\mathrm{m^2s^{-1}}$. If we take (for the purposes of argument) $U = 1\,\mathrm{ms^{-1}}$ and $L = 1\,\mathrm{m}$, this implies that $Pe \sim 10^5$ or equivalently that the time for diffusion through distance L, $10^5\,\mathrm{s}$, is 10^5 times greater than the time for advection through distance L, $1\,\mathrm{s}$. The effects of molecular diffusivity are on this basis expected to be very weak. However, the effects of molecular diffusion cannot be neglected entirely, since this would rule out any molecular mixing (so, for example, two species A and B released in different regions of the flow could never come together to react).

A crucial point here is that Pe depends on the assumed length scale, L, which might be regarded as externally imposed, by the flow geometry, or by the inherent length scales in the flow. But in fact, for the above estimates of terms in Eq. (3.2) to apply, L must be the length scale of the concentration field. An important property of many (but not all) fluid flows is that the evolution of the chemical concentration field as predicted by Eq. (3.2) tends to reduce systematically the actual length scale of the concentration field, l, until the timescale for diffusion is comparable to that for advection. The reduction in scale is achieved by the stirring process, and when l is small enough for diffusion to be effective the stirring is followed by mixing.

3.3. *Stretching in Linear Flows*

More precise insight into the stirring process can be obtained by considering a simple model in which the scale of the concentration field is much less than the scale of the velocity field. On the scale of the concentration field the velocity field can be approximated by a Taylor expansion. The first (constant) term may be removed by transforming to a frame of reference moving with the local flow velocity, leaving a velocity field that is a linear function of space,

$$\mathbf{u}(\mathbf{x}, t) \simeq \mathbf{A}(t) \,.\, \mathbf{x} \tag{3.4}$$

where $A(t)$ is the local velocity gradient tensor $\nabla\mathbf{u}$ with components $\partial u_i/\partial x_j$. Note that if the flow is incompressible then $\nabla.\mathbf{u} = 0$ implying that the trace of the tensor $\mathbf{A}$ must be zero for each t. Incompressibility turns out to be a good approximation for atmospheric flows that are important for transport and mixing.

This leads to a simplied evolution equation for a line element $\mathbf{l}(t)$, i.e., the line joining two nearby marked points,

$$\frac{d\mathbf{l}}{dt} = \mathbf{A}(t) \,.\, \mathbf{l} \tag{3.5}$$

which may be derived from Eq. (3.1) by considering two nearby solutions $\mathbf{X}_1(t)$ and $\mathbf{X}_2(t) = \mathbf{X}_1(t) + \mathbf{l}(t)$. The corresponding equation for scalar concentration is that

$$\frac{\partial\chi}{\partial t} + (\mathbf{A}(t).\mathbf{x}).\nabla\chi = \kappa\nabla^2\chi. \tag{3.6}$$

$\mathbf{A}(t)$ is velocity gradient tensor following a fluid particle. It is therefore the time history of this tensor following the flow that determines the stretching process and its coupling to mixing.

Both equations are significant simplifications over their analogues for general flow, but even so solving them for general $\mathbf{A}(t)$ is not straightforward. It is useful to consider the simplest possible case of two-dimensional flow where the velocity field $\mathbf{u}$ is a linear function of space $\mathbf{u} = \mathbf{A} \,.\, \mathbf{x}$ with $\mathbf{A}$ constant in time. ∇u is therefore constant in space and time (and is therefore constant following fluid particles). There are three possible sorts of behavior for these flows, illustrated by the following three examples.

The first is steady "pure-strain" flow $\mathbf{u} = (\Gamma x, -\Gamma y)$, where Γ is constant. Then

$$\mathbf{A} = \begin{pmatrix} \Gamma & 0 \\ 0 & -\Gamma \end{pmatrix}$$

and it follows that the solution of Eq. (3.5) is

$$\mathbf{l}(t) = (l_1(t), l_2(t)) = (l_1(0)e^{\Gamma t}, l_2(0)e^{-\Gamma t}).$$

Thus, in this case, unless the initial direction of $\mathbf{l}$ is perfectly aligned with the y-axis, which is the compression direction for this strain field, $|\mathbf{l}|$ increases exponentially with t, and $\mathbf{l}$ becomes more and more closely aligned with the x-axis, which the stretching direction.

The second example is steady unidirectional shear flow $\mathbf{u} = (\Lambda y, 0)$ where Λ is constant. Then

$$\mathbf{A} = \begin{pmatrix} 0 & \Lambda \\ 0 & 0 \end{pmatrix}$$

and $\mathbf{l}(t) = (l_1(t), l_2(t)) = (l_1(0) + l_2(0)\Lambda t, l_2(0))$, implying that $|\mathbf{l}|$ increases linearly with t.

The third example is the rotational flow $\mathbf{u} = (-\Omega y, \Omega x)$ where Ω is constant. Then

$$\mathbf{A} = \begin{pmatrix} 0 & -\Omega \\ \Omega & 0 \end{pmatrix}$$

and $\mathbf{l}(t) = (l_1(t), l_2(t)) = (l_1(0)\cos\Omega t - l_2(0)\sin\Omega t, l_2(0)\cos\Omega t + l_1(0) \sin\Omega t)$, implying that $|\mathbf{l}|$ stays constant with time — the vector $\mathbf{l}$ simply rotates at angular velocity Ω.

The above three examples illustrate the three possible sorts of behavior for a line element in a two-dimensional flow that is a linear time-independent function of space. In this two-dimensional case the behavior is determined by $\det \mathbf{A}$. If $\det \mathbf{A} < 0$ then $\mathbf{l}$ increases exponentially with time, if $\det \mathbf{A} = 0$ then $\mathbf{l}$ increases linearly with time and if $\det \mathbf{A} > 0$ then $\mathbf{l}$ oscillates with no systematic increase in time. In a corresponding three-dimensional flow there are similar possibilities, though the criteria are more complicated.

Some insight into the case where $\mathbf{A}$ is time-dependent can be obtained by considering the case of a pure strain that is randomly varying in time — a simple case is where the magnitude of the strain rate is constant, equal to Γ, say, but the axes of strain randomly change direction after a time interval δ. The average effect of the stretching over each time interval δ may be calculated by noting that the effect is equivalent to that of a strain field with stretching axis in fixed direction, e.g., as considered above, but acting on a line element that is randomly oriented at the beginning of the time interval. If the line element is initially given by $(\cos\theta, \sin\theta)$ then the effect of the strain field acting over time δ is to deform the line element to $(e^{\Gamma\delta}\cos\theta, e^{-\Gamma\delta}\sin\theta)$. Anticipating that increase in length will be exponential in time, it is useful to consider the change in $\log|\mathbf{l}(t)|$ over time δ which is given by

$$\log\{|\mathbf{l}(\delta)|/|\mathbf{l}(0)|\} = \frac{1}{2}\log\{e^{2\Gamma\delta}\cos^2\theta + e^{-2\Gamma\delta}\sin^2\theta\}. \tag{3.7}$$

The average value of this quantity, obtained by integrating with respect to θ from 0 to 2π and then dividing by 2π is $\frac{1}{2}\log\{\frac{1}{2}(1+\cosh 2\Gamma\delta)\}$ and the average rate of stretching s over many time intervals δ is therefore $s = \frac{1}{2}\delta^{-1}\log\{\frac{1}{2}(1+\cosh 2\Gamma\delta)\}$. A key quantity in determining the size of s is the product $\Gamma\delta$. When $\Gamma\delta \ll 1$, i.e., the direction of the strain field changes on a time that is much less than the inverse strain rate Γ^{-1}, the above expression reduces to $s \simeq \frac{1}{2}\Gamma^2\delta = \Gamma \times \frac{1}{2}\Gamma\delta$. s is therefore much less than the stretching rate for the steady strain field. When $\Gamma\delta \gg 1$, i.e., the direction of the strain field changes on a time that is much greater than the inverse strain rate Γ^{-1} the corresponding expression is $s \simeq \Gamma - \delta^{-1}\log 2$. s in this case is therefore close to the stretching rate for the steady strain field, but slightly reduced, as a result of the fact that each time the strain field reorients, it takes some time for the line element to align in the stretching direction.

The important point is that for all values of $\Gamma\delta$ there is exponential stretching, even though the average strain field at any fixed point is apparently zero (in the sense that at the beginning of each time interval δ is as likely to be aligned in the compression direction of the strain field as in the stretching direction). The conclusion is that exponential stretching of line elements is something rather robust which does not depend, for example, on steadiness of the strain field. More sophisticated mathematical models of a randomly varying strain field can be formulated, but the general property of all such models is that if the strain has correlation time τ and magnitude Γ then the stretching rate $s \sim \Gamma\min\{\Gamma\tau, 1\}$. Note, recalling Eq. (3.5) as an example, that τ is the correlation time, of the strain, following a fluid element, sometimes known as the Lagrangian correlation time. (See also the chapter "A Brief Introduction to Vortex Dynamics and Turbulence" in this volume for a discussion of exponential stretching in three-dimensional isotropic turbulence.)

So far we have not mentioned the dynamics of the flow — it is, of course, the dynamics that determines the time evolution of $\mathbf{A}$. In a turbulent flow the time evolution might be modeled by a random function — "random straining model" implying exponential increase in length with time. We might conclude that in complex flows exponential stretching is "usual." Cases such as the steady shear flow are "unusual." For "most" $\mathbf{A}(t)$, $|\mathbf{l}(t)|$ increases exponentially with time, i.e., as $e^{\lambda t}$ where λ may be time-dependent but does not decrease or increase systematically with time. Indeed for a given fluid line element $\mathbf{l}$ it is useful to define

$$\lambda = \frac{1}{t} \log \left\{ \frac{|\mathbf{l}(t)|}{|\mathbf{l}(0)|} \right\} \tag{3.8}$$

as a measure of stretching rate.

There are of course parts of atmospheric flows where, at least on limited timescales, exponential stretching does not hold and the behavior is more like that of steady shear flow. These include the interior of long-lived coherent eddies (which might be eddies in the turbulent boundary layer or larger-scale flows such as hurricanes or extratropical cyclones) or strong jets (such as the subtropical jet).

3.4. *The Relation Between Stretching and Mixing*

To emphasize the implications of material line lengthening and relative dispersion for stirring and mixing, it is useful to consider the evolution of a small material surface (assumed smaller than the length scale on which the velocity field varies) that is initially a sphere (or, in two-dimensions, a small material contour that is initially a circle). The tendency of line elements to stretch implies that the sphere is deformed into an ellipsoid, at least one axis of which systematically increases in time. In an incompressible flow the volume of the sphere remains constant with time, therefore the systematic increase in length of one axis is inevitably accompanied by the systematic decrease in length of another axis. This is a manifestation of the scale reduction that leads to mixing. In a compressible flow there is no absolute constraint on the volume of the sphere, but nonetheless it is the case that in almost all flows the density will not systematically reduce, implying again that one axis must systematically reduce in length. The geometry of the ellipsoidal material surface becomes more complicated when its maximum dimension becomes as large as the length scale on which the velocity field varies. The surface is then strongly distorted and folded as different parts of the surface sample very different velocity gradients.

How is this picture affected by molecular diffusivity? It was noted earlier that the relative size of advective and diffusive terms is $Pe = UL/\kappa$. In the local view expressed by Eq. (3.6) there is no obvious velocity scale and it is best to consider the typical value, S, of the velocity gradient, implying an alternative definition $Pe = SL^2/\kappa$ and a length scale $(\kappa/\lambda)^{1/2}$ above which advection dominates and below which molecular diffusion dominates. Following the picture presented above, a circular patch of tracer of radius $r_0 \gg (\kappa/\lambda)^{1/2}$ will stretch and thin into an ellipse until its minor axis $r_0 e^{-\lambda t} \sim (\kappa/\lambda)^{1/2}$. After this the minor axis decreases no further (since the broadening effect of diffusion is balanced by the narrow effect of continued stretching) but the major axis continues to increase as $r_0 e^{\lambda t}$. Since area increases as $e^{\lambda t}$ the typical tracer concentration must reduce as $e^{-\lambda t}$ (as the patch mixes with its environment). When $r_0 e^{\lambda t}$ becomes comparable to the length scale on which the velocity field varies, the elliptical patch folds back on itself to become a lengthening filament, but the typical width remains as $(\kappa/\lambda)^{1/2}$.

The relevance of deformation of material surfaces or curves to the evolution of the concentration of a chemical species is emphasized by noting that a similar picture holds in backward time. Neglecting the effects of diffusivity for the present, the values of concentration in a small spherical region will be the values that were present in the same material region at the initial time. If that material region is stretched (in backward time) to length scales greater than those on which the concentration varies in the initial condition, then that stretched region, and hence the small spherical region, will contain a wide range of different concentration values. It can therefore be safely assumed that the effect of diffusivity will be to homogenize those values over the small spherical region. These complementary views in forward and backward time are pictured schematically in Fig. 2. The intimate relation between relative dispersion, i.e., the separation of nearby particles, and mixing has been exploited in many theoretical studies of the mixing problem.

3.5. *"Type I" and "Type II" Flows*

The stirring and mixing process has so far been described as completely generic. One could equally well be considering the mixing of a smoke plume from a factory into the surrounding boundary layer air, or the mixing into the upper troposphere of boundary layer air that has been lifted in

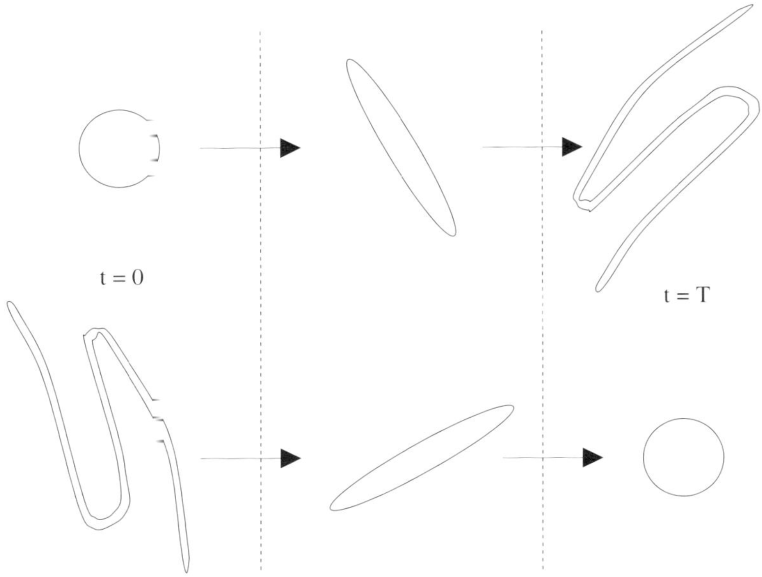

Fig. 2. Schematic of the deformation of two material curves/surfaces. The top panel shows a small circle at time $t = 0$, which is then deformed into an ellipse (while its maximum dimension is less than the characteristic scale of the flow) and then into a more complex structure (as different parts of the curve experience very different velocity fields). The bottom panel shows a small circle at time $t = T$, which originated from a complex filamental structure at $t = 0$. This structure may be obtained by deforming the circle in backward time. The values of chemical concentration inside the circle at $t = T$ are just those sampled by the filamental structure at time $t = 0$. (Note that the two panels do not imply any kind of reversibility — the lower panel corresponds to a particular choice of initial condition that evolves into a circle at time T. If the evolution were continued after time T the circle would stretch and eventually become geometrically complex, much as in the top panel.)

a convective cloud or a convective complex, or the mixing of stratospheric ozone-depleted Antarctic air into mid-latitudes as the polar vortex breaks up in the late spring. These examples range in scales from a hundred meters or so to several thousand kilometers. But the flows that are responsible for stirring and mixing in each of the cases are very different and that has important implications for the stirring and mixing process.

There are two important paradigms for transport and mixing in complex flows. The first, which we might call a "Type I" flow, is exemplified by three-dimensional turbulent flow. The classical Kolmogorov theory of such flow (see the chapter "A Brief Introduction to Vortex Dynamics and

Turbulence") states that energy is put into system at large scale and is dissipated at small scale. A key parameter, indeed the only externally imposed dimensional parameter, is the energy input rate per unit volume ϵ. If the energy injection scale is L and the velocity on that scale is U then ϵ can be estimated as $\epsilon \sim U^3/L$. Dimensional analysis implies that at scale l, the velocity u is of order $\epsilon^{1/3} l^{1/3}$ and therefore that the velocity gradient, and hence the stretching rate, u/l is of order $\epsilon^{1/3} l^{-2/3}$. The stretching rate therefore increases as l decreases. This implies that the velocity gradient has a complex structure in space and time, and also that the tracer field at scale l is dominated by the local stretching characteristics of the flow, so that it too has a complex structure in space and time. Note that the Peclet number at scale l, Pe_l is estimated to be ul/κ is of order $\epsilon^{1/3} l^{4/3}/\kappa$ which is O(1) when l is of order $\kappa^{3/4}\epsilon^{-1/4}$, with the latter scale being that on which molecular mixing occurs. (We assume that this scale is no less than the scale at which energy dissipates, which follows if the diffusivity for the tracer is no less than the kinematic viscosity ν. This assumption is good for most chemical species in the atmosphere. However, it is not good for many common chemical species dissolved in water, so needs to considered carefully for the ocean.) Now consider the time needed for a tracer structure starting with scale L to be deformed so that diffusive mixing becomes important. At any scale l the time to reduce in scale by a factor of 2 can be estimated as $\epsilon^{-1/3} l^{2/3}$ and the time to the mixing scale is therefore estimated by $\epsilon^{-1/3} L^{2/3}(1 + 2^{-2/3} + 2^{-4/3} + \cdots + \kappa^{1/2} L^{-2/3} \epsilon - 1/6)$. Since the series is a geometric series and converges as the number of terms increases, this time is relatively insensitive to κ and well estimated by $\epsilon^{-1/3} L^{2/3}$ which is the eddy turnover time on scale L. This implies that the timescale for molecular mixing (or *micromixing*) starting with a concentration distribution varying on length scale L is similar to, i.e., some modest multiple of, the timescale for advective rearrangement of the concentration field, sometimes called *macromixing* over the length scale L.

The second type of flow, a "Type II" flow, has a smooth structure in space and time, so that the velocity gradient also has a smooth structure. However, this does not imply simplicity for particle trajectories. The mathematical theory of dynamical systems, which when applied to particle motion in smooth flows implies the phenomenon of *chaotic advection*, shows that particle trajectories can be very complex. (What is meant by 'chaotic' is that particle trajectories are quasi-random, in other words that the position of a particle at one time gives little information on its position at some future time. Equivalently, nearby particles separately exponentially in time or stretching of material line elements is exponential in time.) Since these flows are smooth, the local velocity gradient and hence local stretching rate, are well estimated by the large-scale velocity gradient U/L where U and L are defined as above. In this case the time to stretch by a factor of 2 is independent of scale and the consequence is that the time for a tracer structure to be deformed from the large scale L to the scale $(\kappa L/U)^{1/2}$ where molecular diffusion is important is estimated by $(L/U)\log UL/\kappa$, i.e., it depends logarithmically on κ. This implies that the time for micromixing is larger, by a factor $\log UL/\kappa$, than the timescale L/U for macromixing. Furthermore, because the velocity gradient is a smooth function of space, the direction of stretched filaments is also a smooth function of space. Since, when the diffusivity κ is small the filaments are very thin, this implies that the filaments must be locally undirectional. (In a Type I flow on the other hand, the velocity gradient, and hence the direction of stretched filaments, varies strongly on small-scales.)

In the atmosphere the Type I paradigm applies to the active turbulent regions — the planetary boundary layer and actively convecting regions both in the tropics and the extratropics. The Type II paradigm applies to flows dominated by stable stratification and rotation, i.e., the large-scale flow in the troposphere and stratosphere due to synoptic-scale baroclinic eddies (extratropical cyclones and anticyclones), large-scale waves and non-convecting regions of monsoon circulations and tropical cyclones.

3.6. *Stirring and Transport in Quasi-Two-Dimensional Flows*

The previous sections have emphasized the importance of the quasi-horizontal large-scale flow, a Type II flow in long-range atmospheric transport. Therefore it is worth considering simple mathematical models of transport and stirring in such a flow. A convenient idealization is to consider a two-dimensional and incompressible flow. For such a flow the velocity components u and v in the x and y directions, respectively, may be represented in terms of a streamfunction $\psi(x, y, t)$ as $u = -\partial\psi/\partial y$ and $v = \partial\psi/\partial x$. This might be seen as a model of the flow on an single isentropic surface in the atmosphere — this flow would not be exactly incompressible but the implications for transport and mixing are small.

A first idealization might be a steady flow, with $\partial\psi/\partial t = 0$. For such steady flows the streamfunction ψ is constant following fluid particles. This puts a very strong constraint on fluid transport — the streamlines. The contours of ψ are fixed curves and particles can move only along these curves, not across. The streamlines might be regarded as *transport barriers.* Furthermore, in steady flows the stretching of fluid elements, and hence the stirring, is weak, just as the stretching in a linear shear flow was shown to be weak in Sec. 3.3. This can be seen, at least for the case of closed streamlines, by the following argument. Consider a line element $\mathbf{l}(t)$ which at time $t = 0$ has one end at $\mathbf{x}_0$, on the streamline $\psi(\mathbf{x}) = \Psi$, and the other end on the neighboring streamline $\psi(\mathbf{x}) = \Psi + \delta\Psi$, implying that $\mathbf{l}(0).\nabla\psi(\mathbf{x}_0) = \delta\Psi$. Now suppose that the time taken for a particle starting on the streamline $\psi = \Psi$ to move around that streamline once is $T(\Psi)$. After this time one end of the line element will have returned to its original position $\mathbf{x}_0$. However, the other end, on the streamline $T(\Psi + \delta\Psi)$, would take $T(\Psi + \delta\Psi)$ to return, implying that at time $T(\Psi)$ it is displaced by $-T'(\Psi)\mathbf{u}(\mathbf{x}_0) = -\mathbf{l}(0)$. $\nabla\psi(\mathbf{x}_0)T'(\Psi)\mathbf{u}(\mathbf{x}_0)$ from its original position. This displacement is the change in $\mathbf{l}$ over the time $T(\Psi)$. We deduce that $\mathbf{l}(T(\psi)) = M\mathbf{l}(0)$ where the matrix M has elements $M_{ij} = \delta_{ij} - T'(\Psi)u_i(\mathbf{x}_0)\partial\psi/\partial x_j(\mathbf{x}_0)$. The fact that $\mathbf{u}(\mathbf{x}_0).\nabla\psi(\mathbf{x}_0) = 0$ implies that $(M^n)_{ij} = \delta_{ij} - nT'(\Psi)u_i(\mathbf{x}_0)\partial\psi/\partial x_j(\mathbf{x}_0)$ and hence $|\mathbf{l}(nT(\Psi)|$ increases only linearly with n.

If the flow is time-dependent, then the strong constraints on particle transport and on stretching are relaxed, since particles no longer remain on a fixed streamline for all time. There have been many studies over the last 30 years or so of the changes in transport and mixing that occur as a flow changes, through suitable change in 1 or more parameters, from a steady flow to one in which there is strong time-dependence and the resulting behavior is described, for example, Meiss (1992) and Wiggins (1992). A typical example is where a specified time-periodic perturbation is added to a time-independent streamfunction. Studies of these flows are usually "kinematic" in the sense that no attention is paid to whether the flows are dynamically consistent — the velocity field is simply assumed and the transport and mixing properties investigated.

An example of the changes in transport and mixing as a time-periodic component as added to a steady flow is illustrated in Fig. 3. In the underlying steady flow particle trajectories lie along streamlines which are therefore transport barriers. The top panel of Fig. 3 shows the streamlines for the example case. Dots and crosses indicate stagnation points, with dots being *elliptic* (a line element centered at this point in the steady flow would rotate but not be stretched systematically) and crosses being *hyperbolic* (a line element centered at this point in the steady flow will be stretched systematically). When a time-periodic perturbation of very small amplitude is added to the steady flow "most" of the transport barriers corresponding, in the purely steady case, to streamlines are preserved. Between the surviving

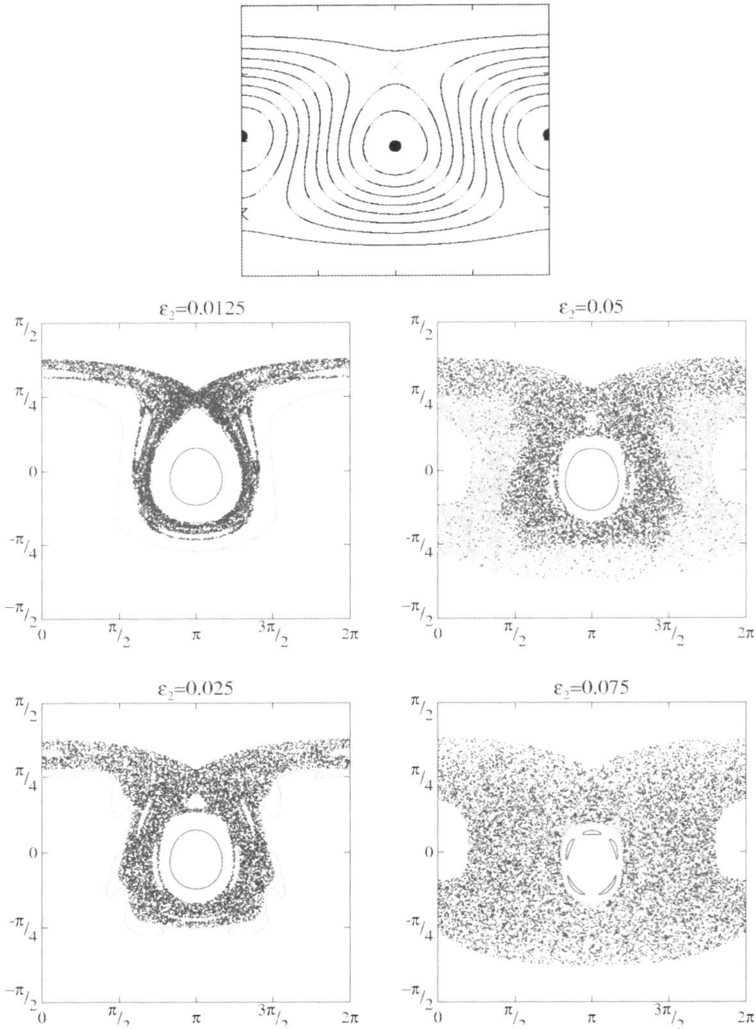

Fig. 3. Variation of transport and mixing in a simple time-periodic flow on a sphere as the unsteady (time-periodic) component is increased. Top panel: Streamlines of the underlying steady flow. Lower panels: Poincaré sections showing points on a given trajectory at time intervals corresponding to the time period of the flow, for different values of ϵ_2 which is the amplitude of the time-periodic component. Three different Poincaré sections, corresponding to three different initial conditions for trajectories, are shown in each case, as light, medium, and dark gray. [*Source*: Shuckburgh and Haynes (2003). See text for further details and explanation. Copyright 2003 American Institute of Physics. Reproduced by permission of American Institute of Physics.]

barriers there are thin regions in which particle trajectories are chaotic and stretching of material line elements is exponential in time, with the thickest regions usually centered on the location of streamlines of the unperturbed flow that pass through hyperbolic stagnation points. This is shown in the middle left-hand panel of Fig. 3, for $\epsilon_2 = 0.125$ with ϵ_2 being the amplitude of the time-periodic component. What are shown are *Poincaré sections* consisting of a large set of points on a given trajectory at time intervals corresponding to the time period of the flow. Three different Poincaré sections, corresponding to three different initial conditions, are shown, as light, middle, and dark gray. The light and middle gray sections have the form of single curves and correspond to transport barriers that have been preserved from the steady case. The dark gray section, which includes the hyperbolic stagnation point in the unsteady flow, fills out a finite area and corresponds to a thin mixing region. As the amplitude of the perturbation increases further barriers disappear and mixing regions increase in thickness. Thus in the bottom left-hand panel of Fig. 3, for $\epsilon_2 = 0.025$ the dark gray section, mapping out a single mixing region, has increased in size from $\epsilon_2 = 0.0125$, but the light and middle gray sections still correspond to single curves and therefore to barriers. Note in particular that the light gray curve corresponds to a barrier that separates the "northern" part of the domain from the "southern" part of the domain. Increasing ϵ_2 further to 0.05, shown in the middle right-hand panel, suggests a significant change, as the light gray section now fills a finite area and therefore corresponds to a mixing region. The light gray and dark gray mixing regions still appear to be distinct, suggesting that they may be separated by a barrier (which would still in effect divide "northern" and "southern" parts of the domain). In fact, increasing the length of the calculation shows that there is no absolute barrier, but it is clear that there is nonetheless significant organization to the transport which inhibits exchange between the "northern" and "southern" parts even if it does not prevent it. Finally, as illustrated in the bottom right-hand panel, increasing ϵ_2 further again, to 0.075, allows rapid exchange between "northern" and "southern" parts so that there is a single large mixing region, filled by both light gray and dark gray sections. Figure 3 illustrates this for a particular example of a flow, but the general pattern of behavior seen here is generic.

Whilst the real large-scale atmospheric flow is certainly not time-periodic, it shares some of important features described above. In particular, there are apparently strong barriers to transport associated with the subtropical jets, with neighboring regions of mixing on both poleward and equatorward sides (see comments on Fig. 1). The precise "cause-and-effect" relation between jets and transport barriers is subtle. For example, in a kinematic model where a systematic jet is added to a simple background eddy field the transport in the cross-jet direction will often be inhibited, i.e., the jet seems to cause the transport barrier. On the other hand, the dynamics of rotating stratified flow is such that the presence of a transport barrier naturally leads to the formation of a jet, i.e., the transport barrier seems to cause the jet. These issues are discussed further in Haynes *et al.* (2007) and Dritschel and McIntyre (2008).

4. Modeling Approaches

There are many highly developed methods of calculating atmospheric chemical fields. Some of these have been motivated by research, e.g., using such calculations as an aid to interpreting chemical measurements in a field campaign or even as a guide to where and when to take observations; whilst others have been motivated by practical concerns, e.g., predicting the effects of accidental chemical release or establishing the origin of industrial pollutants. Most weather forecasting agencies now provide some kind of air quality forecast and there has already been significant progress toward integrating chemical measurements and chemical model calculations in the same way that meteorological measurements and meteorological models have been integrated over the last 40 years (see Geer *et al.*, 2006).

The essence of a chemical calculation is to solve the advection-diffusion Eq. (3.2) for each chemical species, with production and destruction reaction terms on the right-hand side. This first of all requires velocity fields and these may be taken from an observational dataset or else generated by a meteorological numerical model. The former is made possible by the fact that major weather forecasting centers archive global datasets of velocity, temperature, and other quantities, which are generated during the forecast process. These datasets are not "pure" observational data, but are the product of an "analysis" procedure which finds a best fit of the underlying numerical model to the available data. Part of the reason why this procedure is necessary is that observational data is provided at irregular space and time intervals that cannot straightforwardly be inserted into a meteorological numerical model and neither, of course, could it be straightforwardly be used in a chemical calculation. In the case of velocity fields generated by a meteorological model, the chemical calculation may be *off-line*, meaning that the velocity fields are stored at suitable time intervals during the integration of the meteorological model and then later used for the chemical calculation (the term *chemical transport model* (CTM) is often used for the associated numerical model with which the chemical calculation is performed), or *on-line* meaning that the chemical calculations are carried out concurrently with the dynamical calculations required for the meteorological model, which may be a local or global forecast model, or a climate model.

In either on-line or off-line cases the spatial resolution of the velocity field will typically be 10–100 km for current global analysis datasets or global numerical models, or 1–10 km for local or regional numerical models. In some flows, and certainly in Type I flows where there is an active role for small scales, the effects of these small scales must be represented through *parametrization*, i.e., artificial terms in the model equations which represent the effects of small-scale processes. Again the development of suitable parametrizations has been an important part of the development of meteorological models, for weather forecasting or climate, and for chemical calculations those parametrizations must be adapted, or new parametrizations developed. For example, parametrization of the effects of cumulus convection (which occurs on length scales too small to be represented at all in most models) has been a major effort in meteorological modeling since the associated physical processes, such as transport of water vapor and heating by condensation, play a major role on the larger scale. Correspondingly, over the last ten years or so there has been a major effort to extend cumular parametrizations to include transport of a wider class of chemical species (water vapor is a chemical species, but it is peculiar in that its concentration is strongly limited by temperature) and to include representation of cloud-processing effects such as removal of water-soluble species by precipitation. (see Tost *et al.* (2010) for recent discussion). Of course there are many other processes that need to be included in models, with representation of emissions, either natural or anthropogenic, being particularly important (see Monks *et al.* (2009) for a recent review of the broad subject of global and regional air quality including modeling).

In meteorological modeling the dynamical equations are almost invariably solved by Eulerian techniques, where the functions describing the various dynamical quantities are represented on a fixed spatial grid (or there is some equivalent representation of such a fixed set of basis functions) and a corresponding approximation to the governing equations is solved. However, in solving for chemical fields there remains a genuine choice between the Eulerian approach based on Eq. (3.2), and the Lagrangian approach based on Eq. (3.1) following the trajectories of air parcels and at the same time solving the chemical equations for the evolution of the concentrations of different species in the air parcel. The Lagrangian approach, at least in its simplest form, regards air parcels as isolated from their surroundings, and therefore implicitly neglects the molecular mixing effects of the diffusive terms on the right-hand side of Eq. (3.2). But Eulerian approaches tend to overestimate these mixing effects, essentially because they cannot represent variations in chemical concentrations on scales smaller than the grid scale. As has been discussed in Sec. 2, generation of small scales is an essential part of the route to molecular mixing and for the foreseeable future Eulerian models will not be able to resolve the scales at which mixing actually occurs.

Note that parametrizations of unresolved processes can be introduced into the Lagrangian approach just as they can into the Eulerian approach. For example, the effect of small-scale turbulence in the boundary layer, which in an Eulerian approach might be represented by parametrized flux terms, perhaps augmented diffusive fluxes or some suitable generalization, can in the Lagrangian approach be represented adding random displacements to the trajectories (Stohl *et al.*, 2005). The same sort of approach can represent encounter with convective clouds (Forster *et al.*, 2007).

One of the great practical advantages of the Lagrangian approach is simplicity, both practical and conceptual, but this has to be balanced against the artificiality of the "no mixing assumption".

5. Examples

The following section will describe some examples of recent combined observational and modeling studies of atmospheric chemistry which highlight the different approaches used, both in observation and modeling.

5.1. *The 2000 ACTO Campaign — Combining Chemical Measurements and Backward Trajectory Calculations*

The ACTO (Atmospheric Chemistry and Transport) aircraft campaign was based in Prestwick, Scotland, during May 2000. Instruments on a C-130 aircraft of the UK Meteorological Research Flight were used to make measurements to the northwest of Scotland of various meteorological parameters, plus concentrations of chemical species including ozone and carbon monoxide. The Methven *et al.* paper (2003) discusses results from the campaign and gives some nice examples of the sort of work that it is now

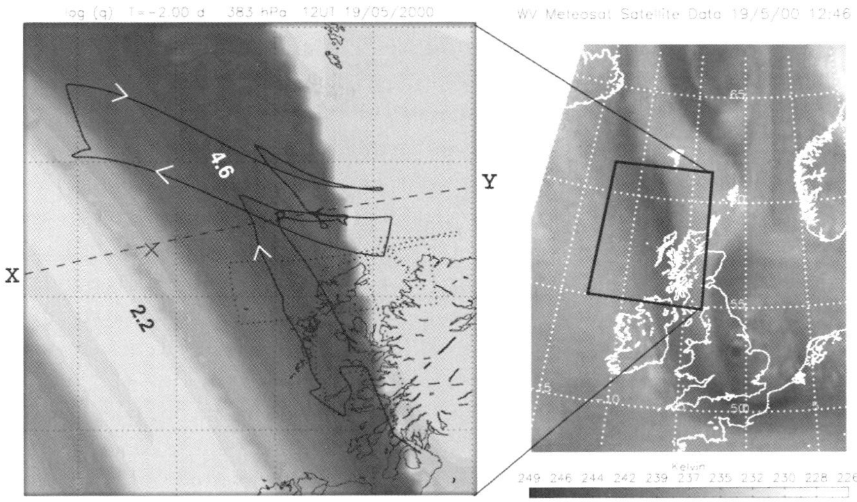

Fig. 4. The right panel shows Meteosat water vapor channel brightness temperature at 12:46 UT, 19 May 2000. Dark shading indicates dry air which originates from the stratosphere (often called a *dry intrusion*). The left panel shows a RDF3D simulation of specific humidity at 12 UT, zooming in on the flight domain. The darkest shading is for $\log(q) < 4.6$; the lightest shading is for $\log(q) > -2.2$. The bold dotted line is the aircraft flight track, and the bold solid line is the same track shifted to be relative to the air at 12 UT. The arrows show the direction of flight. (The dashed line XY marked a particular great circle section and is not relevant in these notes.) [*Source*: Methven *et al.* (2003). Copyright 2003 American Geophysical Union. Reproduced by permission of American Geophysical Union.]

possible using a combination of chemical data and trajectory calculations based on velocity fields from global meteorological datasets (as described in the previous section).

Figure 4 shows (right-hand panel) satellite-measured brightness temperature in the upper troposphere which gives an estimate of water vapor concentrations. Dark colors correspond to dry air, so it can be seen that a filament of dry air runs north–south along the west coast of Scotland, the Irish Sea and southwest England. This dry filament has been drawn out of the lowermost stratosphere, essentially along the 315 K (and neighboring) potential temperature surfaces (recall Fig. 1).

The left-hand panel of Fig. 4 is a reconstruction of the water vapor concentration on the 383 hPa pressure level, corresponding roughly to the center of the vertical layer sampled by the satellite instrument.

The reconstruction has been performed using a technique known as *Reverse Domain Filling (RDF) Trajectories*. This technique uses trajectories integrated backward in time from points distributed across the domain of interest — in this case points distributed in the horizontal on the 383 hPa pressure level. Note that whereas forward trajectory calculations from a region give information on where the air parcels in that region will be transported in future, backward trajectory calculations given information on the origin of those air parcels, i.e., where they have come from in the past, (Recall Fig. 2): Reading the top part from left to right corresponds to forward trajectories from the circular region. Reading the bottom part from right to left corresponds to backward trajectories from the circular region.) A backward trajectory approach therefore allows construction of a field of chemical concentration from the corresponding field known at some earlier time (specifying a suitable initial condition). The value of the concentration at any point is simply determined by following a backward trajectory from that point to the earlier time and then setting the value equal to that of the known concentration field at the position of the backward trajectory at that earlier time. (In practice the known concentration field will be defined on a spatial grid and will therefore have to be interpolated to the position of the backward trajectory which will be unlikely to fall exactly on a grid point.) In the case shown in Fig. 4 the relevant chemical species is water vapor and the known concentration field is taken, just as the velocity fields for the trajectory calculations, from global meteorological analysis datasets, in this case from the European Center for Medium Range Weather Forecasting (ECMWF) (taken two days before the time of the reconstruction). There is close correspondence, in both thickness and orientation, between the observed (right-hand panel) and predicted (left-hand panel), indicating the success of the backward trajectory approach, or more specifically the RDF approach, in this case.

For the case of Type II flows, the backward trajectory approach potentially gives the possibility of predicting chemical concentration features on length scales that are significantly smaller than either the length scale on which the velocity field is resolved or that on which the initial chemical concentration field are resolved. The former follows because the flow is Type II and therefore advection and stretching are dominated by large (and therefore resolved) scales. The latter follows on the principle that an initial concentration field may be of large-scale (and therefore resolved) and then subsequently deformed by the flow to give much smaller-scales (as depicted by the top part of Fig. 2). Thus the dry filament shown in Fig. 4

might have arisen from deformation over the previous two days of a feature that was initially of much larger-scale. Of course, in practice the chemical concentration field at the time at which the initial condition is applied is never completely large-scale. Also, the choice of the time at which the initial condition is applied (two days before the observation for the case shown in Fig. 4) is somewhat arbitrary. If the time difference between initial condition and prediction is small then advective effects will be modest and there will not be time for the development of filaments from initially large-scale features. If the time difference is many days then the process of scale-reduction and filamentation shown in the top part of Fig. 2 will give a concentration field with strong variation at very small-scales. But in reality the variation in the concentration field at small-scale will be limited by molecular diffusion (probably enhanced by the effects of small-scale Type I three-dimensional turbulent flow). A further difficulty is that small errors in the trajectory calculation imply errors in the position of filamentary structures and the thinner the filament, the more significant these errors appear.

Figure 5 shows selected back trajectories from the region of the aircraft measurements. The trajectories separate in backward time (recall the lower part of Fig. 2) suggesting that the air sampled in the measurement regions converged from a large range of locations. In fact, from the trajectories three clearly distinct regions can be identified from where this air originated. One region (A) is in the mid-troposphere in the Eastern Atlantic, another region (S) is the lower stratosphere over central Canada, and another (E) is the lower troposphere over central Europe.

Features corresponding to these different regions of origin can also be clearly identified in Fig. 6 which shows a time series of measured quantities from one particular flight, together with results from backward trajectory calculations to the points (in space and time) along the flight path. In each panel the thicker line is the measurement and the thinner dashed and solid lines respectively correspond to the backward trajectory reconstruction or to a related *air-mass average* reconstruction (see Methven *et al.* (2003) for further detail) which is based on using potential temperature (more precisely equivalent potential temperature which allows for the effect of latent heating due to condensation) and water vapor to specify air mass properties. For (a) pressure and (b) potential temperature the initial condition for the backward trajectory reconstruction comes from ECMWF data. Pressure is, of course, the concentration field of a chemical species and which does not obey Eq. (3.2), so the "reconstructed" quantity in this case is the pressure

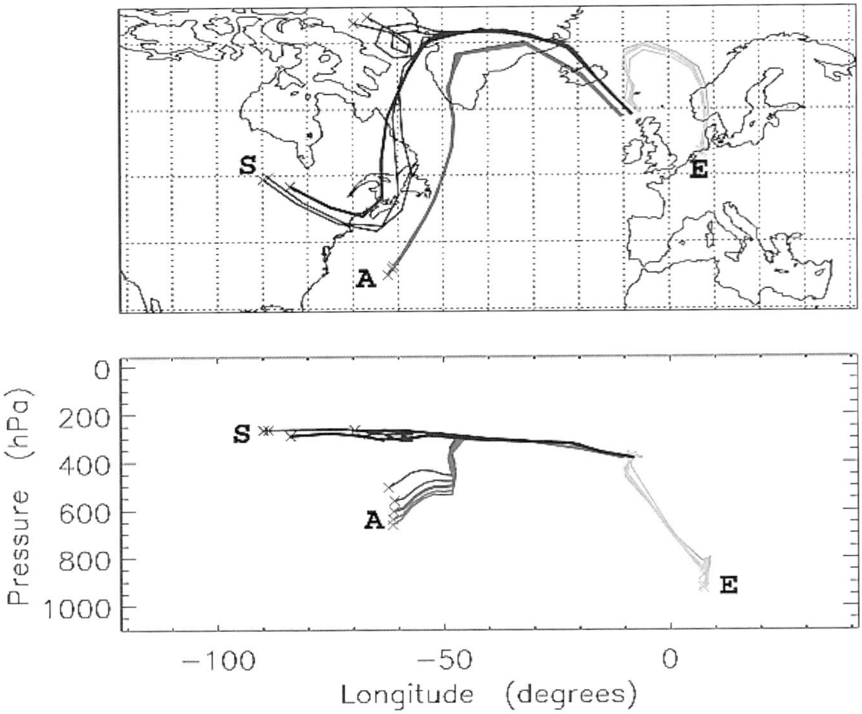

Fig. 5. Three-day back trajectories from region of observation, shown in longitude-latitude in upper panel and longitude-pressure in lower panel. [*Source*: Methven *et al.* (2003). Copyright 2003 American Geophysical Union. Reproduced by permission of American Geophysical Union.]

two days earlier of an air parcel arriving at a given point on the flight track. On the other hand, potential does satisfy Eq. (3.2), so the "reconstructed" can be compared directly against the observation. For (e) ozone and (f) carbon monoxide the initial condition comes from an Eulerian global chemical transport model. In this case the thin dotted line shows the back-trajectory prediction and the thin solid line shows the prediction according to a calculation that integrates the chemical reaction equations along the trajectory. The difference between solid and dotted lines in each case therefore corresponds to the amount of chemical production (or destruction if negative) along the trajectory.

Air from each of these regions has a different chemical signature. That from region A is moist and relatively low in ozone (O_3) and carbon

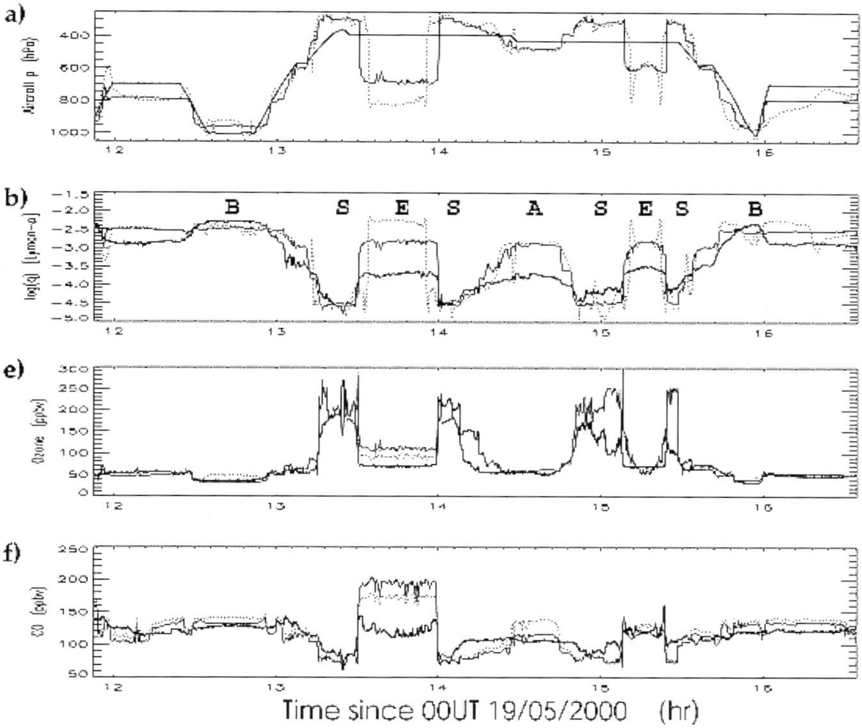

Fig. 6. Time series of observations (thick lines) along the ACTO flight on 19 May 2000 compared to results from trajectory simulations. (a) Pressure and (b) log(q), where q is concentration of water in gkg^{-1}. Thin dotted lines show values interpolated from global meterological data to the trajectory origins at 12 UT, 17 May 2000. Solid lines show an "air-mass average" of these modeled values. "A", "E" and "S" denote air masses that, according to the back-trajectory calculations, have apparently come from identified source regions (see text). (c) and (d) not shown. (e) Ozone and (f) carbon monooxide concentrations. Dotted line is air-mass average at origin of trajectories. Solid is prediction of chemical model integrated along the trajectory. [*Source*: Methven *et al.* (2003). Copyright 2003 American Geophysical Union. Reproduced by permission of American Geophysical Union.]

monooxide (CO). That from region S is dry, high in O_3, and low in CO. That from region E is similar to that from A in that it is moist (both regions are in the troposhere) but different in that it is relatively polluted and therefore high in CO and in O_3 (some of which is likely to have formed through photochemical production as air moves from region E to the region of measurement). Comparison with the detailed chemical fields measured by the aircraft, as shown in Fig. 2, shows that the positions and chemical characteristics of the different air masses are generally well predicted by the back trajectory calculation, even, in many cases, down to small-scale features. For example, the thin dotted line in (b), the simple back trajectory reconstruction of water vapor concentration, is in qualitative agreement with the measurement, but the filament boundaries are in slightly different locations. This is a manifestation of the displacement error mentioned above. Additionally, in the features labeled A and E the predicted water vapor concentration is much larger than the measured concentration. This is because in reality condensation has occurred in these air masses as they have moved toward the measurement point over the previous two days. The use of the air-mass average removes these displacement errors, since the predicted quantity is now specified as a function of the measured quantity.

5.2. *"Around the World in 17 Days" — Transport of Smoke From Russian Forest Fires (Damoah et al., 2004)*

In May 2003 forest fires in southeast Russia gave rise to smoke plumes which extended very large distances across the northern hemisphere and were clearly detected by several different satellite instruments. Indeed the plumes, which were advected by the eastward winds in the northern hemisphere upper troposphere, could be traced all the way around the globe and back to Scandinavia and Eastern Europe. Whether or not these sorts of plumes from high latitude fires, which can sometimes penetrate the lower stratosphere, have a significant effect on regional-scale chemical distributions is still open to question, but they certainly provide a good opportunity to test modeling skill in atmospheric transport and chemistry. Damoah *et al.* (2004) report satellite observations of the May 2003 plumes and show that a numerical model can successfully predict their evolution over the 17 days or so taken for transport around the globe.

The model used by Damoah *et al.* (2004) is a trajectory model (FLEXPART), similar to that used in the Methven *et al.* work (2003) reported previously. FLEXPART (e.g., Stohl *et al.*, 2005) has been widely used to study many different aspects of atmospheric chemistry and transport. It uses velocity fields from global meteorological analysis datasets. The version used by Damoah *et al.* (2004) includes a parametrization of small-scale three-dimensional turbulence, incorporated by adding stochastic fluctuations to the analysis velocity fields, and also a parametrization of convection, incorporated by adding random displacements to the trajectories, with the probability distribution for the displacement set by convective mass fluxes (Forster *et al.*, 2007). The smoke plumes in this model are represented by starting trajectories at the locations and times of the fires, with each trajectory representing the path of an air parcel containing a specified mass of carbon monoxide and then defining the smoke distribution at subsequent times by the spatial density of the air parcels. The positions of the fires were detected using the MODIS (Moderate-Resolution Imaging Spectroradiometer) fire product which identifies hot spots. The mass of CO emitted in one day from a hot spot was assumed to be proportional to the area of that hot spot as identified during that day.

Figure 7 shows the total column CO tracer as simulated by the FLEXPART calculation. The left-hand panels show the results of a calculation based on winds from an ECMWF dataset. The right-hand panels show the results of a corresponding calculation based on winds from the Global Forecast System (GFS) of the National Center for Environmental Prediction in the U.S. Generally the two calculations are in good agreement with each other, giving confidence in the wind data which, whilst based on largely

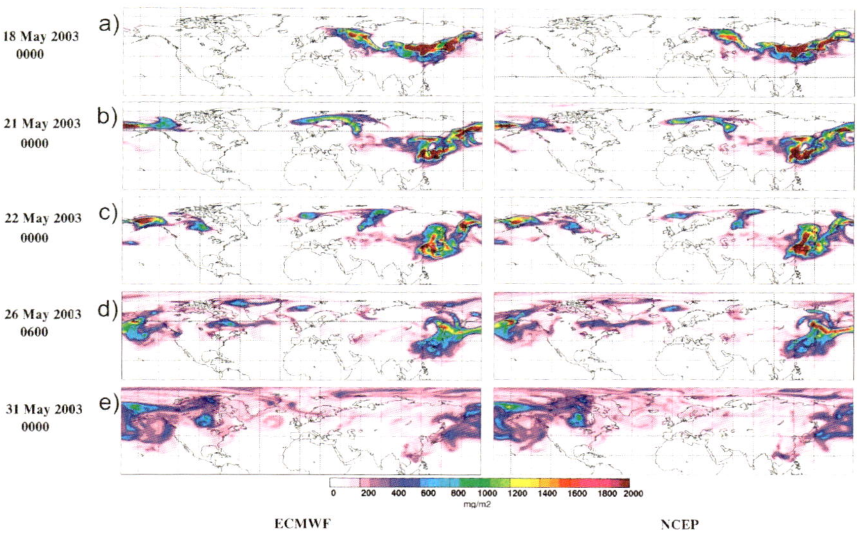

Fig. 7. Total CO tracer columns from FLEXPART simulations using ECMWF data (left column) and GFS data (right columns) on (a) 18 May 2003 at 00 UTC; (b) 21 May at 00 UTC; (c) 22 May at 06 UTC; (d) 26 May at 06 UTC; and (e) 31 May at 00 UTC, respectively. [*Source*: Damoah *et al.* (2004). Reproduced by permission of European Geophysical Union.]

the same observational data, is processed completely independently between the two cases. Panel (a) shows that the CO tracer arising from the fires separates into two patches, one of which is advected northwest towards Scandinavia and the other of which is advected eastward over Japan and then further over Canada and back toward northern Europe. It is this second patch which is the main focus of attention in Damoah *et al.* (2004). The evolution shown in Fig. 7 is, of course, a further demonstration of the stirring and mixing processes described in Sec. 3. In this case smoke emissions over a limited spatial region, but for a relatively long time period, lead to a large patch of the CO tracer, which is subsequently advected by the large-scale flow. The fact that the tracer reaches regions remote from the source region demonstrates transport, the deformation of the region containing the tracer demonstrates stirring, and the fact that the different parts of the initial patch are transported to different locations demonstrates dispersion. The general decrease in peak tracer concentrations and the fact that the region occupied by the tracer appears to increase suggests mixing, but whether or not this mixing is an artificial mixing implied by the FLEXPART approximation to the evolution of a continuously distributed tracer or a real mixing that would be consistent with small-scale observations of the evolving aerosol is harder to tell.

More detail is shown in Figs. 8 and 9 which show both the calculated CO tracer, using ECMWF winds, and also images from the Sea-viewing Wide Field Sensor (Sea WiFS) instrument on the Sea Star satellite, which detects aerosol (i.e., smoke particles in this case). More detail is shown in Figs. 8 and 9, which show rather good agreement between the calculated CO tracer and the Sea WiFS aerosol. In Fig. 8 there is a large region of aerosol over western Alaska, including the Aleutians. In Fig. 9 there is aerosol extending from the North Sea across Denmark and into the rest of Scandinavia. The conclusion here is that the FLEXPART model has significant skill in predicting the spread of the aerosol over many thousands of kilometers and over a period of several days.

5.3. *"Observational and Modeling Analysis of a Severe Air Pollution Episode in Western Hong Kong" (Fung et al., 2005)*

This third example is concerned with a severe air pollution episode in the western part of Hong Kong in late December 1999 (Fung *et al.*, 2005). Consistent with the emphasis elsewhere in these lecture notes, Fung *et al.*

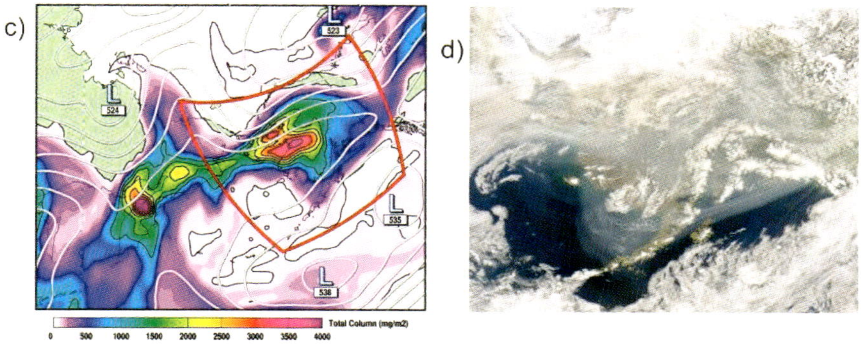

Fig. 8. (c) FLEXPART ECMWF CO tracer columns over the Bering Sea and adjacent regions with superimposed contours of the 500 hPa geopotential surface, based on GFS analyses, contour interval 5 dam, on 22 May 2003 at 00 UTC. Green areas represent land surface, oceans are white. The red rectangle shows approximately the area shown in panel (d). (d) SeaWiFS image showing smoke over Alaska at 23 UTC on 21 May; whitish colors are snow, ice, and clouds, whereas the blue-gray indicate smoke. [*Source*: Damoah *et al.* (2004). Reproduced by permission of European Geophysical Union.]

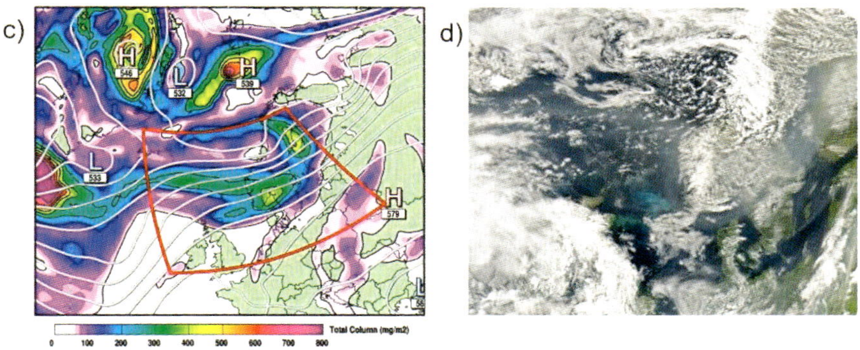

Fig. 9. (c) FLEXPART ECMWF CO tracer columns over the northeast Atlantic, Europe, and Greenland with superimposed contours of the 500 hPa geopotential surface, based on GFS analyses, contour interval 5 dam, at 27 May 15 UTC. (d) Image of SeaWiFS sensor showing smoke over Scandinavia on 27 May 2003 at 12:54 UTC. [*Source*: Damoah *et al.* (2004). Reproduced by permission of European Geophysical Union.]

argue that this episode was not due to emissions within Hong Kong itself, but due to biomass burning about 100 km to the east.

Figure 10 shows concentration of nitrogen dioxide NO_2 at two different observing stations in Hong Kong. The solid line shows Tung Chung

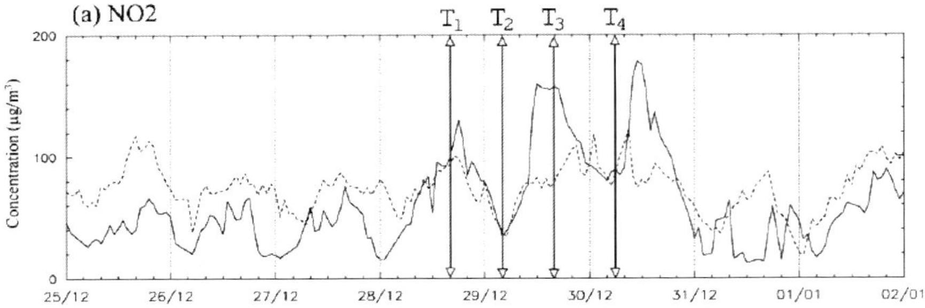

Fig. 10. Concentration of nitrogen dioxide (NO_2) at Tung Chung station (solid line) and Causeway Bay station (dashed line) from 25 December 1999 to 2 January 2000. The pollution episode which lasted for three consecutive days (28–30 December 1999) falls during the middle of this period. Daily highs and lows of NO_2 concentration during the episode are indicated by times T_i ($i = 1, 2, 3, 4$). Note that T_1 corresponds to panel (b), T_2 to panel (c), and T_3 approximately to panel (d) of Fig. 11. [*Source*: Fung *et al.* (2005). Copyright 2005 American Geophysical Union. Reproduced by permission of American Geophysical Union.]

station which is in western Hong Kong, in an open area on the outskirts of the territory. The dashed line shows Causeway Bay station, which is in the most densely occupied central urban area. It is therefore not surprising that nitrogen dioxide concentrations at Tung Chung are generally lower than those at Causeway Bay. However, Fig. 10 shows that for periods during 28 to 31 December, nitrogen dioxide concentrations at Tung Chung significiantly exceed those at Causeway Bay, and indeed exceed those observed at Causeway Bay at any time during the eight-day period shown. Peak concentrations at Tung Chung are considerably greater than values considered to be potentially harmful to human health.

Fung *et al.* (2005) use a variety of evidence to argue that the high levels of NO_2 observed at Tung Chung result not from the local emissions but from the burning of vegetation in a region about 100 km to the north of Hong Kong. This includes information on the chemical composition of the polluted air, but the most straightforward are satellite images which show the presence of fires and the resulting smoke and haze which extends across the western half of Hong Kong, where Tung Chang is located, but which is much less apparent over the eastern part where Causeway Bay is located.

The previous Secs. 5.1. and 5.2., have discussed successful application of transport calculation based on wind fields from large-scale meteorological datasets. An important point made in Fung *et al.* (2005) is that when considering local variation in pollutant concentration in a region of complex topography such as Hong Kong, the usefulness of transport calculations will be strongly limited by the spatial and temporal resolution of the wind field, and that provided by the large-scale meteorological datasets is unlikely to be sufficient. Therefore the approach taken by Fung *et al.* (2005) is to use velocity fields predicted by a regional-scale meteorological model to drive a particle-based transport calculation. The regional-scale model is based on a set of four nested domains, with the spatial resolution becoming finer from the outermost domain to the innermost (varying from a horizontal grid size of 40.5 km for the outermost domain to 1.5 km for the innermost domain). Even for the innermost domain it is accepted that turbulent velocity fluctuations at unresolved scales are potentially important and the effect of these is represented by adding random fluctuations to the velocity field used to advect the particles. (This approach is well developed for Type I turbulence which is what is being represented here.) The parameters for the random fluctuations are set in part by the predicted characteristics of the sub-grid-scale parametrized turbulence in the meteorological model.

Figure 11 shows calculated distributions of particles released in four different locations. The red and green particles are released respectively, at 100 m and 200 m in the region to the north of Hong Kong where the fires occurred. The light and dark blue particles on the other hand are released to the west of Hong Kong to highlight the existence of a convergence region in the horizontal flow, which is argued to favor trapping of pollutant species and therefore to contribute to the large concentrations observed at Tung Chung. Panels (a), (b), and (d) show daytime conditions where the complex pattern of the blue particles reveals the complicated circulation caused by land-sea contrasts and topographic effects. Other diagnostics show that at these times there is transport of pollutant downwards of pollutant species from the top of the planetary boundary layer. At night time, on the other hand, as shown in panel (c) the pattern of circulation is much simpler and the red and green particles are simply advected to the south and southwest without any tendency for trapping.

The study in Fung *et al.* (2005) therefore shows convincingly that remote sources of pollution can be very important for air quality in urban areas, but also that, particularly in a region such as Hong Kong, local details of the flow must be included in any transport calculation in order to capture the spatial and temporal variations in concentration of pollutant species.

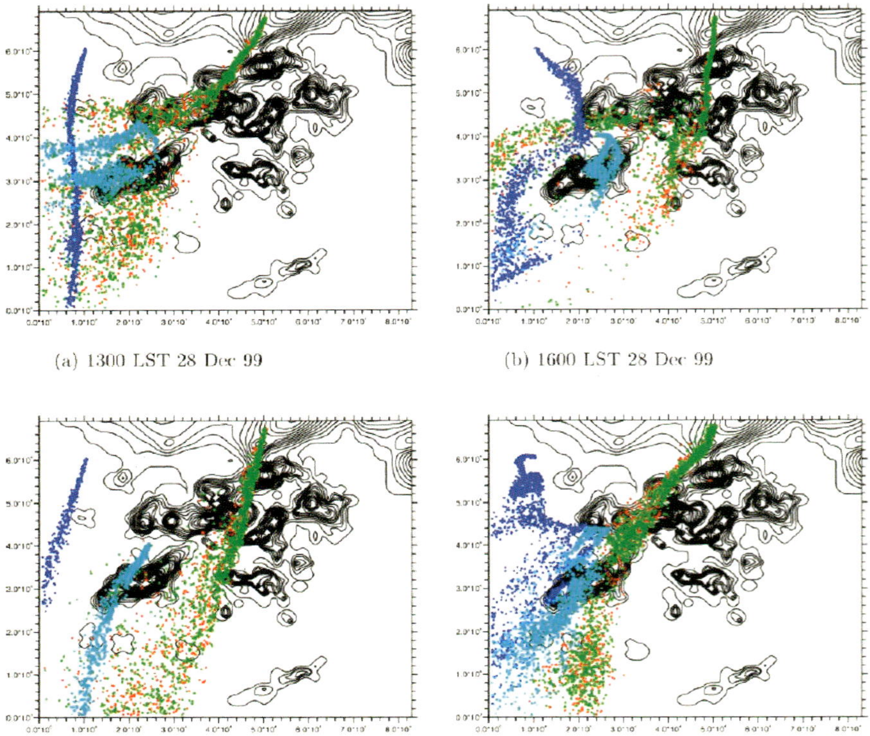

(a) 1300 LST 28 Dec 99

(b) 1600 LST 28 Dec 99

(c) 0400 LST 29 Dec 99

(d) 1500 LST 29 Dec 99

Fig. 11. Tracer plumes, indicated by particle positions, at different times during the pollution episode. (See text for further detail.) The contours correspond to topographic height. Tick marks on x and y axes are at intervals of 1 km. [*Source*: Fung *et al.* (2005). Copyright 2003 American Geophysical Union. Reproduced by permission of American Geophysical Union.]

6. Conclusion

There are many different aspects of atmospheric physics and chemistry that are relevant to understanding and predicting air quality. These notes have focused on the fluid dynamics of transport and mixing, in particular, while giving an impression of how this topic relates to the broader question of how to use atmospheric models to understand observations and to make useful predictions. The reader is reminded of Akimoto (2003) and Monks *et al.* (2009) as articles that give a much broader perspective.

It is certainly the case that in the past weather forecasting and climate prediction on the one hand, and air quality on the other, have been regarded as separate issues: The strong relation between them is now accepted. Therefore (e.g., see Monks *et al.*, 2009) it needs to be accepted that future strategies for minimizing climate impact, e.g., by reducing emissions of long-lived greenhouse gases, and for improving air quality, e.g., by changing engine technology, need to be considered together, since there are trade-offs — some options that are beneficial for one are adverse for the other. It is also the case that prediction of future air quality not only needs to take account of likely changes in emissions, but also the fact that those emissions are in a background atmosphere that is different from our current atmosphere. Finally, some aspects of climate change themselves have air quality implications. These include, for example, the possibility of more summer heatwaves. A study of the effects of the 2003 European heatwave over the UK (Lee *et al.*, 2006) suggests that not only did the associated anticyclonic circulation and elevated temperatures lead to high values of ozone from remote sources, but also that increased emissions of ozone precursors from vegetation as a result of the high temperatures might also have significantly enhanced local ozone concentrations. Another larger-scale effect might be that the effects of climate change on the coupled troposphere-stratosphere system are such that the transport of ozone from stratosphere to troposphere increases, implying an increase in background concentrations of ozone in the troposphere (Zeng and Pyle, 2003). This means that the maximum ozone concentrations arising in polluted regions will also typically increase and therefore, potentially, that concentrations regarded as potentially harmful to health will be encountered more frequently.

References

Akimoto, H. (2003). Global air quality and pollution, *Science* **302**, pp. 1716–1719.

Amann, M., Derwent, D., Forsberg, B., *et al.* (2008). *Health risks of ozone from long-range transboundary air pollution* (World Health Organization Regional Office for Europe, Copenhagen). [Online] Available at http://www.euro.who.int/data/assets/pdf_file/0005/78647/E91843.pdf.

Damoah, R., Spichtinger, N., Forster, C., *et al.* (2004). Around the world in 17 days — hemispheric transport of forest fire smoke from Russia in May 2003, *Atmos. Chem. Phys.* **4**, pp. 1311–1321.

Derwent, R., Simmonds, P., O'Doherty, S., *et al.* (2006). External influences on Europe's air quality: baseline methane, carbon monoxide and ozone from 1990 to 2030 at Mace Head, Ireland, *Atmos. Environ.* **40**, pp. 844–855.

Dritschel, D. and McIntyre, M. (2008). Multiple jets as PV staircases: the Phillips effect and the resilience of eddy-transport barriers, *J. Atmos. Sci.* **65**, pp. 855–873.

Forster, C., Stohl, A. and Seibert, P. (2007). Parameterization of convective transport in a Lagrangian particle dispersion model and its evaluation, *J. Appl. Meteorol.* **46**, pp. 403–422.

Fung, J. C. H., Lau, A. K. H., Lam, J. S. L., *et al.* (2005). Observational and modeling analysis of a severe air pollution episode in western Hong Kong, *J. Geophys. Res.* **110**, p. D09105.

Geer, A., Lahoz, W., Bekki, S., *et al.* (2006). The ASSET intercomparison of ozone analyses: method and first results, *Atmos. Chem. Phys.* **6**, pp. 5445–5474.

Haynes, P., Poet, D. and Shuckburgh, E. (2007). Transport and mixing in kinematic and dynamically consistent flows, *J. Atmos. Sci.* **64**, pp. 3640–3651.

Holton, J. R. (2004). *An Introduction to Dynamic Meteorology*, 4th edn. (Academic Press, San Diego).

Holton, J., Haynes, P., McIntyre, M., *et al.* (1995). Stratosphere-troposphere exchange, *Rev. Geophys.* **33**, pp. 403–439.

Houghton, J. (2002). *The Physics of Atmospheres*, 3rd edn. (Cambridge University Press, Cambridge).

Lee, J. D., *et al.* (2006). Ozone photochemistry and elevated isoprene during the UK heatwave of August 2003, *Atmos. Environ.* **40**, pp. 7598–7613.

Meiss, J. (1992). Symplectic maps, variational principles and transport, *Rev. Mod. Phys.* **64**, pp. 795–848.

Methven, J., Arnold, S. R., O'Connor, F. M., *et al.* (2003). Estimating photochemically produced ozone throughout a domain using flight data and Lagrangian model, *J. Geophys. Res.* **108**, p. 4721.

Monks, P., *et al.* (2009). Atmospheric composition change — global and regional air quality, *Atmos. Environ.* **43**, pp. 5268–5350.

Newell, R., Thouret, V., Cho, J., *et al.* (1999). Ubiquity of quasi-horizontal layers in the troposphere, *Nature* **398**, pp. 316–319.

Royal Society (2008). Ground-level ozone in the 21st century, policy document 15/08 edn. [Online]. Available at: http://royalsociety.org/Ground-level-ozone-in-the-21st-century-future-trends-impacts-and-policy-implications-/).

Shuckburgh, E. and Haynes, P. (2003). Diagnosing transport and mixing using a tracer-based coordinate system, *Phys. Fluids* **15**, pp. 3342–3357.

Stohl, A., Forster, C., Frank, A., *et al.* (2005). Technical note: the Lagrangian particle dispersion model FLEXPART version 6.2, *Atmos. Chem. Phys.* **5**, pp. 2461–2474.

Tost, H., Lawrence, M., Bruehl, C., *et al.* (2010). Uncertainties in atmospheric chemistry modelling due to convection parameterisations and subsequent scavenging, *Atmos. Chem. Phys.* **4**, pp. 1931–1951.

Vallis, G. K. (2006). *Atmospheric and Oceanic Fluid Dynamics* (Cambridge Univ. Press, Cambridge, UK).

Wiggins, S. (1992). *Chaotic Transport in Dynamical Systems* (Springer-Verlag, German).

Zeng, G. and Pyle, J. A. (2003). Changes in tropospheric ozone between 2000 and 2100 modeled in a chemistry-climate model, *Geophys. Res. Lett.* **30**, p. 1392.

Chapter 7

Extreme Rain Events in Mid-Latitudes

Gerd Tetzlaff, Janek Zimmer, and Robin Faulwetter
Institute for Meteorology, University of Leipzig
Stephanstr. 3, 04103 Leipzig, Germany
tetzlaff@uni-leipzig.de

Extreme precipitation (rain) and subsequent flooding is a major environmental hazard. Methods have been developed, mainly on a statistical basis, to estimate maximum probable or even maximum possible precipitation. Higher resolution weather forecast models, with complex representation of the physical processes involved, can be used for calculation of maximum precipitation. This requires input data with known probabilities, but also allows for estimation of the sensitivities of relevant parameters. Orographic (mountain) structures are a major influence in the formation of very high precipitation. The model applied here is used to study the sensitivity of orographic influence. The long-term climatic development of heavy precipitation and floods in Europe during recent centuries will be used to illustrate these numerical studies.

1. Motivation

Rain is one of the key phenomena characterizing weather and climate, both on a global scale and locally. In weather and climate, rain is described by frequency of occurrence, together with its scale in time and space. Major deviations from long-term average conditions, i.e., extreme events, may have adverse effects, and often cause disasters. To reduce these adverse effects and thus to prepare better for the occurrence of rare events, a good understanding of the underlying physical processes is necessary. A statistical treatment requires a data base of sufficient quality and length, often not available. To estimate the size of extreme rain events, physical modeling at least allows us to perform sensitivity studies.

Over the last two decades the number of disasters has increased from year to year (Rodriguez *et al.*, 2009). Weather-related phenomena were responsible for most of these disasters, a major proportion of these being river floods, a consequence of heavy rains. To quantify the adverse effects of disasters, three scaling parameters have been suggested (Guha-Sapir *et al.*, 2004): The number of fatalities, the material damage, and the number of people affected, whether by homelessness, health, or evacuation, the last of these being most widely used for comparison of disasters on the global scale.

Globally the number of people falling victim to floods has increased by 40% in the last two decades. This overall increase is mainly due to an increase of small-scale flood events (Hoyois *et al.*, 2007), while the larger ones actually contributed very little because such events are relatively rare. The countries affected were found all around the globe, including countries in mid-latitudes. In the following we focus on extreme rain events in the mid-latitudes extending over large areas.

The availability of estimates of the upper physical limit of the amount of rain that may fall in any given river catchment area is relevant for the design of flood protection and preparedness. In most cases, series of rain observations on their own do not allow such estimates. To fill this gap the concept of PMP, probable maximum precipitation, was introduced some time ago (WMO, 1973; DVWK, 1997). This is based on statistical methods, combining maximum observed rain with maximum water vapor content of the atmosphere. For mid-latitude conditions, PMP estimates distinguish between large-scale rain events with a typical horizontal length scale of the order of 1000 km, and convective scale events extending from below 10 km to 100 km. The dominant weather processes in the mid-latitudes belong to the larger length scale. To estimate maximum rains on this scale two methods will be used. The first takes the information available from the global climatic setting, deducing maximum rains from this limited information. The second applies a full-scale three-dimensional model and calculates the sensitivities of the maximum rains to the parameters involved.

2. Climatic Setting

The sun supplies the Earth with a continuous flow of radiation energy, which drives the weather processes and the formation of all the rain. At the top of the atmosphere, on a surface perpendicular to the solar radiation, the average solar energy flux density amounts to about 1370 Wm^{-2}. The amount of absorbed solar energy decreases from equator to pole, because of the spherical shape of the Earth. In the simplest situation when the sun lies in the Earth's equatorial plane, the radiation flux density at the poles drops to zero, and the shaded half of Earth does not receive any solar radiation at all. Some variations of the solar radiation are associated with the orbital parameters of the Earth, and are not considered here. Within the atmosphere and at the surface, about 70% of the total incident radiation is absorbed and heats the Earth.

The Earth itself radiates to space and on average sends back the absorbed solar radiation, thus maintaining long-term energy equilibrium. The average energy budget of the Earth in the northern hemisphere shows a surplus area south of about 30°N, and an approximately equal deficit area between 30°N and the pole. At the same time the long-term average of temperature also remains constant in both these zones. To maintain this situation, a horizontal energy flux from the equatorial surplus zone to the polar deficit zone is required. The global temperature distribution shows an average temperature decrease from equator to pole. In the mid-troposphere, temperature differences between the equatorial and the polar region amount to about 15 K in northern summer, 35 K in northern winter, and 24 K averaged over the year.

The annual average of the vertically integrated columnar tropospheric water vapor content amounts to about 45 kgm^{-2} close to the equator, and about 12 kgm^{-2} in the polar region, with a global average of 25 kgm^{-2}. The liquid water content of the atmosphere is most important when creating rain. However, the absolute columnar liquid water content is about two to three orders of magnitude less than that of water in the vapor state.

Water vapor is brought into the atmosphere from evaporation and/or transpiration at the surface. A phase change between water vapor and liquid water consumes or releases energy. The latent heat of condensation is about $2.5 \cdot 10^6$ Jkg^{-1}. Averaged over time and over the Earth's surface, rain and evaporation are in equilibrium. However, locally and for short periods there is no such equilibrium. From theoretical estimates, the amount of evaporation cannot exceed a value of about 7 mmday^{-1} in horizontally homogeneous conditions.

There is no rain without clouds. Both rain and clouds consist of condensed water vapor. On average, about 0.3% of the atmosphere consists of water vapor. The Clausius–Clapeyron relationship describes the exponential increase of the saturation water vapor pressure E as a function of temperature. As a rule of thumb E doubles with a 10 K increase of temperature, or changes by 20% for a 3 K change of temperature.

Rain consists of drops of typical diameter 2 mm, and terminal fall velocity of about $6\,\text{ms}^{-1}$. The global annual average rainfall is close to 1000 liters per square meter of the Earth's surface. This is equivalent to 1000 mm of rain, the most widely used unit for rain. Rain is the only rechargeable supply of fresh water and therefore of prime importance to humanity and to the biosphere.

In mid-latitudes, rain falls on average about 5% of the time in any year. Globally averaged, the amount of rain is 2.7 mmday^{-1}, with a range from 0 to about 30 mmday^{-1}. In the mid-latitudes, negative effects have to be expected when rainfall is more than about 50 mmday^{-1}. Values greater than 100 mmday^{-1} occur rather frequently; the world maximum that has been recorded is 1900 mmday^{-1} (Wiesner, 1970). An amount of 100 mmday^{-1} is equivalent to an energy flux of 3000 Wm^{-2}, more than double the maximum incoming solar radiation at the top of the atmosphere. Neither the atmospheric water vapor reservoir nor the supply from local evapotranspiration can account for such rains.

In addition, there is no transport of liquid water from the Earth's surface into the cloud layer levels of the atmosphere. There is also no relevant reservoir of liquid water in the atmosphere. In most clouds with about 10^4 to 10^6 water droplets per liter, the liquid water content does not exceed $0.5\,\text{gm}^{-3}$, small in relation to the rains mentioned even when integrated over the whole troposphere. The water vapor content in lower cloud levels generally exceeds the liquid water content by a factor of 10 to 100. Thus the supply of water vapor for rain formation needs horizontal water vapor transport into the raining area.

3. Horizontal Energy Transport in the Atmosphere

Transport of energy from the equatorial surplus zone to the polar deficit zone occurs mainly within the atmosphere. The poleward energy fluxes are largest at the latitude of the equilibrated energy budget, i.e., at about 30°N. Further north, the flux gradually decreases. The basic formulation of transport Tr consists of three components: The meridional transport velocity v, the density of the transporting medium ρ, and the gradient of the transported energy Ψ:

$$Tr(\lambda, t) = \rho v \frac{d\Psi}{dy}, \tag{3.1}$$

Tr (in $\text{Wm}^{-2}\text{m}^{-1}$) across a fixed latitudinal circle and integrated with respect to height over the whole troposphere is a function of longitude λ

and time t. In atmospheric conditions the transportable forms of energy (per unit mass, and measured in Jkg^{-1}) are the potential energy P, the sensible heat H, the latent heat L, and the kinetic energy K (Eqs. (3.2)–(3.5)). To achieve effective transport it is necessary to have a gradient of the transported energy along a trajectory.

$$P = gz \tag{3.2}$$
$$H = c_p T \tag{3.3}$$
$$L = L'q \tag{3.4}$$
$$K = \frac{v^2}{2}, \tag{3.5}$$

where g is the gravitational acceleration, z the height, c_p the specific heat at constant pressure, L' the latent heat of condensation, q the specific humidity (i.e., mass of water vapor per unit mass of air), and v the meridional velocity. These and all other dimensional variables are assumed to be measured in SI units unless stated otherwise. Kinetic energy is irrelevant for energy transport (Oort, 1971) and is therefore neglected in the following:

The cubic expansion coefficient of air provides the basis for a linear increase of potential energy with temperature. The same increase applies to sensible heat. Latent heat energy depends on q, and saturation vapor pressure is a function of temperature alone. Therefore, all three forms of energy depend on temperature, which is therefore the prime variable determining the total energy content of an air parcel.

To get the average transport across a circle of latitude, we need to average the transports of potential energy, sensible heat and latent heat over this circle and over time. We can split Tr into its average and fluctuating contributions:

$$Tr = \overline{Tr} + Tr'. \tag{3.6}$$

Assuming the density fluctuations to be small compared to the mean value (e.g., Oort, 1971), the transport averaged over longitude λ and time t can be presented as $\overline{v\Psi_{\lambda,t}}$ (as the result of two consecutive averaging operations):

$$\overline{v\Psi_{\lambda,t}} = \overline{(\bar{v} + v'_\lambda + v'_t)(\bar{\Psi} + \Psi'_\lambda + \Psi'_t)}. \tag{3.7}$$

The index λ denotes averaging over a full latitudinal circle, t a long-term average (i.e., 30 years). Overbar is the average, prime the deviation from the average. Hence, v'_λ is the deviation from the longitudinal average of the meridional velocity and v'_t the deviation from the temporal average.

This can then be summarized to three terms:

$$\overline{v\Psi_{\lambda,t}} = \overline{v_{\lambda,t}}\,\overline{\Psi_{\lambda,t}} + \overline{[(\bar{\Psi}_t)'_\lambda(\bar{v}_t)'_\lambda]_{\lambda,t}} + \overline{[v'_t\Psi'_t]_{\lambda,t}}. \tag{3.8}$$

The first term here describes the long-term average of energy transport uniform round a circle of latitude. Transport of this kind in one layer of the atmosphere must be compensated by a counter-transport at another level; the axis of the resulting meridional circulation is horizontal. The second and the third terms describe energy fluxes that arise from deviation from the average on a latitudinal circle, for stationary and unstationary conditions, respectively. Flow across one part of the circle has to be compensated by counter-flow across another part, implying rotation around a vertical axis in the form of stationary and propagating tropospheric waves.

The meridional circulation does not extend beyond about 30°N (e.g., Hartman, 1994). Momentum balance and observed zonal flow are in agreement there, reaching about $70\,\mathrm{ms}^{-1}$ in the upper part of the troposphere. North of this latitude, in mid-latitudes, atmospheric flow and energy transport are dominated by waves. To provide poleward energy transport, these waves must be baroclinic, i.e., pressure and density surfaces must intersect.

4. Rain Making

In the atmosphere an upward moving air parcel follows a dry adiabatic lapse rate (Eq. (4.1)), as derived from the first law of thermodynamics:

$$\frac{dT}{dz} = -\frac{g}{c_p}, \tag{4.1}$$

where T is temperature, z is the geometric height, g is the gravitational acceleration, and c_p is the specific heat of air.

The numerical value of the dry adiabatic lapse rate is close to $-1\,\mathrm{K}$ $(100\,\mathrm{m})^{-1}$. Moving an air parcel upward requires buoyancy. The upward motion has to happen simultaneously in a column of some vertical extent, maximally the whole troposphere. Rain can only occur when such upward motion allows the parcel to reach the dew point temperature, i.e., the temperature at which water vapor condenses onto small condensation nuclei, thus producing small droplets and/or ice crystals. In areas of further cooling or lifting, these droplets grow until they reach the size of raindrops.

Water droplets or ice crystals grow most efficiently when both phases coexist. This is because the saturation water vapor pressure over ice is smaller than that over water, supplying ice crystals with the water molecules evaporated from the water droplets. This allows a fast growth, increasing the terminal fall velocity of the particles. While they fall, these particles can even accelerate their growth rate. They collide with slower falling droplets and particles, finally forming raindrops. In order to produce a significant quantity of rain, this process must be sustained for some time, typically several hours. The simplest geometrical setting is that in which all processes necessary to form rain happen in a horizontally homogeneous column with vertical orientation.

As already mentioned, the specific humidity q describes the proportion by mass of water vapor in an air parcel:

$$q = 0.622 \frac{e}{p - 0.378e} \approx 0.622 \frac{e}{p}, \tag{4.2}$$

where 0.622 is the ratio of the gas constants of water vapor and dry air, p the air pressure, and e the water vapor pressure. The specific humidity of an air parcel is conserved when no sinks or sources are present. When an air parcel is lifted to the condensation level, the saturation specific humidity is reached. If the lifting is continued beyond this level, condensation occurs (and so q decreases). The mass of water vapor transported through a certain level, TrW (in $\text{kgm}^{-2}\text{s}^{-1}$), is given by

$$TrW = \rho w q. \tag{4.3}$$

The vertical velocity w can be approximated by $\Delta z/\Delta t$, where Δz is the height through which the air parcel is lifted in time Δt. The rate of production of condensed vapor CO (condensate, in $\text{kgm}^{-2}\text{s}^{-1}$) in the column is then given by the difference Δq between the lower and upper levels of a selected layer of depth Δz:

$$CO = \rho_0 w \Delta q = \rho_0 \frac{\Delta z}{\Delta t} \Delta q, \tag{4.4}$$

where ρ_0 is the mean air density in the layer. The amount of condensate depends linearly on the vertical velocity w while its dependence on Δq is actually more complicated for the following reasons. In the whole layer, the temperature is assumed to be equal to the dew point temperature.

Below the condensation level the dry adiabatic lapse rate applies, but this is not so above the condensation level, because the latent heat released by condensation warms the air. The quantity of Δq influences the temperature at the upper portions of the air layer, providing feedback between T and q at this level. This has to be considered when determining the value of CO as a function of Δq. The vertical temperature gradient within a cloud layer is given by the moist adiabatic lapse rate dT_m/dz, combining the effects of the dry adiabatic lapse rate $-g/c_p$ and the effects of latent heat warming when an air parcel rises:

$$\frac{dT_m}{dz} = -\frac{g\left(1 + \frac{L'q}{RT}\right)}{c_p + L'q\frac{1}{E}\frac{dE}{dT}}. \qquad (4.5)$$

dT_m/dz is in Km^{-1}, R is the specific gas constant for dry air, and the saturation water vapor pressure E is a function of temperature T. This lapse rate depends on the temperature and the dew point temperature at the lower condensation level, thus determining Δq in any layer of depth Δz. Since the specific humidity q is a linear function of saturation water vapor pressure (Eq. (4.2)), the specific humidity depends exponentially on this quantity. The lower the temperature is, determining the saturation specific humidity, the closer the moist adiabatic lapse rate approaches the dry one.

To illustrate the sensitivity of the condensate CO with respect to vertical velocity w and specific humidity Δq some examples are shown in Table 1. The layer is taken to be located between the pressure levels 950 hPa at the lower level, and 850 hPa at the upper. The air density $\rho_0 = 1\ \mathrm{kgm}^{-3}$ is assumed to be constant in the whole layer. A 10 K temperature difference is selected in the examples, because this produces a factor of 2 in the water vapor saturation pressure and also in the specific humidity. The resulting condensate CO does not show the same dependence (Table 1).

Halving q at the lower level increases the moist adiabatic lapse rate (Eq. (4.5)). As a result the temperature lapse rate in the layer changes from ≈ -4.3 K (20°C at the lower level) to -5.3 K (10°C at the lower level). The amount of condensate produced in the layer decreases by only about a quarter. This damps the influence of temperature changes at the lower cloud level. In the mid-latitudes at average condensation level the temperature difference between summer and winter is about 10°C, causing relatively small effects on the condensate production in the lower layers of the troposphere.

Table 1. Condensate in mm/time interval for selected examples of vertical velocity w, T and q at lower condensation level of an air layer with $\Delta z \approx 1000\,\text{m}$, and $\rho_0 \approx 1\,\text{kgm}^{-3}$.

	CO in mm/1 h	CO in mm/day
$w = 0.2\,\text{m/s} \approx 1000\,\text{m}/1.5\,\text{h}$ $T_{950} = 20°\text{C}$ $(\Delta T \approx -4.3\,\text{K})$	1.5	34
$w = 0.2\,\text{m/s}$ $T_{950} = 10°\text{C}$ $(\Delta T \approx -5.3\,\text{K})$	1.2	28
$w = 0.1\,\text{m/s}$ $T_{950} = 10°\text{C}$ $(\Delta T \approx -5.3\,\text{K})$	0.6	14

Mid and upper levels are generally colder, so the dependence of condensate on temperature is different here from that at lower levels. In fact, by lowering the temperature by 10 K in an elevated layer (e.g., 500 hPa), the condensate CO decreases by more than just one quarter. This explains why deep atmospheric lifting (through the entire troposphere) is necessary to produce extreme rain events in the warmer regions.

5. Baroclinic Instability and the Synoptic Scale

Energy transport in mid-latitudes occurs, as described above, in baroclinic waves with a vertical axis of rotation, requiring a poleward decrease of temperature. The total temperature difference between equator and pole varies between 15 and about 35 K in winter. The mid-latitudes extend over about 5000 km, half the distance between equator and pole. Therefore, a horizontal temperature difference of 20 K over this distance may be assumed to drive the waves in these mid-latitudes.

The increase of the geostrophic wind with height is described by the thermal wind equation (e.g., Holton, 1992):

$$\frac{d\mathbf{v_g}}{dz} = \frac{g}{f\overline{T}}\mathbf{k} \times \nabla_{\mathbf{h}} T, \tag{5.1}$$

where $\mathbf{v_g}$ is the geostrophic wind, f the Coriolis parameter (the vertical component of the Earth's angular velocity), $\overline{T}$ the layer-averaged temperature, $\mathbf{k}$ the unit vertical vector, and T the temperature, and other symbols are as defined previously. This equation also serves to quantify baroclinicity.

With the above poleward temperature gradient (and no pressure differences at the surface), this thermal wind equation implies a westerly wind of about 10 ms^{-1} in the mid-troposphere and about 20 ms^{-1} at the top of the troposphere.

The development of the baroclinic waves needed to transport energy polewards is described by the baroclinic instability theory of Charney (1947). Calculations based on this theory (e.g., Holton, 1992) show that baroclinic wave growth starts at a minimum thermal wind speed of about 8 ms^{-1} in the mid-troposphere at a wavelength of about 4000 km. Stronger thermal winds enhance the growth rate of baroclinic waves, and so increase the energy transport.

Rotational motion within the waves, including meridional flow, is strongest near the wave troughs and crests, often developing closed circulations at the surface. The circulation is clockwise (anticyclonic) at the crests and counter-clockwise (cyclonic) in the troughs. The detailed configuration of the cyclones and the anti-cyclones has been well known for a long time (e.g., Bergeron, 1928). The typical south–north extent of a cyclone is about 2000 km, with the temperature gradient not being evenly distributed across the entire cyclone.

The length scale of these waves is called the synoptic scale, as used in mid-latitude weather analysis and forecast. On this scale, the pressure field is hydrostatic to high accuracy (e.g., Holton, 1992). The mass budget of a control volume then gives a simplified form of the equation of continuity:

$$\nabla \cdot \mathbf{v} = \frac{\partial u}{\partial x} + \frac{\partial v}{\partial y} + \frac{\partial w}{\partial z}. \tag{5.2}$$

Here, for simplification, a purely zonal flow is assumed. Thus, with $v = 0$ the above equation simplifies to:

$$\frac{\partial u}{\partial x} = -\frac{\partial w}{\partial z} \quad \text{or as differences}: \quad \Delta \mathrm{u} = -\Delta \mathrm{x} \frac{\Delta \mathrm{w}}{\Delta \mathrm{z}}. \tag{5.3}$$

This equation expresses the fact that any horizontal wind divergence is associated with a vertical wind component.

To arrive at the vertical velocity, Eq. (5.3) can be integrated to give:

$$w = -\int \frac{du}{dx} dz \quad \left(\text{in difference form}: \quad \mathrm{w} = -\frac{\Delta \mathrm{u}}{\Delta \mathrm{x}} \int \mathrm{dz}\right). \tag{5.4}$$

The vertical velocity w thus depends on the horizontal divergence and the depth of the layer. When considering the whole troposphere the effective average value for the divergence is found in the mid-troposphere, close to the altitude of 3 km below which rain is formed most effectively.

When two air masses with different temperature are placed next to each other, the interface cannot be oriented vertically, but must be slanted. The angle of slope was derived by Margules (1906). For most frontal systems, the frontal interface exhibits a slope close to $1:100$. The troposphere is about 10 km high, implying that such a frontal interface typically covers a horizontal extent of about 1000 km, containing a large share of the temperature difference between 30°N and the pole. No other process is available on the synoptic scale to provide greater horizontal temperature gradients.

6. Horizontal Wind Speed and Wind Divergence

To estimate the horizontal wind divergence the difference form of Eq. (5.1) may be applied (assuming only north–south temperature gradients):

$$\Delta u = -\frac{g}{f\bar{T}}\frac{\Delta T}{\Delta y}\Delta z, \tag{6.1}$$

Δu is the zonal wind speed difference through a tropospheric layer of depth Δz, $\bar{T}$ the average temperature of the layer and $\Delta T/\Delta y$ the meridional temperature gradient.

Assuming homogeneous surface pressure and a horizontal temperature difference of 20 K, the resulting wind speed difference Δu is about $25\,\mathrm{ms}^{-1}$ at 3 km height, and about $80\,\mathrm{ms}^{-1}$ at the top of the troposphere.

The integrated vertical velocity w can now be estimated from Eq. (5.4). The plausible maximum divergence occurs when the wind velocity is reduced to zero, resulting in a vertical velocity of $0.25\,\mathrm{ms}^{-1}$, an average for the layer of the troposphere where most rain is produced. At higher altitudes, the temperature and the specific humidity drop, so that the fomation of condensate is small (see Eqs. (4.2) and (4.4)).

7. Propagation Speed of Synoptic Weather Systems

To obtain rain rates at a given location, the typical duration of a rain event is needed. This is determined by the spatial extent of the weather system (see Sec. 5.) and its propagation speed, which must be estimated. For this purpose simple, analytically solvable models of baroclinic instability are considered. Such models are usually based on linearization of the quasi-geostrophic potential vorticity equation around a suitable base flow with suitable boundary conditions.

For example, in the model of Eady (1949), it is assumed that the flow is confined to a channel oriented in the zonal (west–east) direction, with periodic boundary conditions in this direction. Furthermore, it is postulated that the troposphere has a rigid upper lid, that the base flow has a constant density, and that the potential vorticity gradient in the interior between the upper lid and the flat bottom vanishes (see (Pedlosky, 1992)). The last assumption is achieved by postulating that the Coriolis parameter vanishes and that the eastward base flow increases linearly with height from $U_{bottom} = 0$ to U_{top}.

The solution of the Eady model features two so-called "edge-waves" with extrema at the top and bottom (Davies and Bishop, 1994; Faulwetter, 2006). The top wave propagates at a speed $U_{top} + c_{top}$ and the bottom wave at a speed $U_{bottom} + c_{bottom} = c_{bottom}$, where c_{top} is negative and c_{bottom} positive. As the zonal wavelength increases, the vertical extent of the waves increases and the two waves interact more and more with each other, in such a way that their propagation speeds are modified. Above a certain threshold wavelength, this leads to a so-called "phase-locking" between the waves, i.e., the waves propagate simultaneously at a speed

$$c = U_{top} + c_{top} = c_{bottom} = 0.5\ U_{top}. \tag{7.1}$$

At a given zonal wavelength, phase-locking occurs for two configurations: Either the bottom wave lags behind the top wave at a certain phase difference or the top wave lags behind the bottom. In the latter case the wave is tilted westward, a configuration that extracts energy from the base flow leading to amplification in time. It can be shown that the most unstable Eady wave has a zonal wavelength of approximately $2L$, where L is the characteristic horizontal length scale. With $1000\,\text{km} < L < 2000\,\text{km}$ this yields wavelengths between 2000 km and 4000 km for the most unstable waves, in agreement with the estimate given above.

The above estimate $c = 0.5\,U_{top}$ for the propagation speed of baroclinic synoptic systems clearly suffers from the strong assumptions that are the basis of the Eady model. However, it is possible to relax these assumptions. A modified version of the Eady model that allows for a non-zero Coriolis parameter yields $0.4U_{top} < c < 0.5U_{top}$ (Lindzen, 1994). If the constraints of a rigid upper lid — a constant base flow density and zero interior potential vorticity gradient — are relaxed, the Charney model can be obtained (Charney, 1947), which yields propagation speeds that are near to the minimum speed of the zonal base flow, rather than the mean speed as in the Eady model (Pedlosky, 1992).

Hence, according to these simple models, the propagation speed of synoptic-scale baroclinic weather systems is not larger than $0.5\,U_{top}$, which typically yields speeds in the range of $10\,\mathrm{ms}^{-1}$.

8. Conceptual Results for Rain

The above considerations may be summarized as follows. Mid-latitude weather is dominated by baroclinic waves, because they are needed for global energy transport. These waves have a length scale of several 1000 km, feeding on the mid-latitude north–south temperature difference of about 20 K. The maximum concentration of this difference occurs in frontal systems, with an overall horizontal extent of about 1000 km. This allows us to estimate the difference of the wind velocities between the cold and warm air masses, the maximum horizontal divergence, and the vertical velocity. The propagation velocity can be estimated from the baroclinicity of the base flow.

The maximum rainfall in stationary conditions is estimated using Eq. (4.4). This equation assumes strictly vertical motion, all condensate falling vertically as rain. This means that there is no condensate left for liquid water storage in the atmosphere, forming clouds. A mid-latitude maximum rainfall is then estimated with a layer-averaged vertical velocity of 0.25 ms^{-1} and a surface dew point temperature of 20°C. Equation (4.4) then implies rainfall of about 200 mmday^{-1}. The frontal rain band does not extend over the whole troposphere and over the whole front, but shows a breadth of about 300 to 600 km, the cloud reaching from the surface to about 6 km height (see tilted frontal surface mentioned above). If the front propagates at the velocity of the wave and perpendicular to its temperature gradient, the rain reduces to about 65 to 130 mm during such an intense frontal rainfall event.

9. Three Historical Mid-Latitude Extreme Rain Events

The most evident consequence of abundant rain is flooding. In mid-latitudes many people live in flood-prone areas, mostly river valleys. Rivers are fed by rainwaters collected in catchment areas determined by the landscape topography. The further downstream a location is sited, the greater the upstream catchment area from which runoff waters are concentrated. The susceptibility to the rise of water levels beyond threshold values depends on many parameters, and is difficult to quantify. However, protection against such events is standard procedure in order to provide attractive living conditions in river plains. The conflict between wanted use of land close to rivers and unwanted flooding by high water levels is inherent and longstanding; in general it is no solution of the flood problem to suggest abandonment of all flood-prone areas!

There are two kinds of floods, inundations and flash floods, both caused by heavy rainfall. Flash floods occur in areas usually extending over a few km^2 up to several $100\,km^2$. In mid-latitudes they are caused by small-scale, non-hydrostatic weather events, for example, convection. In these events the water levels rise fast, within hours or even minutes, and return to normal within hours. Most of the destruction originates from the sheer mechanical forces of the running water. In many cases these floods occur in more mountainous terrain where river beds slope steeply.

By contrast, inundations occur in large river basins extending over many $1000\ km^2$. They are characterized by a slow rise of water level, occasionally taking days until the peak level is reached. The total duration may be several weeks. The damage is caused by the flood waters remaining for days or weeks. The baroclinic waves on the synoptic scale can bring rain to large river catchment areas of several $10000\,km^2$. Rainfall often exceeds $200\,mmday^{-1}$ in those cases.

It should be noted that river beds are formed by the past history of the flowing river waters and that flood events are an essential part of this history. This is because of the non linear relationship between water level and landscape formation, estimated to follow a one-third to one-fifth power law. Hence it is no surprise to find still today in the terraces of River Main (Germany) traces of the major flood of the year 1342, even though the river bed has been heavily transformed by human activities since then (Bork and Kranz, 2008). The most extreme events have the gravest consequences and demand attention. We present some outstanding examples here by way of illustration.

9.1. *Central Europe: Elbe 2002*

In the summer of 2002 heavy flooding occurred in the River Elbe, whose catchment area extends over about $55000\,km^2$. Along the river, the flood level rose to a peak value in the course of several days, returning to average after about three weeks. The maximum water level reached 9.45 m above the reference value in the city of Dresden, the highest value on record for more than 500 years. The rains causing the flood covered almost all of the catchment area, including the neighboring countries, Poland and the Czech Republic. The bulk of the rain fell on 11 and 12 August. On the whole, the average rainfall over the catchment area was 140 mm. The maximum runoff through the river bed in Dresden at the peak water level was $5800\,\mathrm{m}^3\mathrm{s}^{-1}$. This means a runoff of about 10 mmday^{-1} or 70 mmweek^{-1} when applied to the whole catchment area.

This weather event was exceptional, warm humid Mediterranean air being transported northward (e.g., (Rudolf and Simmer, 2002; Mudelsee *et al.*, 2006)). The baroclinic wave also propagated northward into regions east of the areas of heavy rainfall. On 12 August, the flow of warm, humid air finally came from the north (forming a curve around the northern fringes of the cyclone), directed toward the Erzgebirge mountain range which falls within the River Elbe catchment area. Baroclinicity shows up in the eastern parts of Germany (Fig. 1, left), where the isotherms and the isobars cross each other. The scale of this baroclinic area has the extent of the synoptic scale. Rains causing the flood exceeded the monthly average rain by a factor of about 2 (Fig. 1, right). The average rainfall in the Elbe catchment area was 140 mm. Near the Erzgebirge mountains' crest the rainfall reached nearly 400 mm. As a consequence, some of the

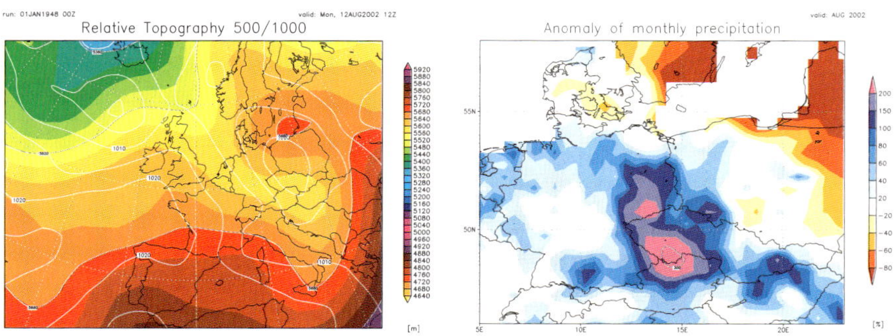

Fig. 1. Characterization of the synoptic situation causing heavy rain in the catchment area of the River Elbe. The relative topography depends on the temperature of the layer between the 1000 hPa and 500 hPa pressure levels; the isobars at the surface describe the flow field. The most baroclinic part of a wave is located in major parts of eastern Germany and the neighboring countries, causing heavy rains there. Right: Monthly rain in central Europe for the month of August 2002. More than 80% of the flood-causing rains in Austria, the Czech Republic, and the eastern parts of Germany occurred on 11 and 12 August. (Data: NCEP Reanalyses; GPCC monthly precipitation.)

Elbe contributary catchments areas suffered from heavy flash flood events. On the other hand, the synoptic scale of the rain event is emphasized by the southern extension of the areas with large rainfall, actually extending into other catchment areas, which also experienced flooding (Rudolf and Simmer, 2002).

9.2. *England 2007*

The 2007 floods were the product of several rainfall events, starting in June and ending only toward the end of July. They caused several floods in different parts of England. Most of the events affected more than 60000 km^2, almost half the country. The floods were caused by synoptic-scale events, but also in some locations by a series of convective events. For the month of July 2007, the deviation from the average July rainfall was more than 200% in the flooded areas (Fig. 2, right). The most significant rainfall event occurred on 20 July when, in one day, about twice as much rain fell compared to an entire average month. The maximum values observed reached about 120 mm of rainfall in one day, and that in terrain with rather small terrain height differences. The rains were caused by a synoptic-scale weather system, propagating across the country from south to north. The front was already rather old, but nevertheless active, clearly visible in the baroclinicity (Fig. 2, left).

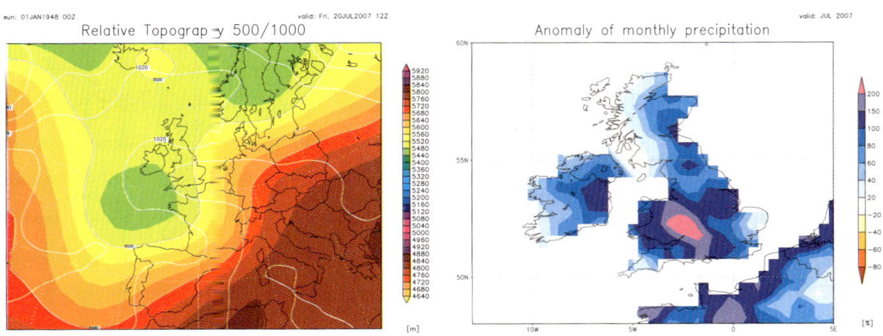

Fig. 2. Left: The weather situation shows a synoptic-scale weather system with the frontal system extending from east to west over England (low values of surface pressure). In the flood affected areas the daily rain amounts reached some 80% of the total monthly rains observed in July 2007. Right: Monthly deviation from the long-term monthly average over Great Britain and Ireland for July 2007. In parts of England about 300% of the long term July rains were observed (an anomaly of 200%).

The frontal system extended from east to west, with England situated at the tip of the frontal process.

9.3. *Mississippi 1993*

The Mississippi floods of 1993 provide another good example of synoptic-scale rain. The floods affected an area of almost one million square kilometers in the Midwest of the United States. Heavy rain and flooding occurred from May to September with repeated rain events and subsequent flood waves. The sequence of the rains was such that the flood levels rose stepwise with each rain event, because the waters of the preceding rain event had not run off through the river bed. The flood period ended only in October, some areas having been continuously flooded for more than four months. Overall, the 1993 Midwest flood was among the most damaging natural disasters ever to hit the U.S.A, with damage exceeding 15 billion dollars, and 50 fatalities. Many people evacuated for months (Larson, 1997).

The weather of 4 July 1993 is chosen to show the type of rainfall-inducing pattern. Figure 3 (left) shows a cyclone situated over the northwestern part of the Great Plains. On its eastern flank a southerly baroclinic flow developed, with frontal systems bringing large amounts of rain. The baroclinicity is quite strong, both on the east side of the cyclone with warm air advection, and on the west side with cold air advection. The

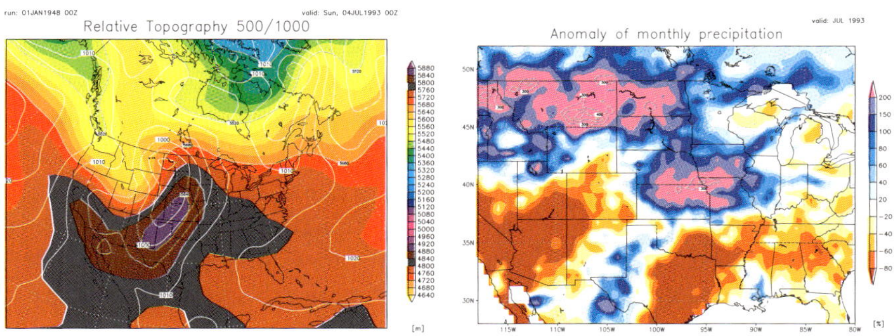

Fig. 3. Left: A rain-bringing cyclone located over the Missouri river catchment area, feeding the central Mississippi river. The synoptic-scale extent and the frontal character are apparent. Right: The monthly rainfall in the central and upper Mississippi river catchment areas surpasses about three times the average monthly amount of rain with more than 500 mm per month, of which about 150 mm of rain fell on the 20 July 1993.

maximum daily rain reached values of about 150 mm, close to the maximum as estimated above, at the same time being close to the average monthly amount. Figure 3 (right) shows the deviations from long-term July rain. In the most severely hit areas the monthly rainfall totals reached four times the long-term value. At the same time it is clear that there were several events with heavy rainfall.

10. Orographic Precipitation Modeling

Comparable to lifting processes taking place at frontal surfaces, orographic obstacles (i.e., mountains) act as a source for lift if the upstream air is forced to flow over the obstacle. Provided the air mass contains a sufficient amount of moisture, this lifting can result in condensation and hence rainfall or snow as described above.

The intensity of orographic precipitation enhancement depends crucially on three main ingredients:

- the orographic lifting w_{oro} proportional to the wind speed U perpendicular to the ridge/hill crest and the terrain slope dH/dx;
- the stability of the inflowing air mass; and
- the vertical profile of specific humidity q and hence the relative humidity rH.

The vertical velocity induced by orography can be computed from the lower boundary condition

$$w_{oro} = U\frac{dH}{dx}. \tag{10.1}$$

The magnitude of w_{oro} is therefore easily defined at the surface, but the vertical structure $w(z)$ depends on several non-trivial factors. In most real cases, the orography-induced lifting vanishes at some height above the surface. Although about one-third of the column's moisture is contained within the lowest 1500 m, the remaining two-thirds play an important role when it comes to extreme events. The maximum precipitation intensity is therefore critically dependent on the degree of vertical decay $w^*(z)$ of the surface-induced lifting w_{oro}.

The vertical profile of atmospheric moisture content, represented by the specific humidity q and the relative humidity rH, influences both the intensity and the horizontal distribution of orographic rainfall. On the one hand, the specific humidity profile determines the amount of condensate released during the condensation process (see Eq. (4.4)). On the other hand, the degree of saturation given by the relative humidity is essential for the condensation level height and for the static stability of the air mass. Lower condensation levels (or higher relative humidities) allow the formation of clouds farther upstream of the crest, thus increasing the time for particles to form. Moreover, the vertical extent of the air lifted by orography and experiencing condensation increases as well. Because of the lowered stability of a saturated air mass compared to unsaturated air, the suppression of orographic lifting by mountain waves will be less pronounced.

The atmospheric stability can be described with several parameters, the most important for orographic precipitation being the moist Brunt–Väisälä frequency (BVF, after (Lalas and Einaudi, 1974)):

$$N_m^2 = \frac{g}{T_0}\left(\frac{dT}{dz} - \frac{dT_m}{dz}\right)\left(1 + \frac{L'q}{RT}\right) - \frac{g}{1+q_w}\frac{dq_w}{dz}, \tag{10.2}$$

with the moist adiabatic lapse rate dT_m/dz (Eq. (4.5)) and the total water content $q_w = q + q_c$. At higher values of N_m the atmosphere is more stable. Increased stability suppresses lifting not only for thermodynamic reasons (since warmer air aloft "generates" negative buoyancy), but also enhances the formation of mountain waves (see Smith, 1979 among others). These waves typically propagate vertically while their axis of phase is tilted upstream. The degree of this tilting defines the upper limit of the orographic lifting, because the lifting zone is overlayed by sinking air within the lee branch tilted upstream (see Fig. 4 in Sec. 10.2.).

The moist BVF approaches zero in the case of neutral stratification, that is a saturated air mass with the temperature lapse rate being equal to the moist adiabatic lapse rate dT_m/dz. In this situation, air flowing over an obstacle can rise freely and without the formation of mountain waves. This is supported by numerical experiments over small hills and under undisturbed ambient conditions (Miglietta and Rotunno, 2005). For taller obstacles, non linear effects such as the drag exerted by the pressure difference between the windward side and the lee (called "form drag") tend to re-establish a wave-like vertical velocity pattern including tilt. Moreover, the creation of a saturated air column with $dT/dz = dT_m/dz$ requires some sort of lifting before reaching the orography. Lifting is mostly associated with a frontal system, which in turn requires at least some stability in the vertical. A neutrally stratified air mass will therefore be limited to small spatial and/or temporal "windows," with the rest of the air mass being in the state often called "near-neutral."

10.1. *Non-Hydrostatic Numerical Modeling: The Meso-Scale Numerical Weather Prediction Model COSMO*

Since non linear effects present great analytical difficulties, it is necessary to turn to numerical models in order to determine the vertical velocities over orography. Numerical models also allow for the inclusion of microphysical processes such as the formation, growth, and evaporation of hydrometeors via physical parametrizations. As an example of state-of-the-art numerical weather prediction (NWP) models, the COSMO model (originally developed by the German Weather Service) is presented here. A detailed description is given in Doms and Schaettler (2002).

This is a non-hydrostatic numerical weather prediction model that uses a fully compressible formulation of the primitive equations. As it is designed as a limited area model, it is possible to use a grid of equidistant cells in the horizontal plane. To ensure this, the model equations are actually formulated on a rotated geographical coordinate system which virtually "moves" the model domain toward the equator. The vertical coordinate is of the generalized terrain-following type. An extensive package of physical parametrizations is provided to represent subgrid-scale processes such as turbulence, radiation or microphysics.

10.1.1. *Basic equations*

As opposed to coarse-grid hydrostatic models which diagnose vertical motion from an equilibrium condition, a non-hydrostatic set of equations involves prognostic treatment of the three-dimensional distribution of the pressure perturbation (or equivalently density). Thus, an additional equation of mass is coupled with the prognostic equations of momentum, heat and water species:

$$\rho \frac{d\mathbf{v}}{dt} = -\nabla p + \mathbf{g}\rho - 2\mathbf{\Omega} \times (\rho \mathbf{v}) - \nabla \cdot \underline{\mathbf{t}} \tag{10.3}$$

$$\rho \frac{de}{dt} = -p\nabla \cdot \mathbf{v} - \nabla \cdot (\mathbf{J_e} + \mathbf{R}) + \varepsilon \tag{10.4}$$

$$\rho \frac{dq_x}{dt} = -\nabla \cdot \mathbf{J_x} + I_x \tag{10.5}$$

$$\frac{d\rho}{dt} = -\rho \nabla \cdot \mathbf{v}, \tag{10.6}$$

where the index x can be d, v, l, or f, representing dry air (d), water vapor (v), liquid water (l), and frozen water (f) such as snow or ice, respectively.

In order to satisfy the conservation of mass in the total volume, the sum of the specific masses of all constituents must equal unity, and the sum of sources and sinks I_x must be zero; similarly for the sum of the diffusion fluxes $\mathbf{J_x}$.

The symbols used in Eqs. (10.3)–(10.6) represent the following variables:

ρ ... density,
$\mathbf{v}$ ... (horizontal wind) velocity vector,
t ... time,
p ... pressure,
$\mathbf{g}$ ... acceleration of gravity,
$\boldsymbol{\Omega}$ ... angular velocity of the Earth's rotation,
$\underline{\mathbf{t}}$ ... stress tensor due to viscosity,
e ... specific internal energy,
$\mathbf{J_e}$ ... diffusion flux of internal energy (heat flux),
$\mathbf{R}$ ... flux density of solar and thermal radiation,
ε ... kinetic energy dissipation due to viscosity,
q_x ... water constituents,
$\mathbf{J_x}$ ... diffusion flux of constituent x,
I_x ... sources and sinks of constituent x.

The derivative operator $\rho\frac{d\psi}{dt}$ is equivalent to $\frac{\partial(\rho\psi)}{\partial t} + \nabla\cdot(\rho\mathbf{v}\psi)$, which would be required to express the set of equations in flux form.

The set of Eqs. (10.3)–(10.6) requires some simplification before the model can be easily integrated in time. Moreover, to obtain a closed set of prognostic equations, the fluxes of the water constituents $\mathbf{J_x}$, their rates of phase change (incorporated in the source terms I_x), the sensible heat flux $\mathbf{J_s}$ (included in the internal energy term $\mathbf{J_e}$), the radiative flux $\mathbf{R}$, and the stress tensor $\underline{\mathbf{t}}$ have to be known. This is achieved by several physical parametrization schemes which attempt to provide a grid-scale average value of unresolved subgrid-scale processes.

10.1.2. *Physical parametrizations*

Mesoscale numerical models with typical grid spacings of $\sim$1$-$10 km do not resolve physical processes that act on smaller-scales (spatial and temporal). These processes include all molecular interactions such as radiation, microphysics, and molecular transfer processes. Furthermore, microscale phenomena such as turbulent fluxes cannot be treated explicitly as well. The same applies for convective motion developing in unstable environments if the horizontal model grid spacing exceeds approximately 2–4 km. All of those subgrid-scale processes have to be expressed in terms of their grid-scale averages; this is realized by a package of physical parametrizations.

10.1.3. *Microphysics*

Microphysical processes play an important role during the life cycle of precipitation. Every kilogram of water vapor which is cooled below its dew point (or equivalently, lifted above the condensation level) will undergo microphysical changes before it finally falls to the ground as rain or snow. After the particles have formed by nucleation of individual water vapor molecules (the condensation process), they grow to sizes of a few hundred micrometers or more due to diffusion and coalescence with other particles. After reaching a certain mass threshold, the particles cannot continue to remain in suspension, and they fall to the ground as precipitation. Even at this stage, microphysical changes continue to occur due to evaporation of drops or crystals while falling through unsaturated air below the cloud base.

All of these processes have to be represented with balance equations which are connected via conversion rates of the individual water constituents (vapor, cloud water, cloud ice, rain, snow, hail). These equations involve numerous empirical constants which have to be determined experimentally or at least have to be estimated. The reader is referred to the COSMO model's manual for details.

A major issue of microphysics is to account for the formation time of particles. The ambient conditions (temperature, relative humidity, turbulence, and condensation nuclei) affect the nucleation and growth rates, and therefore influence both location and intensity of precipitation.

Another important aspect of precipitation physics within the model is the drift of the particles with the wind. In the COSMO model and most other high-resolution models, precipitation drift is taken into account (also called the "prognostic treatment" of precipitation).

Parametrizing microphysical processes remains a challenge since most of the processes involved are either poorly understood or only examined in laboratory experiments to date. This is especially true for environments involving the ice phase — affecting practically all clouds at upper levels of the troposphere. However, for extreme rain events, the effects of microphysics (delayed particle formation, evaporation, etc.) are negligible in a first approximation. For conditions of strong uplift in a warm atmosphere, it is legitimate to consider only the instantaneous conversion of water vapor surplus into liquid water. The condensation rate is then given by Eq. (4.4), which assigns the amount of water vapor Δq exceeding the saturation specific humidity q_{sat} to condensate. This approach is called "saturation adjustment" and is the basic part of the microphysics package. Nearly all other processes tend to reduce the condensation rate.

10.1.4. *Convection — parametrized vs. explicit*

Lifting that acts on scales larger than a few model grid cells will finally result in saturation, followed by condensation, which can be captured by the microphysical parametrization. This type of precipitation formation is referred to as grid-scale or resolved precipitation, taking place in fronts or over orography.

All updrafts which are roughly equal to or even smaller than the size of a single model grid cell will not lead to saturation of sufficiently large magnitude within the cell. Thus, subgrid-scale condensation due to convective updrafts will be underestimated by the model. Besides a poor precipitation distribution, this can have a negative impact on the model's evolution of the flow, since convective instability will be able to grow in absence of a vertical mixing process such as convection.

A parametrization of subgrid-scale (convective) precipitation fills this gap and acts to reduce static instability. Given suitable conditions, i.e., an unstable environment and a certain threshold of moisture convergence, the parametrization will trigger convective up- and downdrafts within individual grid cells. The air column will then produce rain/snow/hail at the rate of moisture convergence into the cell until the convergence falls below a certain threshold. This technique is commonly employed by mass-flux-scheme parametrizations (see Tiedtke, 1989).

For very high-resolution models, whose horizontal grid spacing does not exceed approximately 2 to 4 km, the convective motion can be adequately resolved, making convective parametrization obsolete. The COSMO model at 2.8 km grid-cell size falls within this range.

10.1.5. *Turbulence*

Atmospheric turbulence contributes significantly to the flux of heat, momentum and moisture between the surface and the free atmosphere. In the

COSMO model, the parametrization of those fluxes can be handled either on a fully three-dimensional basis in small-scale simulations, or assuming horizontal homogeneity (so that turbulence affects only vertical transport) in coarser-grid simulations.

A traditional way to handle subgrid-scale turbulence is to treat it as a diffusive mechanism. The rate of turbulent mixing is then computed from the mixing quantities $M_{\mathbf{v}}$, M_T and $M_{q,x}$. In a modified set of model Eqs. (10.3)–(10.6), these quantities M_Ψ are related to the divergence of the turbulent fluxes F_Ψ of momentum, heat and moisture:

$$\rho\, M_\Psi = -\nabla \cdot \mathbf{F}_{\mathbf{\Psi}}. \tag{10.7}$$

By applying a parametrization based on K-theory for the turbulent flux $\mathbf{F}_\psi$ of the variable ψ,

$$\mathbf{F}_\psi = -\underline{\mathbf{K}}_\psi \cdot \nabla\psi, \tag{10.8}$$

these can be expressed by the gradient of the variable ψ and a diffusion coefficent $\underline{\mathbf{K}}_\psi$ which is itself a defined constant for horizontal and vertical fluxes of heat and moisture, $K_{h,h}$ and $K_{h,v}$, while for momentum it is set to $K_{m,h}$ and $K_{m,v}$. The diffusion coefficients K_m and K_h are stability-dependent and are usually estimated through a mixing length approach (see (Blackadar, 1962)).

We should note that turbulence can be parametrized in much more complex ways (e.g., by three-dimensional closure techniques based on turbulent kinetic energy) which are beyond the scope of the present treatment.

10.2. *Sensitivity Studies of Orographic Precipitation Using the COSMO Model*

The COSMO model has been used in sensitivity studies to investigate the structure of the vertical velocity induced by orography. By way of example, we describe here a situation in which a bell-shaped ridge is placed in a northerly flow, in the absence of any synoptic- or meso-scale disturbances. In this way it is possible to trace the vertical velocity pattern in the vicinity of the obstacle without superimposed lifting and/or sinking belonging to fronts or other inhomogeneities.

A bell-shaped ridge of height $H = 800$ m and with the half width $a = 20$ km was chosen in order to represent the basic geometry of a typical low mountain range resembling the Erzgebirge mountains in eastern Germany.

The atmospheric flow was established by different upstream vertical temperature profiles, but in every case being horizontally homogeneous. The Coriolis force has been included in all of the runs to better capture the nonlinear effect of "blocking" (see this section) which is reduced if the Coriolis force is considered (Pierrehumbert and Wyman, 1985).

The ridge is oriented perpendicularly to the incident flow, so that the orographically induced vertical velocity w_{oro} at the surface can be computed from Eq. (10.1). This value is reached in almost all of the numerical simulations which show no significant blocking of air on the windward (upstream) side. Depending on the static stability of the air mass — varied through the upstream vertical temperature gradient — w_{oro} vanishes more or less quickly with increasing height, as seen in Fig. 4 (left).

The upstream tilting of the mountain wave increases with increasing stability (reduced temperature gradient), while the vertical wavelength decreases. Thus, the upward directed branch of orographic lifting will not reach the middle and upper troposphere in stably stratified air masses.

Even in near-neutral flow conditions (thick contours in Fig. 4), with a saturated moist adiabatic temperature profile upstream, the ascending branch reaches heights of only a little more than 4 km above the surface before it is displaced by the first descending wave trough above that height. This is in agreement with other numerical studies of near-neutral flows (Miglietta and Rotunno, 2005) over orographic obstacles of this size. Only

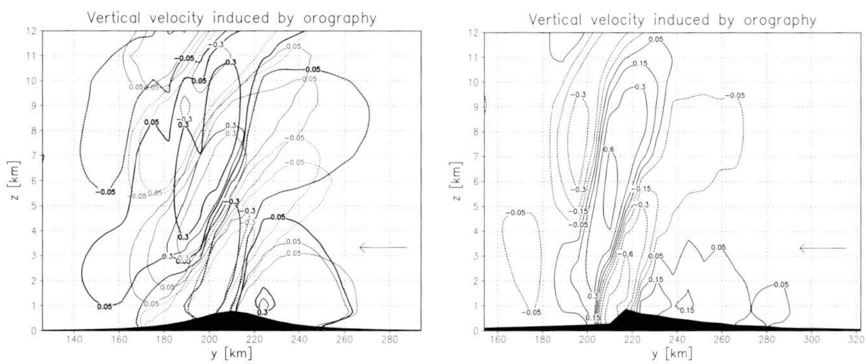

Fig. 4. Vertical velocity induced by orography as simulated by the COSMO numerical model. Left: Variation of the upstream static stability; thin lines represent stable conditions and medium thickness reduced stability; thick lines mark the case of near-neutral stratification. Right: Near-neutral flow over an idealized ridge resembling the Erzgebirge mountains. Flow is from the right (north) with $U = 15\,\mathrm{ms}^{-1}$ in both figures.

for very small hills does the mountain wave tilt vanish, allowing nearly undisturbed rising on the windward slope. The non linear interaction of the flow with orography, made up of form and wave drag, will induce waves as long as the incoming air mass is not convectively unstable. Such unstable flow regimes are treated in numerous recent studies (e.g., (Miglietta and Rotunno, 2009; Kirshbaum and Durran, 2004)), but are not discussed further here.

The horizontal wind speed of the impinging air is directly proportional to the orographic lifting near the surface. Thus, doubling the wind speed would double the magnitude of lifting. Again, there are restrictions that complicate this simple relation. One of these is the degree of blocking of air on the windward side. The blocked air mass creates a kind of an air cushion that can be considered as the "new orography." The incoming air will rise over the cushion instead of directly following the slope of the terrain. As a result, the induced precipitation is generated further upstream of the mountain crest and is also weaker due to the reduced effective slope.

If this blocking is strong, the air will likely flow round the mountain instead of over it. This behavior is made easier during periods of weak wind and/or taller mountain height, but it is also influenced by the static stability since colder air near the surface cannot climb up the hill as easily as would warmer air. Precipitation due to blocked air may extend tens of kilometers upstream of the main slope where pure orographic lifting could not explain its intensity.

Different orographic shapes also exhibit modified mountain wave patterns, which further modify the lifting process. An asymmetric ridge profile with moderate windward slope and steeper slope on the lee side enhances the lee branch of the wave, which in turn reduces the upper limit of orographic lift. This can be explained by the increasing drag resulting from the amplified downward motion in the lee. Although the two ridges in Fig. 4 posess similar windward terrain slopes (in terms of w_{oro}), the orographic lifting is more shallow over the asymmetric ridge. The orographic rain rate in this example drops from $4\,\mathrm{mmh}^{-1}$ (millimeters per hour) over the bell ridge to slightly above $2\,\mathrm{mmh}^{-1}$ over the idealized Erzgebirge ridge. However, not all of this decrease is attributed to the asymmetry; some is attributed to the non-uniform windward slope, maximal near the crest, and also to the enhanced blocking; the direct comparison is not shown here.

In any case, for the purpose of estimating maximum orographic precipitation, it is necessary to assume a gradual decay of the surface-induced vertical velocity with increasing height. If no scaling is applied to w_{oro}, the precipitation intensity is not representative of "classic" upslope motion of stable or neutral air, but falls into the regime that is influenced by convection. Although the spectrum of vertical motion can be quite complex for different conditions, a simple scaling with an upper limit of orographic lifting in the middle troposphere gives reasonable results.

10.3. *Estimating Maximum Orographic Precipitation*

As opposed to statistical approaches combining observed precipitation extremes with maximized observed atmospheric conditions (PMP), the technique presented in this section predicts the maximum possible (orographic) precipitation for given climatic conditions. This is advantageous for mountainous regions with low density of observing stations or with only short recording periods. On the other hand, the method is limited to this particular lifting process.

The estimation described in the following assumes a near-equilibrium between moisture supply and moisture conversion due to orographic uplift. By using a simple diagnostic maximum precipitation model after Tetzlaff and Raabe (1999), the horizontal and vertical distributions of the rainfall production rate can be visualized for various profiles of orography as well as for varying upstream atmospheric conditions. Since the model is designed for stationary flow (not changing in time), the computational cost is minimal because only one integration time-step is needed for each grid cell (as opposed to the prognostic formulation of complex numerical weather prediction (NWP) models). This allows very high-resolution computations with grid spacings of less than 1 km.

The model requires the vertical decay rate of orographic lifting — as discussed in the previous section — as a critical external parameter since it is not designed to compute the complex mountain wave dynamics. In the following, the surface-induced vertical velocity w_{oro} is scaled with the exponential function

$$w(z) = w_{oro}\, w^*(z) = w_{oro}\, exp\left[-\frac{z}{H_{scale}}\right], \qquad (10.9)$$

where $H_{scale} = 4000\,\text{m}$ is the height at which w is decreased by a factor $1/e$. This profile is similar to that seen in Fig. 4, but with somewhat higher extent at mid-tropospheric levels since this is a maximization approach.

Precipitation is formed when parcels of air rise beyond the condensation level, become saturated and continuously release the surplus of water vapor as condensate (Eq. (4.4), applied for each model layer over each grid cell). The condensation process is assumed to be instantaneous, and the drift of falling particles with the mean horizontal wind is included in the model. The terminal fall velocity of precipitation particles is adopted from Sinclair (1994), with the maximum fall velocities of rain and snow being 7 and $2\,\mathrm{ms}^{-1}$, respectively.

For the given flow over a bell-shaped ridge described in Sec. 10.2, the diagnostic model predicts a maximum rain rate of about $4.5\,\mathrm{mmh}^{-1}$ some kilometers upstream of the mountain crest (Fig. 5). The influence of precipitation drift can be clearly seen: It shifts the precipitation maximum towards the crest and it reduces the maximum value. The same flow configuration treated with the COSMO model produces about $4\,\mathrm{mmh}^{-1}$. This

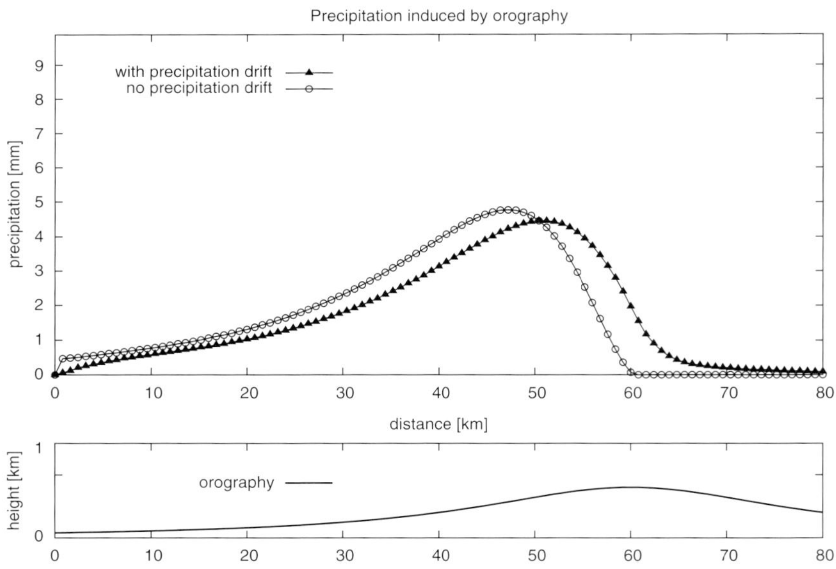

Fig. 5. Orographic precipitation rate [mmh^{-1}] over a bell-shaped mountain of height $H = 800$ m and half width $a = 20$ km, computed with a diagnostic maximum precipitation model. The air flow is from the left at $U = 15\ \mathrm{ms}^{-1}$. The precipitation rate is shown for the cases with and without horizontal drift of the drops/crystals.

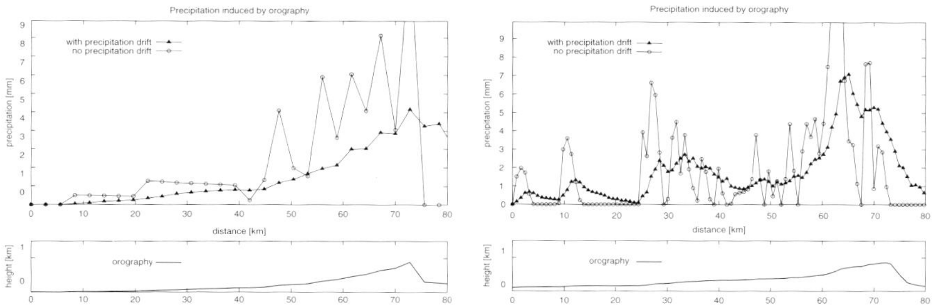

Fig. 6. Orographic precipitation rate [mmh^{-1}] along the northern slope of the Erzgebirge mountains. Left: Using the COSMO model's orography at 2800 m grid spacing. Right: using a higher resolution orography at $dx = 800\,\text{m}$. The air flow is from the left at $\text{U} = 15\,\text{ms}^{-1}$. The precipitation rate is shown for the cases with and without horizontal drift of the drops/crystals.

suggests that the chosen configuration does not suffer significant reduction from small-scale or nonlinear effects or due to air flowing around the obstacle.

For the more realistic orographic profile shown in Fig. 6, the diagnostic model predicts higher rain rates than the numerical model. Depending on the resolution of the underlying orography, the diagnostic model computes 5 and 7 mmh^{-1} for grid spacings of 2.8 and 0.8 km, respectively. The rainfall in the COSMO model hardly exceeded $2\,\text{mmh}^{-1}$, but with the maximum located upstream of the steepest slope. This demonstrates the complications that can arise from complex dynamic interactions such as those discussed in Sec. 10.2.

The flood event of August 2002 in central Europe affected large portions of the Erzgebirge mountains in eastern Germany. Near the crest, more than half of the total rainfall of about $300\,\text{mmday}^{-1}$ was generated by orographic enhancement within the strong northerly flow. The rainfall intensities given in Fig. 6 (right) match the average orographic fraction of the observed amounts (see Zimmer *et al.*, 2006). The (simple) diagnostic model is able to predict the upper limit of the orographically induced precipitation in cases of non-blocked flows if a valid assumption regarding the vertical velocity profile can be applied.

The orographic lifting in the above examples was on the order of $w_{oro} \approx 0.1..0.5\,\text{ms}^{-1}$. Given a suitable arrangement of fronts, the contributions of frontal lifting ($w \approx 0.1..0.2\,\text{ms}^{-1}$) and orographic lifting may add. That is why extreme precipitation events in mid-latitudes are mainly found on the slopes of mountain ranges which are crossed by propagating frontal systems. Among these are the coastal parts of the Rocky Mountains of North America or the Norwegian mountain range in Northern Europe. On occasion, the orographic lifting can reach values of ≈ 1 ms^{-1} over several hours, when strong winds transport saturated masses of air toward the mountain crest. Frontal systems alone cannot supply such an intense lifting over a comparable area and time period. In this way, orographic lifting on the steepest slopes can produce rain rates of about $20\,\mathrm{mmh}^{-1}$.

11. Convective Precipitation

The term "convection" commonly refers to a process involving vertical motion due to static instability. Instability is generated by vertical temperature gradients exceeding the adiabatic lapse rate. A parcel of air becomes positively buoyant if it is warmer than the surrounding air mass while it rises. In unsaturated air, the temperature gradient has to be stronger than the dry adiabatic lapse rate $dT/dz \approx -1$ K $(100\,\mathrm{m})^{-1}$ (Eq. (4.1)). Dry convection is usually limited to the atmospheric boundary layer during times of strong solar radiation and is often associated with rising thermals of surface-warmed air.

If the air becomes saturated (i.e., above the condensation level), positive buoyancy emerges from the release of latent heat due to condensation. This is the case if the moist adiabatic lapse rate dT_m/dz (Eq. (4.5)), which the saturated parcel is following upward, is smaller than the ambient temperature lapse rate. If the temperature difference is integrated over the entire column in which the deviation is positive, the resulting quantity is a measure of the convective available potential energy (CAPE):

$$CAPE = \int g \frac{T_{par} - T_{env}}{T_{par}} dz, \tag{11.1}$$

where T_{par} and T_{env} represent parcel and environmental temperature.

In an idealized framework, the potential energy represented by $CAPE$ can be converted completely into kinetic energy during the ascent of the parcel. Because internal energy (sensible and latent heat, H and L) is already incorporated in $CAPE$, the balance equation for energy simplifies to:

$$0 = P + K = P + \frac{U^2}{2}. \tag{11.2}$$

Inserting $CAPE$ for potential energy P and w for U, the maximum (vertical) velocity for moist convection is given by

$$w_{max} = \sqrt{2CAPE}. \tag{11.3}$$

The climatology of $CAPE$ varies greatly with latitude as it depends strongly on the available temperature and moisture near the surface (a warmer temperature yields weaker moist adiabatic lapse rates, allowing greater positive deviations of the rising parcel). During the warm season, $CAPE$ can reach values as much as 2000 to 4000 $\mathrm{Jkg^{-1}}$ in the mid-latitudes, while in some parts of the globe it can exceed 6000 $\mathrm{Jkg^{-1}}$, for example during early monsoon in Southern Asia.

Applying Eq. (11.3) for $CAPE = 2000\ \mathrm{Jkg^{-1}}$ results in a maximum vertical velocity of more than 60 $\mathrm{ms^{-1}}$. Recalling the typical vertical motion within baroclinic lows on the order of 0.1 $\mathrm{ms^{-1}}$, this suggests devastating rainfall production rates around 2400 $\mathrm{mmh^{-1}}$ (according to Eq. (4.4), including the entire column). Fortunately, there are (at least) four limiting factors that restrict convective processes of that strength to short temporal and spatial scales. These are:

- The restoring force of non-hydrostatic pressure perturbations (growing rapidly with increasing diameter of the updraft);
- the limited buoyancy in case of wide updrafts (because buoyancy emerges from the temperature difference over a limited horizontal distance);
- the supply of unstable air feeding the convective updraft; and
- the fact that the rising air is accelerated during its ascent, so that w_{max} is reached only near the top of the cloud, while w is much lower near the surface.

In reality, convective updrafts are further slowed due to the inclusion of unsaturated cooler environmental air ("entrainment") and the mass of the produced condensate ("water loading").

If all the reducing factors are included, the vertically averaged ascent rate of such a convective "turret" can reach 10 $\mathrm{ms^{-1}}$ for diameters of up to a few kilometers. This translates to a rain rate of roughly 7 mm per minute in a warm and unstable air mass, a value which is occasionally observed during intense (not only tropical) showers. The horizontal extent and duration of such events are limited by the restraining effects mentioned above (see Zimmer, 2008), for details on a diagnostic approach of maximum convective precipitation).

The great diversity of shapes and sizes of convection complicates the estimation of areal coverage and mean intensity of convectively dominated systems, in contrast to the approach in Sec. 8. While individual convective cells will provide the maximum convective precipitation over areas of a few to some tens of square kilometers, organized convection as in Mesoscale Convective Systems (MCS) will have the same effect for much larger areas, but at reduced intensity.

11.1. *Mesoscale Convective Systems*

Mesoscale convective systems (MCS) consist of numerous individual convective cells, each living no longer than one hour, but contributing to the larger-scale lifting. Through the existence of a main region of converging air near the surface, a sequence of individual cells supports the longevity of the system by converting convective available potential energy (CAPE) into upward motion.

Unlike a baroclinic synoptic-scale vortex, an MCS develops some distance away from the main frontal boundary, an area which still needs to be supportive for large-scale ascent. Due to the vast diversity of convective arrangement and intensity under different synoptic forcing, MCSs can take on various shapes and sizes. One of the most important types is the so-called "squall line," consisting of a leading line of convective cells and followed by a trailing region of stratiform precipitation (e.g., (Houze, 1997)).

The horizontal extent can reach several hundreds of kilometers perpendicular to the direction of movement and $\approx$100 km along its path. As those systems effectively convert available moisture into rainfall, they can contribute significant portions to warm-season precipitation in parts of the mid-latitudes.

The lifetime of such a system depends on several factors, such as the synoptic forcing, the ambient conditions (CAPE and wind shear), and the system's intensity itself. An estimate of the lifetime can be attempted using some empirical assumptions concerning size and intensity of large MCSs embedded in the general (westerly) flow. For a balanced mass flux, the upward motion within the precipitating part of the system requires compensating subsidence (see Fritsch, 1975), generally occuring outside of that area. If the subsiding air is assumed to spread across the entire warm quarter of a low-amplitude baroclinic wave minus the rainfall-generating area of the MCS,

$$\frac{L_x}{2}\frac{L_y}{2} - A_{MCS} \approx 1000\,\text{km} \cdot 500\,\text{km} - 100\,\text{km} \cdot 500\,\text{km} = 450000\,\text{km}^2, \tag{11.4}$$

the vertical motion w_{sink} outside of the MCS needs to fulfill the equation

$$A_{MCS}\ w_{MCS} + A_{sink}\ w_{sink} = 0. \tag{11.5}$$

Given strong ascent within the upward branch, $w_{MCS} \approx 0.5\ \text{ms}^{-1}$, corresponding to an average rainfall rate of roughly $20\,\text{mmh}^{-1}$, the subsidence amounts to

$$w_{sink} = -\frac{A_{MCS}\ w_{MCS}}{A_{sink}} \approx -0.05\,\text{ms}^{-1}. \tag{11.6}$$

Since the subsiding air warms the middle and upper levels most (as a consequence of the strongest ascent at those levels), the convective instability represented by CAPE gradually decreases with time. In this example it vanishes after about eight hours, if the initial value is assumed with $2500\,\text{Jkg}^{-1}$ (a typical upper limit for active weather conditions in mid-latitudes).

The above result can be considered a rough estimate, but it will be different in situations of rapid MCS propagation, that is if the subsiding air adjacent to the MCS's rainfall area does not fully modify the inflowing unstable air mass at the front side of the rainfall area ("leading edge"). This would allow longer lifetimes, the same applying for reduced precipitation intensity (hence weakened subsidence) or larger subsidence areas. However, in the above example, the MCS would cross one location in less than three hours, dropping roughly 50 mm of rain over some 10000 square kilometers during its lifetime.

12. Conclusion

The estimates of maximum mid-latitude synoptic-scale rain applying a simple conceptual approach show a maximum of about $200\,\text{mmday}^{-1}$. Often larger daily amounts of rain are observed in mid-latitude weather conditions. Other supporting processes are needed to allow their formation. Orography is frequently responsible for such an enhancement. Smaller-scale events are influenced by convective-scale processes allowing higher rain rates per time on smaller areas. Meso-scale convective processes need synoptic-scale conditions to develop properly, but in addition combine with convective processes, which do not reach the dimensions of the synoptic scale however.

Rainfall induced by synoptic-scale processes is closely coupled to frontal zones. Because these fronts do not cover the same area as the driving synoptic-scale process, the "synoptic-scale rain" overlaps with smaller, meso-scale events in terms of size. Hence, synoptic-scale lifting processes extend over several 10,000 km^2, mesoscale convective systems over not more than a few 10,000 km^2, and convection over some 10 to 100 km^2.

As a rough estimate, the area-average maximum rainfall reaches about 0.1 mm min^{-1} for the synoptic scale in flat terrain, up to 0.3 mm min^{-1} in orographically structured terrain and in mesoscale convective systems, and 7 mm min^{-1} for individual convective cells. The typical maximum duration of the rainfall periods over one location is 24 hours for the synoptic-scale rain, three hours for mesoscale convective systems, and about half an hour for individual convective cells.

References

Bergeron, T. (1928). über die dreidimensional verknüpfende Wetteranalyse, *Geofys. Publ.* **5**, pp. 1–111.

Blackadar, A. K. (1962). The vertical distribution of wind and turbulent exchange in a neutral atmosphere, *J. Geophys. Res.* **67**, pp. 309–3102.

Bork, H. R. and Kranz, A. (2008). Die Jahrtausendflut des Jahres 1342 prägt Deutschland'. *Jber. Wetterau Ges. Naturkunde* **158**, pp. 119–129.

Charney, J. G. (1947). The dynamics of long waves in a baroclinic westerly current, *J. Meteorol.* **4**, pp. 135–163.

Davies, H. C. and Bishop, C. H. (1994). Eady edge waves and rapid development, *J. Atmos. Sci.* **51**, pp. 1930–1946.

Doms, G. and Schaettler, U. (2002). A description of the nonhydrostatic regional model LM. Consortium for Small-Scale Modelling, Deutscher Wetterdienst, Offenbach. http://www.cosmo-model.org/content/model/documentation/core/default.htm.

DVWK (1997). Maximierte gebietsniederschlagshöhe für Deutschland, *Deutscher Verband für Wasserwirtschaft und Kulturbau (DVWK), Wirtschafts- und Verlagsgesellschaft Gas und Wasser*, Bonn, 112 pp.

Eady, E. T. (1949). Long waves and cyclone waves, *Tellus* **1**, pp. 33–52.

Faulwetter, R. (2006). Optimal growth of Eady edge waves, *Meteorologische Zeitschrift* **15**, pp. 423–437.

Fritsch, J. M. (1975). Cumulus dynamics: local compensating subsidence and its implications for cumulus parameterization, *Pure and Applied Geophysics* **113**, 1, pp. 851–867.

Guha-Sapir, D., Hargitt, D. and Hoyois, P. (2004). Thirty years of natural disasters. Report of the centre for Research on the Epidemiology of Disasters (CRED), Mellin (Belgium).

Hartman, D. L. (1994). *Global Physical Climatology* (Academic Press, San Diego).

Holton, J. R. (1992). *Introduction to Dynamic Meteorology*, 3rd edn. (Academic Press, San Diego).

Houze, R. A. (1997). Stratiform precipitation in regions of convection: a meteorological paradox? *B. Am. Meteorol. Soc.* **78**, 10, pp. 2179–2196.

Hoyois, P., Below, R., Scheuren, J. M., *et al.* (2007). Annual disaster statistical review 2006. Report of the centre for Research on the Epidemiology of Disasters (CRED), Mellin (Belgium).

Kirshbaum, D. J. and Durran, D. R. (2004). Factors governing cellular convection in orographic precipitation, *J. Atmos. Sci.* **61**, 6, pp. 682–698.

Lalas, D. P. and Einaudi, F. (1974). On the correct use of the wet adiabatic lapse rate in the stability criteria of a saturated atmosphere, *J. Appl. Meteor.* **13**, pp. 318–324.

Larson, L. W. (1997). The great USA flood of 1993. In *Proc. IAHS Conf. on Destructive Water: Water-Caused Natural Disasters, Their Abatement and Control* (G. H. Leavesley, International Association of Hydrological Sciences, IAHS Publication no. 239, ISBN 1-901502-00-7, IAHS Press, Institute of Hydrology, Wallingford, UK).

Lindzen, R. S. (1994). The Eady problem for a basic state with zero PV gradient but beta non-zero, *J. Atmos. Sci.* **51**, 22, pp. 3221–3226.

Margules, M. (1906). Über Temperaturschichtung in stationär bewegter und ruhender Luft, *Meteorol. Z.* Hann-Volume, pp. 243–254.

Miglietta, M. M. and Rotunno, R. (2005). Numerical simulations of nearly moist neutral flow past a two-dimensional ridge, *J. Atmos. Sci.* **62**, pp. 1410–1427.

Miglietta, M. M. and Rotunno, R. (2009). Numerical simulations of conditionally unstable flows over a mountain ridge, *J. Atmos. Sci.* **66**, pp. 1865–1885.

Mudelsee, M., Börngen, M., Tetzlaff, G., *et al.* (2006). Extreme floods in central Europe over the past 500 years: role of cyclone pathway "Zugstrasse Vb", *J. Geophys. Res.* **109**, D23101.

Oort, A. H. (1971). The observed annual cycle in the meridional transport of atmospheric energy, *J. Atmos. Sci.* **28**, pp. 325–339.

Pedlosky, J. (1992). *Geophysical Fluid Dynamics* (Springer, Berlin).

Pierrehumbert, R. T. and Wyman, B. (1985). Upstream effects of mesoscale mountains, *J. Atmos. Sci.* **42**, 10, pp. 977–1003.

Rodriguez, J., Vos, F., Below, R., *et al.* (2009). Annual disaster statistical review 2008. Report of centre for research on the epidemiology of disasters (CRED), Melui (Belgium).

Rudolf, B. and Simmer, C. (2002). Niederschlag, Starkregen und Hochwasser im Jahr der Geowissenschaften, *Wiss. Mitteil. Inst. f. Meteorol. Univ. Leipzig* **Special Vol.**, pp. 28–44.

Sinclair, M. R. (1994). A diagnostic model for the estimating of orographic precipitation, *J. Appl. Meteor.* **33**, pp. 1163–1175.

Smith, R. B. (1979). The influence of mountains on the atmosphere, *Adv. in Geophys.* **21**, pp. 87–230.

Tetzlaff, G. and Raabe, A. (1999). Räumliche und zeitliche Verteilung maximaler Niederschläge, in *Extreme Naturereignisse und Wasserwirtschaft– Niederschlag und Abfluss, Internationales Symposium, Bayrisches Landesamt für Wirtschaft, Informationsheft*, Vol. 5, pp. 57–64.

Tiedtke, M. (1989). A comprehensive mass flux scheme for cumulus-parameterization in large-scale models, *Monthly Weather Review* **117**, pp. 1779–1799.

Wiesner, C. J. (1970). *Hydrometeorology* (Chapman and Hall Ltd, London).

WMO (1973). Manual for estimation of probable maximum precipitation, *Operational Hydrology* Report No. 1, WMO-No. 332, Geneva, Switzerland, 190 pp.

Zimmer, J. (2008). Estimation of maximum convective precipitation: an idealised model approach, *Wiss. Mitteil. Inst. f. Meteorol. Univ. Leipzig* **42**, pp. 33–41.

Zimmer, J., Raabe, A. and Tetzlaff, G. (2006). Quantification of topographic effects on predicted precipitation in the Erzgebirge, *Wiss. Mitteil. Inst. f. Meteorol. Univ. Leipzig* **37**, pp. 125–136.

Chapter 8

Dynamics of Hydro-Meteorological and Environmental Hazards

A. W. Jayawardena

International Centre for Water Hazard and Risk Management (ICHARM) under the auspices of UNESCO
Public Works Research Institute, Tsukuba, Japan
hrecjaw@hkucc.hku.hk

An overview is presented of the physical and biological factors that cause disasters and of their relationships in quantitative terms to the outcomes of these disasters. The chapter begins with an introduction to the atmosphere, which is the starting point of all hydro-meteorological disasters, including the different processes and links that lead to precipitation. The relationship between precipitation and runoff, or floods, including their forecasting techniques is described. The chapter also covers the types and causes of water-related environmental disasters. A quantitative description of mixing processes by Fickian diffusion and by convective dispersion is given. The governing equations and simplifications for conservative and non-conservative types of pollutants, point and non-point sources of pollution, reaction kinetics for non-conservative pollutants, and modeling approaches, are presented. As the health of a waterbody is measured by the dissolved oxygen concentration, an introduction to the oxygen sag curve in rivers is also given.

1. Introduction

Disasters can be broadly classified as natural or human induced. The former type is difficult if not impossible to prevent whereas the latter type is preventable. In terms of the cost and damage induced by various types of natural disasters, "water-related disasters" by far exceed those by any other natural disaster. In this context, water-related disasters include all types of floods, land and mud slides, storm surges, tsunamis, tidal waves, debris flow, avalanches, droughts, and all types of cyclones. In addition to such geophysical disasters, water-related biological disasters such as epidemics and endemics also take a significant toll in terms of human lives. Human-induced disasters include various types of pollution, accidents, and wildfires, among others. In the modern world, pollution of the water environment is a major environmental disaster in many regions, with some places reaching irreversible conditions. The objective of this chapter is to highlight the causes and mechanisms of such disasters, explore how they can be modeled, and predict the outcomes of impending disasters with a view to mitigate their effects. A better understanding of the initiation and fate of any disaster is important for taking preventive and mitigative actions.

Natural disasters have taken place from time immemorial. In the past, biotic populations living under natural conditions and in harmony with nature were able to live with disasters by adapting their lifestyles or by changing their habitats. With exponential increase in human population and increasing urbanization, natural conditions no longer exist in many places. With increased population density and high value added infrastructure, the impacts have increased manifold.

The definition of a disaster depends upon the agency or organization that collects and disseminates data. There is a wide variation in the criteria used for inclusion in databases. One of the comprehensive databases on disaster information is the Emergency Events Database (EMDAT), which is located in the University Catholic Louvain, Belgium (http://www.emdat.be), and which is updated regularly. They define an event as a disaster if there have been more than 10 deaths or more than 100 people displaced, or if the government of the affected country has declared a state of emergency and asked for international assistance.

According to a report (Adikari *et al.*, 2008) by the International Center for Water Hazard and Risk Management (ICHARM) based on data compiled by EMDAT, there have been 3,050 incidents of flood disasters during the period 1900–2006 causing economic damage to the extent of some US$342 billion. During the same period, there have been 2,758 incidences of windstorm disasters causing US$536 billion worth of damage. Figure 1 illustrates the trends for water-related disasters on a three-year period basis. The numbers of people who lost their lives have been in excess of 6.8 million and 1.2 million, respectively, for flood and windstorm disasters. These two types of disaster alone accounted for over 56% of all natural disasters in that period. Of the 1,000 worst natural disasters in terms of the number

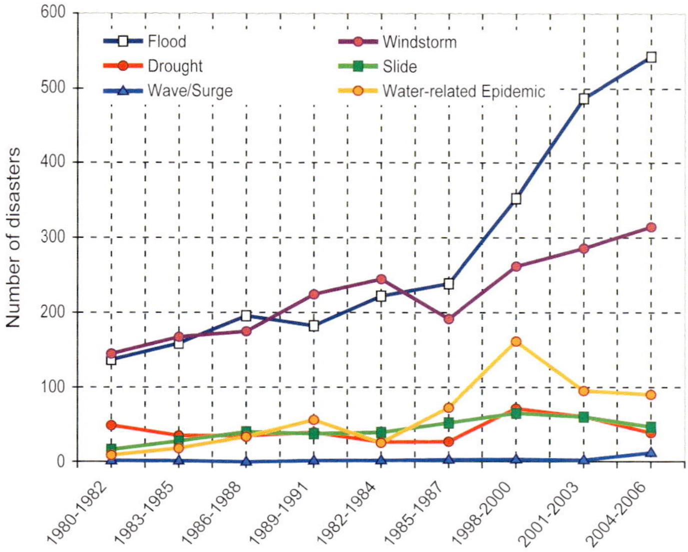

Fig. 1. Trends in different types of disasters.

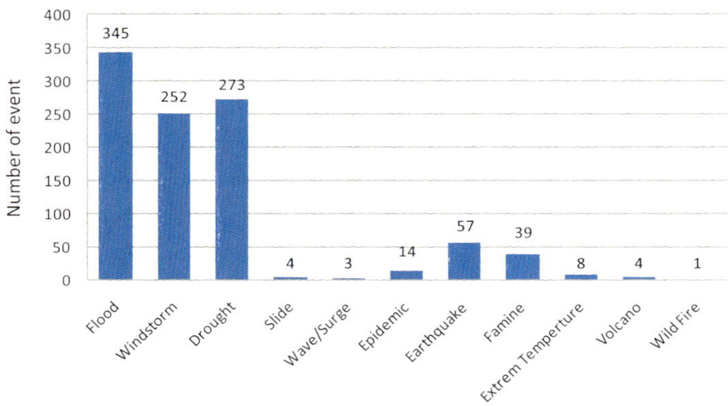

Fig. 2. Statistics of the 1000 worst disasters in the period 1900–2006.

of human casualties that occurred during 1900–2006, floods accounted for 345, windstorms for 252, and droughts for 273 (Fig. 2). All these facts and figures illustrate the importance of hydro-meteorological disasters. It is also important to note that it is not only the numbers of disasters that are increasing but also the number of people affected, because of the migration of people into areas with better economic prospects.

Environmental disasters are mainly human-induced and are therefore preventable. There are disasters caused by actions over a long period of time, as well as those caused by accidents. In the water sector, these include the pollution of waterbodies including rivers, streams, lakes, and reservoirs, groundwater as well as coastal waters. The problem of understanding water pollution involves the study of the fate and transport of any pollutant introduced (deliberately or by accident) into a waterbody. In general the fate and transport are governed by principles of fluid mechanics. However, in real life, it is often difficult to quantify the problem in terms of fluid mechanics without making assumptions and simplifications. Some approaches for particular cases are described in the following sections.

Except for tsunami and tidal waves, all the hydro-meteorological disasters are caused by rain or snowfall. Drought, which is the lack of sufficient rainfall, can lead to a shortage of water for agriculture, industry, and domestic use that in turn can lead to a disaster if it continuously prolongs for long periods of time. Under such conditions, the quality of water deteriorates resulting in undesirable microorganisms that can cause diseases to grow, thereby causing water-borne diseases, which in uncontrollable situations may lead to epidemics. Thus rainfall can be considered as the triggering cause of almost all water-related disasters.

2. Hydro-Meteorology

2.1. *Weather*

2.1.1. *Weather charts*

Usually wind speeds are plotted in weather charts in knots (1 knot = 1.15 mph = 1.85 km/hr = 0.5 m/s). The effect of wind is categorized according to the Beaufort scale (1806) which, in simplified form, is given in Table 1. A front is a narrow zone of transition between air masses of contrasting physical properties. They include stationary fronts which remain stationary over a certain area, cold fronts in which the cold (denser) air is moving into warm (lighter) air, warm fronts in which warm (lighter) air is replacing the cold (denser) air by overrunning, and occluded fronts in which cold fronts that travel twice as fast as warm fronts eventually catch up and merge.

Table 1. The Beaufort scale for wind effects.

Force	Specifications for use on land	Equivalent mean wind speed 10 m above ground (Knots)	(m/s)
0	Calm; smoke rises vertically	0	0
1	Light air; wind direction shown by smoke drift, not by vanes	2	1.0
2	Light breeze; wind felt on face; leaves rustle; vanes move	5	2.57
3	Gentle breeze; leaves and small twigs moving; light flags lift	9	6.3
4	Moderate breeze; dust and loose paper lift; small branches move	13	6.68
5	Fresh breeze; small leafy trees sway; crested wavelets on lakes	19	9.77
6	Strong breeze; large branches sway; telegraph wires whistle; umbrellas difficult to use	24	12.3
7	Near gale; whole trees move; inconvenient to walk against	30	15.4
8	Gale; small twigs break off; impedes all walking	37	19.0
9	Strong gale; slight structural damage	44	22.6
10	Storm; seldom experienced on land; considerable structural damage; trees uprooted	52	26.7
11	Violent storm; rarely experienced; widespread damage	60	30.8
12	Hurricane; at sea visibility is badly affected by driving foam and spray; sea surface completely white	>64	>32.9

Jet streams are easterly winds (speeds > 100 knots) flowing round the entire hemisphere from west to east in the form of a meandering river. In the tropics the core of the jet stream is located at about 13 km (150 mb). In the extratropical latitudes (20–40°) it is located at around 12 km.

2.1.2. *Atmospheric properties*

An "air parcel" refers to a small volume of air, which has uniform temperature, pressure, humidity, density, etc. It may expand, contract as it moves but the matter contained within it remains constant (Fig. 3). It is similar to the "control volume" concept.

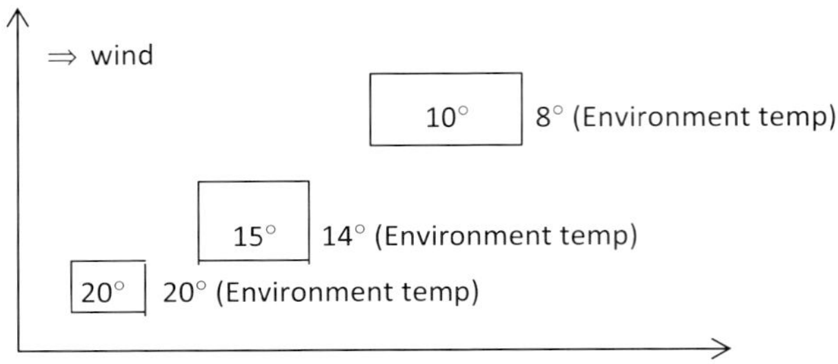

Fig. 3. An ascending air parcel.

The "lapse rate" refers to the temperature gradient with respect to altitude. There are three types of lapse rates in meteorology. The dry adiabatic lapse rate refers to the temperature gradient when there is no heat added or taken away from the atmospheric process. In this case the change in temperature is caused by the change in pressure (expansion and contraction). The dry adiabatic lapse rate Γ_d is given by

$$\Gamma_d = -\frac{dT}{dz} = \frac{g}{c_p}, \quad \frac{T_1}{T_2} = \left(\frac{p_1}{p_2}\right)^{(\gamma-1)/\gamma}, \tag{2.1}$$

where g is the gravitational acceleration and c_p is the specific heat capacity at constant pressure. A parcel of air flowing over a mountain can be adiabatic. The approximate value of the dry adiabatic lapse rate is $9.8°\mathrm{Ckm}^{-1}$. The environmental (ambient) lapse rate is the actual temperature gradient that exists in the environment. It can take a wide range of values. When the moist unsaturated parcel of air rises it will at some altitude reach saturation and condensation will result. This adds the latent heat of condensation to the thermodynamic process resulting in a decrease of the lapse rate to a value of approximately $6°\mathrm{Ckm}^{-1}$. The lowered lapse rate is referred to as the saturated or moist lapse rate.

When phases change, heat must be added or taken away without any change of temperature. The processes of changing phases from solid to liquid and liquid to gas are endothermic or energy absorbing. The reverse processes of changing phases from gas to liquid and liquid to solid are exothermic or energy releasing. Latent heat of vaporization (or condensation) is the amount of heat needed to be added (or released) to change phase from liquid to vapor (or vapor to liquid) and has a value of

$2500.78\,\text{kJkg}^{-1}$ at $0°\text{C}$. The rate of change of latent heat of evaporation with absolute temperature is equal to the difference between the specific heat at constant pressure of the vapor and the specific heat of liquid:

$$\frac{dL_{ev}}{dT} = -2.3697\,\text{kJ/kg°C} \tag{2.2}$$

Therefore, the latent heat of evaporation at $100°\text{C}$ is approximately $2263.81\,\text{kJkg}^{-1}$ (≈ 540 Calg^{-1}). Latent heat of melting, or fusion (which is equal to the latent heat of freezing) is the amount of heat needed to be added (or released) to change phase from solid to liquid (or liquid to solid) and has a value of $334\,\text{kJkg}^{-1}$ at $0°\text{C}$. Some substances undergo phase changes from solid to gas or vice versa without going through the intermediate stage. Latent heat of sublimation is the amount of heat needed to be added (or released) to change phase from solid to vapor (or vapor to solid) and has a value of $2834\,\text{kJkg}^{-1}$ at $0°\text{C}$ (≈ 680 Calg^{-1}). Frost formation is an example of deposition, which is the reverse of the process of sublimation.

If a parcel of air lifted to a certain height returns to its original level when released, then the condition is stable. If it remains at that height the condition is neutral. On the other hand, if it continues to rise further when released, the condition is unstable. These three conditions can be explained with respect to the lapse rates. Instability can occur if the parcel of air is warmer, or, if the parcel of air contains more water vapor than dry air (molecular weight of water vapor is less than that of dry air in the ratio 18 : 29). The former condition is maintained when the atmospheric lapse rate exceeds the dry adiabatic lapse rate. It is also possible for vertical lifting to take place in a stable environment when the surface temperature is very high, e.g., over forest fires, chimneys, explosions, etc.

2.1.3. *Energy in the atmosphere*

Energy in the atmosphere is composed of solar energy, terrestrial energy, and tidal energy. The latter two types are small compared to solar energy. Solar energy comes from the Sun mostly in the form of short-wave radiation (visible, ultraviolet, and infrared rays are all at the short wave end of the spectrum). The solar radiation which is received at the surface of the Earth is partly reflected back into the outer space as long-wave radiation. Of the 1380 Wm^{-2} (solar constant) of energy received at the top of the atmosphere, only about 350 Wm^{-2} is received on average at the Earth's surface. The energy that is absorbed by the Earth's surface is used to heat the unsaturated air in contact by conduction which then gets lifted by convection, orographic, or frontal mechanisms.

Energy utilized in the atmosphere comes from two sources — the heat content of rising air and the heat released by water vapor when condensing to form clouds. The first source is indirectly from solar radiation. In a typical thunderstorm of approximately 5 km in diameter, there may be 500,000 tons of condensed water. In producing these droplets, a quantity of energy equivalent to about 3.5×10^8 kWh would have been released. If the air is dry, relatively small quantities of energy are available. Energy in the atmosphere is dissipated mainly as kinetic energy in various wind systems (lightning also discharges some amount of energy). A comparison of the approximate orders of magnitude of the energy of various wind systems is given in Table 2. Intense vortices in the atmosphere can be taken as signs that the atmosphere is unstable and has high moisture content. The total power of a system is difficult to ascertain because only part of it can be experienced at a time. By any standard, the weather systems in the atmosphere are very powerful. The amount of energy input required to develop such systems is even larger. The difference is dissipated in overcoming friction and heating the air inside and outside the system. The energy of the systems in general is spread over a large area. Therefore, the destructive effect is not apparent when compared to, for example, that of a nuclear bomb. On the other hand, a tornado is concentrated around a smaller area within a radius of about 100 m and therefore the effects are explosive.

Table 2. Order of magnitudes of energy in wind systems (* in desert regions when the ground is heated to very high temperatures).

System	Kinetic energy
Gust	<1
Dust devil*	10
Tornado	10^4
Thunderstorm	10^6
Hurricane	10^{10}
Cyclone	10^{11}
Nagasaki bomb	10^7
Hydrogen bomb	10^{10}

Fig. 4. Water vapor in the atmosphere.

2.1.4. *Water vapor in the atmosphere*

The amount of water vapor contained in the atmosphere (Fig. 4) is a function of several factors such as the availability of a source of moisture, place, temperature, elevation, etc. It is measured by the relative humidity, which is defined as the ratio of the amount of moisture in the air to the amount needed to saturate the air at the same temperature. The saturation vapor pressure (SVP) ranges from about 5 mb (at 0°C) to about 50 mb (at about 32°C).

The "precipitable water" is the total amount of water in a column of air. It is the maximum possible precipitation under total condensation (very rare). However, it gives no indication of the actual precipitation because the air is always in motion and a column when depleted of its moisture will be replaced with more moisture from adjacent columns. Considering a column of unit area of moist air, it can be shown that the total mass of water vapor (m_w) between two pressure levels p_1 and p_2 is given by

$$m_w = -\frac{1}{g}\int_{p_1}^{p_2} (\rho_w/\rho)dp, \tag{2.3}$$

where ρ_w is the water vapor density (= absolute humidity $= m_w/V$) and m_w is the mass of water vapor in volume V, ρ, the density of unsaturated air $\rho = (m_w + m_a)/V$; $\rho > \rho_w$. The pressures are related to the elevations by $p = -\rho g z$ (z measured positive upwards). The ratio ρ_w/ρ is called the specific humidity and is <1.

A water vapor particle undergoes various phases and physical changes before precipitation takes place. The water vapor must first be carried to upper levels where expansion and cooling take place. When the temperature has reached the dew point, condensation will take place releasing the latent heat of condensation to an otherwise adiabatic process. Cloud formation will take place with nucleation around impurities in the water vapor. Droplets coalesce with other droplets forming raindrops which are large enough to cause precipitation.

Normally, if the air is pure, condensation will occur only when the air is greatly supersaturated (taking more water vapor than saturation value). However, impurities present in the atmosphere act as nuclei around which water vapor in normal saturated form condense. The two main types of nuclei are hygroscopic particles having affinity for water vapor upon which condensation begins before saturation (mainly salt particles from the oceans), and non-hygroscopic particles that need some degree of supersaturation (e.g., dust particles, smoke, ash, etc.). Condensation nuclei range in size from a radius of $10^{-3}\,\mu$m to $10\,\mu$m. The average raindrop size is in the range 500–4000 μm (μm is a micron and $1\,\mu\text{m} = 10^{-6}\,\text{m}$). Once cloud droplets are formed, they may grow depending on atmospheric conditions. There are several theories that explain the growth of cloud droplets. However, not all clouds produce precipitation. Small clouds on hot days disappear as a result of evaporation. Large drops are formed by condensation of water droplets on ice crystals or by the collision of droplets with ice crystals. This means that the rain producing clouds must extend to the region where ice crystals are formed (about 5 km). Falling crystals continue to grow both through condensation and the capture of liquid droplets. They change into rain after entering air in which the temperature is above freezing. It is also possible that raindrops may be formed at temperatures above freezing, by the mixing of warm and cold droplets. The warm droplets evaporate and condense on cold droplets. Showers produced by this method are usually light.

2.2. *Atmospheric Circulation*

2.2.1. *Forces in the atmosphere*

The forces in the atmosphere include the gravitational force which is directed towards the center of the Earth, the pressure gradient force, which in the vertical direction acts upward to balance the gravitational force, and in the horizontal direction acts from high to low pressure, the frictional force, and the Coriolis force. The vertical pressure gradient near the ground is about $100\,\text{mbkm}^{-1}$ whereas the horizontal pressure gradient is about $1\,\text{mb}/100\,\text{km}$ at ground level. The horizontal pressure gradient is important in producing wind. The frictional force acts in the direction opposite to that of motion. It is significant only near the ground (up to about 1 km). Coriolis force is a fictitious force (or, acceleration) introduced into the Newtonian equation of motion to make it valid for a rotating frame of reference since atmospheric motions are measured from a frame of reference on Earth which is rotating and therefore is accelerating.

In the case of a particle moving with velocity $\mathbf{u}$ relative to a frame of reference rotating with angular velocity $\boldsymbol{\Omega}$, the Coriolis acceleration can be shown to be equal to $2\boldsymbol{\Omega} \times \mathbf{u}$. Coriolis acceleration is always normal to the direction of $\mathbf{u}$ but may be either to the left or to the right of $\mathbf{u}$ depending on the direction of rotation of the frame of reference. In the Northern Hemisphere, winds are deflected to the right; in the Southern Hemisphere it is to the left. At a point on Earth in the Northern Hemisphere at latitude ϕ, the angular velocity $\boldsymbol{\Omega}$ of the rotation of Earth can be resolved into two components: $\Omega \sin\phi$ along the local vertical, or z-axis, and, $\Omega \cos\phi$ along the poleward horizontal, or y-axis.

2.2.2. *Equations of motion*

Newton's second of law of motion is (Force $\mathbf{F}$ = mass, $m \times$ acceleration $\mathbf{a}$)

$$\mathbf{F} = m\mathbf{a}. \tag{2.4}$$

The acceleration term "$\mathbf{a}$" consists of the acceleration relative to Earth and the Coriolis acceleration (and a centrifugal acceleration term which we will neglect):

$$\mathbf{a} = \frac{D\mathbf{u}}{Dt} + 2\,\boldsymbol{\Omega} \times \mathbf{u}. \tag{2.5}$$

The forces involved are pressure gradient, gravitational, and frictional. Then, for a unit mass, the equations of motion (Navier–Stokes equations) are:

$$\frac{Du}{Dt} = 2v\Omega\sin\phi \quad -2w\Omega\cos\phi - \frac{1}{\rho}\frac{\partial p}{\partial x} \quad +F_x \tag{2.6a}$$

$$\frac{Dv}{Dt} = -2u\Omega\sin\phi \quad - \frac{1}{\rho}\frac{\partial p}{\partial y} \quad +F_y \tag{2.6b}$$

$$\frac{Dw}{Dt} = 2u\Omega\cos\phi \quad - \frac{1}{\rho}\frac{\partial p}{\partial z} \quad +F_z - g \tag{2.6c}$$

where u, v, and w are velocities in the east (x-axis), north (y-axis), and the local vertical (z-axis) directions, respectively and F_x, F_y, and F_z are the frictional forces per unit mass. The general equations of motion (Eq. 2.6) can be simplified to represent different scales of motion.

2.2.3. *Synoptic scales of motion*

The approximate order of magnitude of the various elements of the equations of motion applicable to the synoptic scale can be summarized as follows:

Length (Horizontal)	L	1000 km	10^6 m
Length (Vertical)	H	10 km	10^4 m
Time	t	1 day	10^5 s
Pressure change (Horizontal)	Δp	10 mb	10^3 Pa
Pressure change (Vertical)	p	1000 mb	10^5 Pa
Air density	ρ		1 kgm^{-3}
Earth's angular velocity	$\Omega(= 7 \times 10^{-5})$		10^{-4} rads^{-1}
Acceleration due to gravity	g		10 ms^{-2}
Wind speed (Horizontal)	u, v		10 ms^{-1}
Wind speed (Vertical)	w		10^{-1} ms^{-1}
Acceleration (Horizontal)	u/t, v/t		10^{-4} ms^{-2}
Acceleration (Vertical)	w/t		10^{-6} ms^{-2}
Coriolis acceleration	ΩV		10^{-3} ms^{-2}
Horizontal pressure gradient	$\Delta p/L$		10^{-3} Pam^{-1}

Ignoring the friction terms F_x, F_y, and F_z, an order of magnitude analysis of Eqs. (2.6a)–(2.6c) gives

$$2v\Omega\sin\phi - \frac{1}{\rho}\frac{\partial p}{\partial x} = 0 \tag{2.7a}$$

$$-2u\Omega \sin\phi - \frac{1}{\rho}\frac{\partial p}{\partial y} = 0 \tag{2.7b}$$

$$-\frac{1}{\rho}\frac{\partial p}{\partial z} - g = 0. \tag{2.7c}$$

Eqs. (2.7a) and (2.7b) are called the "geostropic equations" and Eq. (2.7c) the "hydrostatic equation".

2.2.4. *Small-scale motion*

The approximate order of magnitude of the various elements of the equations of motion applicable to the small scale can be summarized as follows:

Length (Horizontal & Vertical)	L	10 km	10^4 m
Minimum timescale	t		10^3 s
Pressure change (Horizontal)	Δp	1 mb	10^2 Pa
Wind speed (Horizontal)	u, v		$10\,\text{ms}^{-1}$
Angular velocity of Earth	Ω		$10^{-4}\,\text{s}^{-1}$
Wind speed (Vertical)	w		$0.5\ \text{ms}^{-1}$
Acceleration (Horizontal)	$u/t;\ v/t$		$10^{-2}\,\text{ms}^{-2}$
Acceleration (Vertical)	w/t		$10^{-3}\,\text{ms}^{-2}$
Pressure gradient (Horizontal)	$\Delta p/L$		$10^{-2}\ \text{Pam}^{-1}$

An order of magnitude analysis, neglecting higher order terms in Eqs. 2.6(a)–(c) gives

$$\frac{Du}{Dt} = -\frac{1}{\rho}\frac{\partial p}{\partial x} \tag{2.8a}$$

$$\frac{Dv}{Dt} = -\frac{1}{\rho}\frac{\partial p}{\partial y} \tag{2.8b}$$

$$g = -\frac{1}{\rho}\frac{\partial p}{\partial z}. \tag{2.8c}$$

In polar co-ordinates, Eqs. (2.8a) and (2.8b) transform to (v_θ is the tangential velocity)

$$\frac{v_\theta^2}{r} = \frac{1}{\rho}\frac{\partial p}{\partial r} \tag{2.9}$$

which is the equation for a forced vortex. It gives a balance of pressure gradient and centrifugal forces. Small-scale phenomena such as tornadoes and waterspouts are described by this equation. In a small-scale phenomenon such as a tornado, the velocities are of the order of $50\,\text{ms}^{-1}$ within a radius of about 100 m. The resulting pressure gradient therefore is of the order of 25 mb/100 m, which is very powerful and destructive.

2.2.5. *General Circulation Models (GCM's)*

Early general circulation models of the atmosphere were developed to solve the governing equations at $5^\circ \times 5^\circ$ latitude-longitude grid ($\cong$550 km $\times$ 550 km). With the advent of high speed and super computers, the grid size has now been considerably reduced. If a $1^\circ \times 1^\circ$ grid is chosen, there would be 129,600 nodes. The governing equations consist of the following:

Equations of motion

Equations (2.6a), (2.6b) and (2.6c) in the x, y, and z directions

Continuity equation

$$\frac{\partial \rho}{\partial t} + \frac{\partial}{\partial x}(\rho u) + \frac{\partial}{\partial y}(\rho v) + \frac{\partial}{\partial z}(\rho w) = 0 \tag{2.10}$$

Thermodynamic equation (First law of thermodynamics)

$$dQ = c_v dT + pdV \tag{2.11}$$

where Q is the heat source/sink per unit mass; c_v is the specific heat of air at constant volume; T is the absolute temperature; and V is the volume.

These equations, or their simplified forms, together with the boundary and initial conditions constitute the basis for GCM's. For a solution to be obtained the frictional force term $\boldsymbol{F}$ and the heat source/sink term Q must be expressed as functions of the unknown variables u, v, w, p, ρ and T. Any two of the three thermodynamic variables p, ρ and T would describe the system completely.

2.3. *Weather Systems*

2.3.1. *Scales of meteorological phenomena*

Various atmospheric phenomena have varying magnitudes both in space and time (Fig. 5). Although there can be variations of an order of magnitude in the same phenomenon, the timescales give a guide to predict the scales of influence of these phenomena. From Fig. 5, it can be seen that it is not possible for a single thunderstorm to affect a large area such as China or USA and that it will not last more than a day. The most important scale for weather is the synoptic scale, or weather map scale. It includes atmospheric phenomena with typical horizontal scales of 800–8000 km.

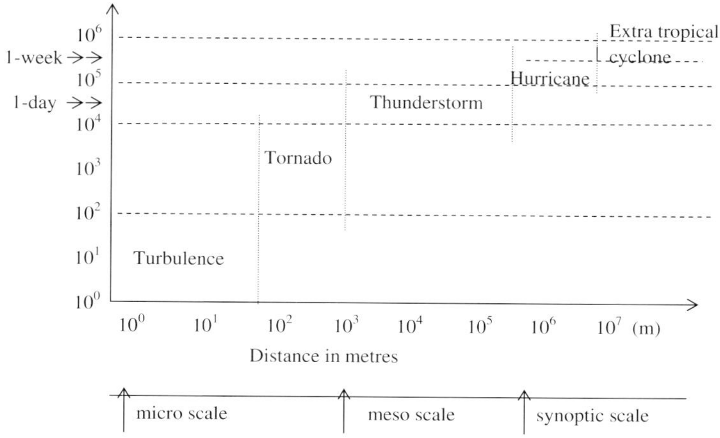

Fig. 5. Scales of meteorological phenomena. (Note: The Earth's circumference, which is approximately 40,000 km, is at the extreme end of the above scale.)

2.3.2. *Monsoons*

The word "monsoon" has its root in the Arabic word *mausim* which means "season." Considering the region in the south and southeast of Asia with the south Asian mountains as a natural boundary, i.e., approximately 35°N–25°S and 30°W–170°E, the monsoon is characterized by a reversal of wind direction between January and July of at least 120°. The monsoons consist of two seasonal circulations — a winter outflow from a cold continental anti-cyclone and a summer inflow into a continental heat low (cyclone), i.e., surface winds flowing persistently from oceans to continents in summer and just as persistently from continents to oceans in winter. The summer winds blowing from the oceans are warm and moist whereas the winter winds blowing from the continents are dry and cool. There is a corresponding change in the surface pressure gradient and in prevailing weather. The important features of northern summer monsoons are: (i) surface pressure (Low on land; high on oceans); (ii) pressure in the upper troposphere (high on land; low on oceans); (iii) zonal wind in the lower troposphere (westerlies on land; easterlies on oceans); (iv) zonal wind in the upper troposphere (easterlies on land; westerlies on oceans); (v) meridional wind in the lower troposphere (southerly on land; northerly on oceans); (vi) tropospheric mean temperature (warm on land; cold on oceans); (vii) total moisture (humid on land; relatively dry on oceans); and (viii) rainfall (much larger on land than in the trade wind belt on oceans).

Monsoons bring large amounts of rainfall. The world's highest recorded annual rainfall of 26,470 mm, and an average annual rainfall of about 12,000 mm, was in Cherrapunji (25°15′N, 91°44′E) in northeast India, which also has a monthly record of 9300 mm. This rainfall is brought about by the Southwest Monsoon. In India, 70% of the rainfall takes place during the Southwest Monsoon (June–September). In Sri Lanka, the Southwest Monsoon which comes in summer during the period April–September is called "Yala" and the Northeast Monsoon which comes in winter during the period October–March is called "Maha."

The driving forces in monsoon winds are the pressure gradients between large land masses and the ocean. It can be thought of as a convective motion generated by differential heating of the land and the oceans. The swirl introduced to wind by the rotation of the Earth is also a contributing factor. The differential heating is caused by the differences in the specific heats of the oceans and the land masses. The specific heat (energy required to raise the temperature by 1°C) of water is twice that of dry soil. Solar energy received on land heats only a few meters of the Earth's sub-surface; much of the energy goes into heating the air. For the oceans, it is quite the opposite; less energy is available for heating the air. The effective heat capacity of the ocean is very much larger than that of land.

In summer, the rise in temperature over the oceans is less than the rise in temperature over land. The mean summer temperature over the oceans is about 5–10°C less than on land at the same latitude. In winter, large heat storage in the oceans leads to higher temperatures in the oceans. Westerly winds at the lower levels and easterly winds at the higher levels generate the convective motion. The reversal takes place at about 6 km elevation. Monsoon arrival is gradual and starts in June. They last from two–four months. In the Indian sub-continent, an extensive anti-cyclone dominates above the monsoon winds. In mid-latitudes, the pressure gradient force and the Coriolis force balance each other. At low latitudes, the Coriolis force weakens and there is no geostropic balance.

2.4. *Extreme Weather*

2.4.1. *Cyclones*

A cyclone is any circulation around a low pressure center regardless of size and intensity. While rotating about the axis, they also move horizontally. They spin (or appear to spin) clockwise in the Southern Hemisphere and counter-clockwise in the Northern Hemisphere, i.e., the same direction as the direction of rotation of the Earth. The main driving force in a cyclone formation is the pressure gradient force which acts from the high pressure to the low pressure region. Tropical cyclones occur in the tropics (23°27′N and S). They originate around 5°–15° latitudes from the equator and are quite common in the Indian and Pacific Ocean parts of the monsoon region. Wind speeds of up to 250 knots at times have been reached. When traveling across continents, they lose energy and die down. Cyclones are usually accompanied by heavy rain. In different regions of the world, tropical cyclones have different names. For example, in North America, they are called hurricanes; in Japan, Northern China, South-east Asia, and the North-western Pacific Ocean, they are called typhoons; in the Indian Ocean, they are called cyclones; in the Philippines, they are called baguios; and in Australia, they are called willy-willys. Any storm is a form of cyclone. Tropical cyclone is the proper generic name whereas tropical storm is a less technical term. The World Meteorological Organization classifies cyclones according to the maximum sustained wind speeds near the center of the cyclone. In the USA, a cyclone with wind speeds in excess of $32.6\,\mathrm{ms}^{-1}$ ($119\,\mathrm{kmhr}^{-1}$) is called a hurricane. Speeds of up to $90\,\mathrm{ms}^{-1}$ ($324\,\mathrm{kmhr}^{-1}$) have been recorded. It has a calm central area called the eye (common to all cyclones). In most cases the surface wind speeds do not usually exceed $67\,\mathrm{ms}^{-1}$ ($241\,\mathrm{kmhr}^{-1}$), but they may occur over a large area. The timescale of a hurricane is of the order of a few days.

2.4.2. *Tornadoes*

Tornadoes are quite common in the USA. They last only for a few minutes but with extreme force. Wind speeds are of the order of 130–180 ms^{-1} (480–640 kmhr^{-1}). Distances affected are of the order of 100–1000 m. Because of the extreme low pressure, no man-made structure can survive a direct hit by a tornado. When tornadoes occur in water, a phenomenon known as a waterspout is formed.

2.4.3. *Thunderstorms*

When the atmosphere is unstable and the moisture content is high, convective cloud development, once started, proceeds at a rapid rate. The cloud air, because of its buoyancy, continues rising. In a very unstable air mass, the rising parcel of air becomes more and more buoyant with altitude. This is because of the temperature decrease with altitude. In some cases the cloud air may be warmer than the environmental air up to the lower layers of the stratosphere. A cloud air ascending at the rate of perhaps 1 ms^{-1} at 1500 m may attain a speed of 25 ms^{-1} at an altitude of 7500 m. In this manner, small clouds become bigger and in turn develop into cumulonimbus clouds — better known as thunderstorms. These extend to altitudes of about 10–20 km. The upper limit of the growth of a thunderstorm is determined by the height of the stratosphere. This is so because the lower layers of the stratosphere are very stable and the temperature gradient at the stratosphere is zero or negative. Once it has reached an altitude where the cloud is colder than the environment, it begins to slow down but will continue upward movement for a few thousand meters because of its momentum.

When the thunderstorm has matured the upward movement takes place at its maximum speed. Because of the growth of precipitation particles (which coalesce and move downward) there is a downward draft of equal magnitude. At this stage heavy rain, electrical effects, and gusts at the surface are common. The lifting of moist low-level air to the high troposphere can take place by three mechanisms: Convectional lifting — when low-level moist air is heated by high surface temperatures caused by solar radiation; orographic lifting — when moist air is forced up by topographical barriers such as mountain ranges; or frontal lifting — convergence of low-level air in the vicinity of cold fronts.

Lightning is another feature of thunderstorms. Electrons from the water droplets accumulate at the base of the cloud. This negative charge induces a positive charge on the Earth's surface below the cloud. A potential gradient of about 1000 $Voltsm^{-1}$ occurs between the cloud and the ground. When this is too large, a discharge of electrons takes place. The rapid heating of the air in the lightning path produces a violent expansion of air which initiates a sound wave propagating outwards at the speed of sound. (Lightning travels at about 10^9 $kmhr^{-1}$, whereas sound travels at about 960 $kmhr^{-1}$). By recording the times between seeing the flash and hearing the sound it is possible to calculate the approximate distance from the place of lightning. Thunderstorms can affect a large area, but will not last more than a day. They bring large amounts of rain. Gustiness and falling temperatures are signs of an approaching thunderstorm.

2.4.4. *Tropical depressions and storms*

Tropical depressions are centers of low pressure which form in the troughs. They produce deep clouds and much precipitation, mainly of the convective type. By classification, wind speeds are less than $17.4\,\mathrm{ms}^{-1}$. Tropical storms are well-developed, low-pressure systems surrounded by strong winds and much rain. By convention, a system qualifies as a tropical storm if winds range from 17.4–$32.6\,\mathrm{ms}^{-1}$ (40–$120\,\mathrm{kmhr}^{-1}$).

3. Hydrology

The two principal processes in the hydrological cycle are precipitation, which deposits the atmospheric water on the surface of Earth, and evaporation, which returns the water on the surface of the Earth back to the atmosphere. Runoff is the outcome of precipitation that can be thought of as an integrator of all catchment processes, which in excessive quantities leads to flooding. Flood disasters on a global scale have been rising in the past few years (Fig. 1) and account for the major share of all natural disasters. Flood damages, direct and indirect, have also been increasing globally. Mitigation of flood damages has now become an essential step toward economic development.

Basically there are two approaches of mitigating flood damages: Structural measures such as construction of storage and detention reservoirs to temporarily store the flood waters, upgrading the hydraulic capacities of drainage networks including natural river courses, and construction of flood defence structures such as levees; and non-structural measures such as early warning systems, flood zoning, flood hazard mapping, and building community awareness. Structural measures are costly and not always environmentally friendly, whereas non-structural measures are less costly and appear to be more favored nowadays. "Adaptation" and "living with floods" have become two widely used slogans in this context. A brief description of the components that constitute an early warning system is given below.

The basic input information that goes into an early warning system for flood damage mitigation is rainfall which can be measured to a high degree of accuracy. Runoff, or river flow, can also be used as basic input information, but it is more difficult and costly to measure. Therefore, runoff is predicted using mathematical models that transform the input rainfall to a corresponding output runoff or stage. Once the impending flood volumes or levels have been predicted, warnings could be issued to the vulnerable areas including information on evacuation routes and locations of shelters. The mode of dissemination of the warning is important for its effectiveness. Various types of communication media could be used, but the message should be unambiguous, unique, and directed from a single authority to avoid confusion. Effective implementation of non-structural measures requires the co-operation and involvement of the community. River discharge (or "stage") prediction becomes an important component of any early flood warning system.

A parameter that is often used in the design of drainage structures is the peak flow which is estimated from rainfall data using empirical approaches. A widely used method in this context is the rational method which relates the peak discharge Q_p to the rainfall intensity I and the catchment area A as

$$Q_p = CIA. \tag{3.1}$$

This formula assumes that the rainfall intensity is uniform over the entire catchment area throughout the duration of the storm and that the duration is longer than the time of concentration of the catchment. The constant C $(0 < C < 1)$ is known as the runoff coefficient. When the time variation of the flow is desired, the discharge hydrograph can be determined by using a number of methods. The unit hydrograph approach, the earliest of such methods, is based upon the concept of rainfall excess. It assumes that the transformation of rainfall excess to direct runoff is linear, and therefore the principles of superposition and proportionality can be used. The difficulty in this approach is how to separate the rainfall excess (or runoff-producing rainfall) from the actual rainfall and direct runoff from actual runoff. With certain assumptions the procedure can be implemented, and the method, due to its simplicity, has stood the test of time.

Other types of rainfall-runoff modeling can be broadly classified into two categories: Data-driven and distributed. Data-driven types include regression methods, stochastic methods, artificial neural networks (ANN), genetic algorithms (GA), and phase-space reconstruction methods, among others. Distributed types are generally physics-based, but some semi-distributed models are conceptual in formulation. Regression models aim to find a regression relationship between the rainfall data and the corresponding runoff data. They are purely statistical in character and do not take into account the processes that transform the rainfall to corresponding runoff. Stochastic models consider the input (rainfall) data and/or output (runoff) data as time series. The time series are decomposed into constituent components such as trends, periodic parts, dependent stochastic parts, and finally the remaining random residual part. Once the structure of the composition of the time series is determined, more samples that will have the same statistical structure could be generated for different random samples of the residual component. Such methods can be used for synthetic data generation as well as for forecasting purposes. Details of stochastic modeling and forecasting are well documented in several textbooks (e.g., Box and Jenkins, 1976; Salas *et al.*, 1980).

Emerging data-driven methods of rainfall-runoff modeling include the application of ANN, GA, genetic programming (GP), and phase-space reconstruction methods. A typical multi-layer perceptron (MLP)-type artificial neural network has a layer of input nodes, one or more layers of hidden nodes, and a layer of output nodes, as illustrated schematically in Fig. 6.

In an MLP-type ANN, the relationship between the input x and the output y can be expressed as

$$y_{pk} = f_k^o \left(\theta_k^o + \sum_{j=1}^{L} w_{kj}^o \left[f_j^k \left(\theta_j^h + \sum_{i=1}^{N} w_{ji}^h \, x_{pi} \right) \right] \right), \tag{3.2}$$

where is y_{pk} the output of the network; f_k^o is the activation function for the output layer; θ_k^o is the bias term for the output layer; w_{kj}^o is the connection weight between the kth hidden node and the jth output node; f_j^k is the activation function at the hidden layer; θ_j^h is the bias term for the hidden layer; w_{ji}^h is the connection weight between the ith input node and the jth hidden node; x_{pi} are the inputs at the input layer; N is the dimension of the input vector; and L is the dimension of the hidden layer. Generally, the activation functions at the hidden and output layers are assumed to be

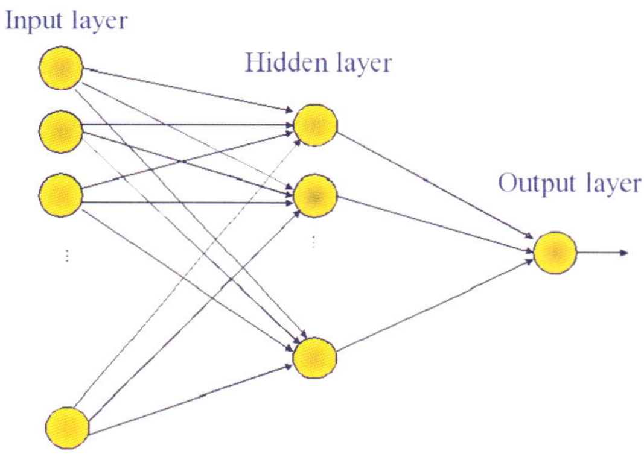

Fig. 6. A multi-layer perceptron (MLP) artificial neural network.

the same. The superscripts h and o refer to the quantities at the hidden and output layers. There are many types of activation functions that can be used but the Sigmoid which take the following forms and which are continuously differentiable are the most popular:

$$f(x) = \begin{cases} \dfrac{1}{1+e^{-rx}} & \text{(logistic type)} \\ \tanh(rx/2) = \dfrac{1-e^{-rx}}{1+e^{-rx}} & \text{(hyperbolic tangent type).} \end{cases} \tag{3.3}$$

In Eq. (3.3), r is the steepness parameter. Once the output for a given set of input values has been estimated, it is compared with the expected output, and the difference, which is the error, is back-propagated to adjust the weights incrementally until a certain stopping criterion is met. The weight adjustment by back-propagation is done according to the back-propagation algorithm which takes the form

$$w_{ij}(k-1) = w_{ij}(k) - \eta \frac{\partial E_p}{\partial w_{ij}} + a[w_{ij}(k) - w_{ij}(k-1)] \tag{3.4}$$

where η, the learning rate and a, the momentum term are user-defined parameters, and E_p is an objective function defined in terms of the error. The network can be fine tuned by this procedure to match with the expected output to any desired degree of accuracy. Theoretical details of ANN can be found in several textbooks (e.g., Haykin, 1999) while applications in hydrology can be found in several research papers (e.g., Jayawardena and Fernando, 1998; Govindaraju, 2000; Jayawardena *et al.*, 2006). An example (Jayawardena, 2009) of the application of ANNs for water level predictions at the Sylhet gauging station (24°42′N; 91°53′E) across Surma River in Bangladesh is shown in Fig. 7.

Distributed models may be of the conceptual type, such as for example, the Xinanjiang model (Zhao *et al.*, 1980; Jayawardena and Zhou, 2000) and the Variable Infiltration Capacity (VIC) model (Wood *et al.*, 1992; Liang *et al.*, 199–; Liang and Xie, 2003; Jayawardena and Ying, 2005), or physics-based type such as, for example, the MIKE-SHE model (Abbott *et al.*, 1986; Refsgaard *et al.*, 1995). The approach to the development of a physics-based model involves the description of the problem, simplification of the problem, definition of a set of governing equations, choice of a set of boundary and initial conditions, identification of the solution domain in space and time, solution of the simplified governing equations subject to the given boundary and initial conditions within the domain of interest using a numerical scheme, calibration, verification, and application. Distributed

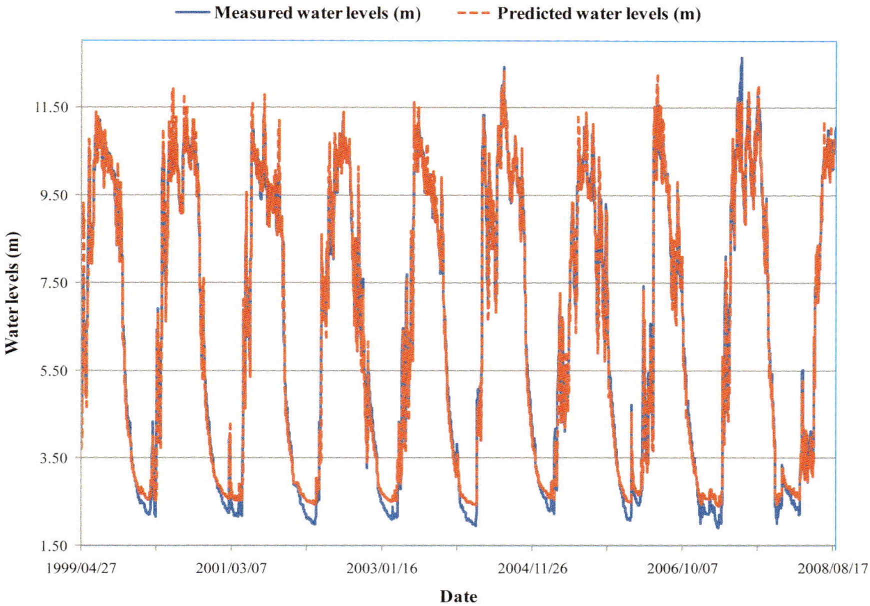

Fig. 7. Time series plot of measured and predicted water level in case of application (calculated by MLP with BP algorithm).

models are more resource intensive and need a great deal more input information than data-driven models. Although, potentially, such models are capable of accommodating spatially varying inputs, outputs, and parameter values, their calibration becomes quite difficult. Because of the interactions among different parameters, very often it is not possible to obtain a unique set of parameter values. A practice that is adopted is to ignore spatial variation of physical and hydraulic parameters and obtain their spatially uniform values by optimization techniques, thereby diluting the meaning of distributed models. Practical applications of such models in a truly distributed manner are still some way ahead.

4. Dynamics of Water-related Environmental Hazards

Unlike hydro-meteorological hazards which cannot be prevented, environmental hazards in general are preventable. Among the many types of environmental disasters, those that are water related by far have the greatest effect on human population. Water related environmental disasters may be caused by pollution in rivers, streams, lakes, reservoirs, coastal bays, groundwater, inland seas, and the open oceans. They may also be caused by the lack of water causing diseases and loss of food production leading to famine. Pollution of waterbodies may take place slowly over a long period of time, or because of accidents, in a short time. To mitigate the consequences of such pollution the first step would be to have a better understanding of the dynamics of the fate and transport of pollutants in waterbodies. Due to the complex nature of the mixing, decay, and transport characteristics of different pollutants under different hydraulic conditions, certain assumptions and simplifications are necessary.

4.1. *Dynamics of Well-mixed Waterbodies*

Water quality systems can be considered under various assumptions. The well-mixed assumption implies that there are no concentration gradients in the horizontal and vertical directions. Concentration is assumed to vary only in the time domain. It is an idealized situation. In practice, well-mixed conditions do not exist in real waterbodies. Nevertheless, the assumption enables an understanding of the gross effects of how a pollutant attains steady-state conditions from an initial state. The well-mixed assumption alone is not sufficient to obtain solutions to the governing equations. Several other assumptions are also necessary. For example, the input waste load can be a constant or time-varying. Similarly, the output can also be either constant or time-varying. The hydraulic parameters such as the flow rate and velocity of the waterbody, as well as the reaction rates, may be considered as time-invariant or time-varying. Finally, the system can be assumed to be either linear or non linear. Non-linear approaches are rarely used for modeling water quality systems because of the difficulties associated with the modeling techniques as well as calibration. They also do not have general applicability. Under a linear assumption, the parameters and inputs may also be considered as constant or time-varying. Different combinations of these assumptions and their variations can lead to a large number of possible modeling systems.

The concentration gradient of a pollutant discharged into a well-mixed waterbody can be obtained from the law of conservation of mass as follows:

$$V\frac{dc}{dt} = W(t) - Qc - kVc, \tag{4.1}$$

where $V[\mathrm{L}^3]$, is the volume of the waterbody, $c[\mathrm{ML}^{-3}]$ is the concentration of the pollutant, $W[\mathrm{MT}^{-1}]$ is the rate of application of the waste load, $Q[\mathrm{L}^3T^{-1}]$ is the net outflow from the waterbody, and k is the decay constant (T^{-1}). The ratio $\frac{V}{Q}$ can be considered as the detention time of the waterbody. Eq. (4.1) can be re-written as

$$V\frac{dc}{dt} + k'c = W(t), \tag{4.2}$$

where $k' = Q + kV$. This is a linear first-order ordinary differential equation and therefore the principles of proportionality and superposition hold. Given the initial condition $c = c_0$ at time $t = 0$, the solution to Eq. (4.2) is of the form

$$c(t) = \frac{1}{V}e^{-\frac{k'}{V}t}\int_0^t W(t)e^{\frac{k'}{V}t}dt + c_0e^{-\frac{k'}{V}t} \tag{4.3}$$

which simplifies to

$$c(t) = c_0e^{-\frac{k'}{V}t}, \tag{4.4}$$

when $W(t) = 0$ and which is the effect of the initial condition. When $W(t) \neq 0$, it can take several forms: Step (constant input for a period of time), periodic, impulse, arbitrary, or stochastic. It can be written as a constant and a variable part as

$$W(t) = \overline{W} + W'(t). \tag{4.5}$$

The solutions to each of these input functions can be obtained as follows.

4.1.1. *Step function input*

If a step input of magnitude $\overline{w}$ is imposed at $t_0 = 0$, the solution, by the principle of proportionality, called the step response, can be shown to be

$$c_u(t) = \frac{\overline{W}}{k'}\left(1 - e^{-\frac{k'}{V}t}\right), \tag{4.6}$$

which as $t \to \infty$, takes the form $c_u \to \frac{\overline{W}}{k'} = \frac{\overline{W}}{Q+kV}$. If $k = 0$ (for a conservative substance), then $c_u = \frac{\overline{W}}{Q}$. The total response (Eq. (4.4) + Eq. (4.6)) is given by

$$c_u(t) = \frac{\overline{W}}{k'}\left(1 - e^{-\frac{k'}{V}t}\right) + c_0e^{-\frac{k'}{V}t}. \tag{4.7}$$

The variation of the total response depends on the relative magnitude of the initial condition with respect to the ultimate steady-state concentration, i.e., whether $c_0 > c_\infty$, or $c_0 < c_\infty$. In the former case, the two concentrations are additive whereas as in the latter case, the total concentration will decrease and attain a new steady state. A typical example of a

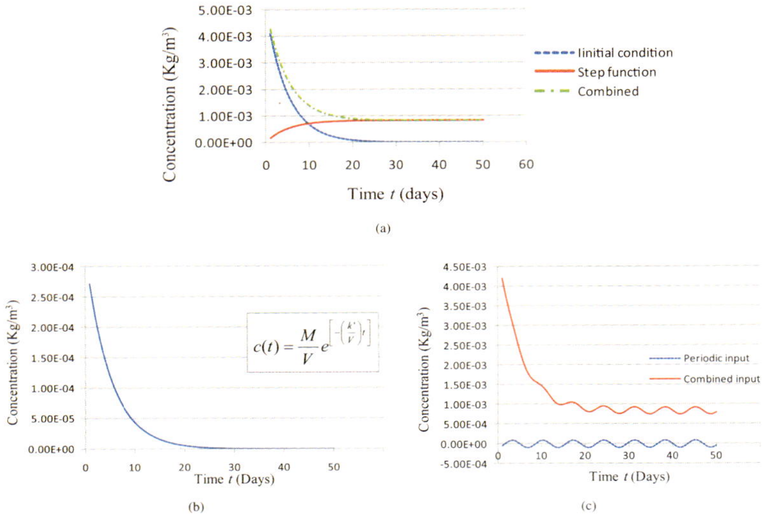

Fig. 8. (a) Effects of initial condition, step function and their combination; (b) response due to an impulse waste input; and (c) response due to a periodic waste input combined with a step function input.

step function input is when a certain amount of waste load enters a waterbody for a fixed period of time. The response to a combination of several step function type waste loads can be easily determined using the principle of superposition. The steady-state and maximum concentrations and the time to attain a specified concentration would be of particular interest. Because of the linear assumption, it is possible to obtain solutions to other forms of $W(t)$. Figs. 8(a)–(c) illustrate the different response functions for the case of a well-mixed waterbody of volume 300 million m^3, an outflow of 0.6 million $\mathrm{m}^3\mathrm{day}^{-1}$, and a decay coefficient of 0.2 day^{-1} with the following input conditions and an initial concentration of 0.005 kgm^{-3}: Fig. 8(a), a step waste input of 50,000 kgday^{-1}; Fig 8(b), an impulse waste input of 100 tons; and Fig. 8(c), a periodic waste input given by the function $W(t) = 50{,}000 + 25{,}000\sin(\frac{2\pi}{7}t)\,\mathrm{kg\,day}^{-1}$.

4.1.2. *Periodic input function*

A periodic input function can be written in the form

$$W(t) = \overline{W} + W_0 \sin(\omega t - a), \tag{4.8}$$

where W_0 is the amplitude [MT^{-1}] of the waste load, a is the phase shift angle in radians measured from $t = 0$ to the beginning of the positive portion of the sine curve, and ω is the angular frequency ($=\frac{2\pi}{T}$, T is the period). Again, because of the linearity assumption, the principle of superposition can be made use of, and therefore only the time-dependent part of the load needs to be considered. The solution for the time-dependent input function $W_0 \sin(\omega t - a)$ is

$$c(t) = W_0 A_m(\omega) \sin(\omega t - a - \theta(\omega)), \tag{4.9}$$

where

$$A_m(\omega) = \frac{\frac{1}{V}}{\left(\left(\frac{k'}{V}\right)^2 + \omega^2\right)^{\frac{1}{2}}} [\mathrm{L}^{-3}\mathrm{T}]; \text{ and } \theta(\omega) = \arctan\left(\frac{\omega}{\frac{k'}{V}}\right).$$

The solution is therefore a function of $A_m(\omega)$ and $\theta(\omega)$ which are in turn functions of ω. The limiting cases are:

When $\omega = 0$ (when $T \to \infty$), $A_m(\omega) = \frac{1}{k'}$; $\theta(\omega) = \arctan(0) = 0$.

When $\omega \to \infty$ (when $T \to 0$), $A_m(\omega) \to 0$; $\theta(\omega) \to \frac{\pi}{2}$.

4.1.3. *Impulse input*

A waste load can be applied at different rates. When the rate of application is very high (implying time of application is very short), it can be approximated by an impulse input which is mathematically equivalent to the Dirac-δ type function. Physically it is represented by the discharge of an amount M (kg) of waste in a very short time. Input function $W(\mathrm{t})$ is then given by

$$W(t) = M\delta(t - t_0). \tag{4.10}$$

In Eq. (4.10), $\delta(t - t_0)$ has the units of [T^{-1}] because $\int_{-\infty}^{\infty} \delta(t - t_0)dt = 1$. If a sequence of impulses are released at different times, then

$$W(t) = \sum_{r=1}^{n} M_r \delta(t - t_r). \tag{4.11}$$

If $t_0 = 0$, then $W(t) = M\delta(t)$. The solution to this case can be shown to be

$$c(t) = \frac{M}{V} e^{-\frac{k'}{V} t}. \tag{4.12}$$

When $M = 1$, the response is referred to as the Impulse Response Function, $I(t)$, the response due to an instantaneous unit load, $I(t) = \frac{1}{V} e^{-\frac{k'}{V} t}$.

4.1.4. *Arbitrary input*

The approach for an arbitrary input consists of approximating the input by a series of finite impulse inputs. The concept and the procedure is the same as that for the unit hydrograph theory.

$$c(t) = \int_{-\infty}^{t} W(\tau)I(t-\tau)d\tau, \tag{4.13}$$

which is the well-known convolution integral.

4.2. *Dissolved Oxygen Systems*

The health of a waterbody can be measured by the amount of dissolved oxygen (DO) which depends upon its temperature, elevation, and salt content. Under pristine conditions, the concentration of dissolved oxygen would be at saturation level, which at 0°C is about 14.6 mg/l, at 30°C about 7.56 mg/l, and at 40°C, about 6.41 mg/l. The dissolved oxygen concentration decreases with decreasing pressure and increasing salt content. In a heavily polluted waterbody, the concentration of dissolved oxygen may become zero. Under such conditions which give rise to a "dead" waterbody, no living species can survive. For fish to survive in a waterbody, the dissolved oxygen concentration must be at least 4–6 mgl^{-1}.

The fluctuation of DO in a waterbody takes place as a result of oxygenation and de-oxygenation. Oxygenation takes place via re-aeration, which is the process of oxygen transfer from the atmosphere to the waterbody through the air/water interface, from tributaries carrying water with higher DO concentration, and by photosynthesis. De-oxygenation takes place via the oxidation of Carbonaceous Biochemical Oxygen Demand (CBOD), Nitrogenous Biochemical Oxygen Demand (NBOD), Sediment Oxygen Demand (SOD), and algal respiration. CBOD refers to the reduction of organic carbon to CO_2 in the presence of micro-organisms such as bacteria, NBOD refers to the biological oxidation of ammonia (NH_3) to nitrates (NO_3^-), and SOD refers to aerobic decay of organic benthic material, which is negligible in flowing water.

When an oxygen-demanding pollutant is released into a waterbody, the dissolved oxygen in the waterbody is depleted. At the same time, a certain amount of re-oxygenation also takes place since the water surface is in contact with the atmosphere. A mass balance for this de-oxygenation and re-oxygenation processes can be written as follows:

$$V\frac{dc}{dt} = O_{in} - O_{out} + (\text{Rate of re-oxygenation})V - (\text{Rate of de-oxygenation})V,$$

where V is the volume of the waterbody [L^3], c, the dissolved oxygen concentration [ML^{-3}] (usually expressed as mgl^{-1}), O_{in} and O_{out}, the rates of external oxygen inflow and outflow [MT^{-1}]. Assuming that there are no external inflows and outflows contributing to the oxygen mass balance, the rate of change of concentration can be written as

$$\frac{dc}{dt} = \left(\frac{dc}{dt}\right)_{\text{decay}} + \left(\frac{dc}{dt}\right)_{\text{re-aeration}}. \tag{4.14}$$

Assuming a first-order decay, the rate of decay (or de-oxygenation) and the rate of re-aeration (or re-oxygenation) can be expressed, respectively, as

$$\left(\frac{dc}{dt}\right)_{\text{decay}} = -k_d L \tag{4.15}$$

and

$$\left(\frac{dc}{dt}\right)_{\text{re-aeration}} = k_r(c_s - c), \tag{4.16}$$

where k_d and k_r are the de-oxygenation and re-oxygenation coefficients, L, the BOD remaining in the water at time t, and c_s, the saturation value of dissolved oxygen concentration, which depends upon the temperature. Equation (4.15) assumes that k_d is the overall de-oxygenation rate that includes both oxidation of settled and soluble BOD. The Committee on Sanitary Engineering Research of the American Society of Civil Engineers (1960) has proposed an empirical equation to relate the saturation dissolved oxygen concentration to temperature which takes the form

$$c_s = 14.652 - 0.41022T + 0.0079910T^2 - 0.0000777774T^3, \tag{4.17}$$

where c_s is in mg/l and T is the temperature in °C. Figure 9 shows the decreasing trend of the saturation value with increasing temperature, according to Eq. (4.17).

Substituting Eqs. (4.15) and (4.16), in Eq. (4.14) gives

$$\frac{dc}{dt} = -k_d L + k_r(c_s - c). \tag{4.18}$$

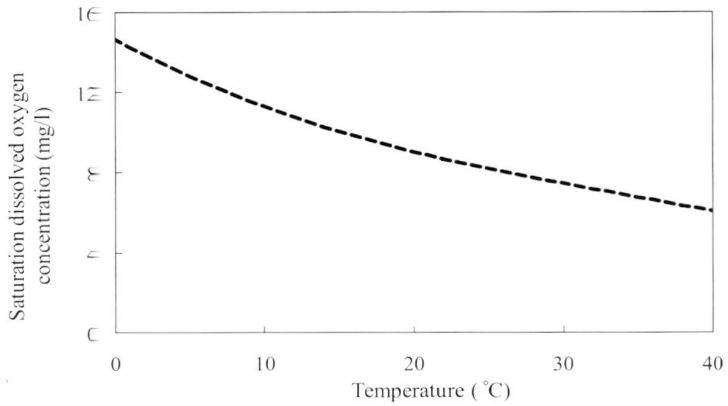

Fig. 9. Variation of saturation dissolved oxygen concentration with temperature.

This is the differential equation that describes the dissolved oxygen concentration variation in a waterbody subjected to BOD loading. In many situations, it is convenient to convert this equation to represent the oxygen deficit D which is defined as $D = c_s - c$. Then Eq. (4.18) becomes

$$\frac{dD}{dt} = k_d L - k_r D. \tag{4.19}$$

Assuming first-order decay for the BOD, this can be written as

$$\frac{dD}{dt} + k_r D = k_d L_0 e^{-k_d t}, \tag{4.20}$$

where L_0 is the ultimate BOD remaining in the water at time t. The solution, which is obtained by using an integrating factor, is of the form

$$D = \frac{k_d L_0}{k_r - k_d}(e^{-k_d t} - e^{-k_r t}) + D_0 e^{-k_r t}, \tag{4.21}$$

or, in terms of the dissolved oxygen concentration,

$$c = c_s - \frac{k_d L_0}{k_r - k_d}(e^{-k_d t} - e^{-k_r t}) - D_0 e^{-k_r t}. \tag{4.22}$$

In river systems, it is often desired to estimate the dissolved oxygen concentration in the downstream direction. This can easily be achieved by converting the time variable to a space variable ($t = \frac{x}{u}$; x is the distance from the outfall, u is the average velocity in the river). Then,

$$c = c_s - \frac{k_d L_0}{k_r - k_d}\left(e^{-k_d \frac{x}{u}} - e^{-k_r \frac{x}{u}}\right) - D_0 e^{-k_r \frac{x}{u}}. \tag{4.23}$$

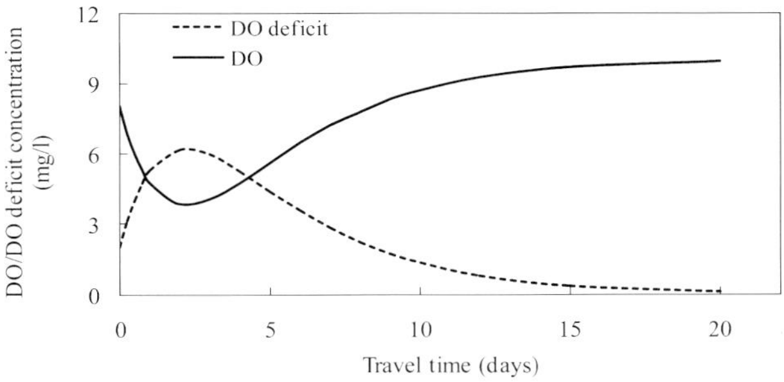

Fig. 10. Dissolved oxygen sag and DO deficit curves (Eq. (4.24) with $k_r = 0.5$/day; $k_d = 0.3$/day; $L_0 = 20$ mg/l; $u = 0.35$ m/s; $D_0 = 2$ mg/l and $c_s = 10$ mg/l)

This solution to the mass balance differential equation was first obtained by Streeter and Phelps (1925), and is referred to as the Streeter–Phelps equation, or the Oxygen Sag Curve. It was first applied to study the water quality in the Ohio River in the USA, and has since then become the basis of many applications of environmental modeling. Implicit in the Streeter–Phelps equation are the assumptions that the flow in the river is non-dispersive, steady-state flow BOD and DO reaction conditions, and the only reactions are de-oxygenation by decay and re-oxygenation by aeration. It should also be noted that the Streeter–Phelps equation is not valid when $k_r = k_d$.

The critical (minimum) oxygen deficit can be estimated by setting $\frac{dD}{dt} = 0$. This occurs when

$$t_c = \frac{1}{k_r - k_d} \ln\left(\frac{k_r}{k_d}\left(1 - \frac{D_0(k_r - k_d)}{k_d L_0}\right)\right). \tag{4.24}$$

Figure 10 shows a DO deficit and DO variation with time for a typical set of parameters for which the time at which the minimum DO deficit occurs is 2.209 days (from Eq. (4.24)). The corresponding DO deficit and DO concentrations are 6.182 mgl^{-1} and 3.818 mgl^{-1}, respectively.

4.3. *Water Quality in Rivers and Streams*

Water quality variation in a river system depends upon many factors such as the hydraulic parameters, presence of tributaries and abstraction points, outfalls of waste material at fixed discharge points, non-point sources of pollution, and whether the system is considered as at steady state or unsteady state. Different conditions lead to different formulations and solutions. The system should therefore be considered under specific assumptions and specific waste input conditions. The simplest is when there is a point source of waste loading in a river which is assumed to be a one-dimensional waterbody.

4.3.1. *Point sources*

The governing equation, or the mass balance equation, is formulated under three major assumptions. First it is assumed that there is no concentration gradient across the width and depth of the river. This is an idealized condition which is justified only after some time (or distance), called the "initial period" (or "mixing length"), has lapsed. After the initial period, complete mixing is assumed to be achieved, at least in theory. The second assumption is that there is no dispersion in the longitudinal direction. This condition is also called the "plug flow system," or "maximum gradient system" or "advective system." There is no mixing of one control volume of water with another control volume. The third assumption is that steady-state conditions prevail.

Three mass balance equations under these assumptions can be written for the flow and the waste material, respectively, as

$$Q_u c_u + Q_s c_s = Q_d c_d \tag{4.25}$$

$$W = Q_s c_s \tag{4.26}$$

$$Q_d = Q_u + Q_s, \tag{4.27}$$

where Q_u, Q_d, and Q_s, respectively, (all in $[L^3T^{-1}]$) are the upstream, downstream, and point source flow discharges; c_u, c_d, and c_s, respectively, (all in $[ML^{-3}]$) are the concentrations of the waste material upstream, downstream, and at the source; and W is waste load $[MT^{-1}]$. From Eqs. (4.25) and (4.27),

$$c_d = \frac{Q_u c_u + Q_s c_s}{Q_d} = \frac{Q_u c_u + Q_s c_s}{Q_u + Q_s}. \tag{4.28}$$

If the upstream waste concentration is zero ($c_u = 0$), then

$$c_d = \frac{Q_s c_s}{Q_d} = \frac{W}{Q_d}, \tag{4.29}$$

which gives the effect of dilution only. This condition can be applied to tributary inflows which bring in waste concentrations. Because of the non-dispersive well-mixed assumptions, the concentrations downstream of the outfall will remain unaltered for a conservative pollutant until an external input of flow or waste material is added or taken away. For a conservative material, such as, for example, total dissolved solids (TDS), chlorides, and certain metals, there is no change in concentration between tributaries or waste inputs. The concentration changes only at a discharge point. It is also assumed that there is no leakage due to seepage. At a discharge point the concentration will undergo a sharp increase or decrease depending upon tributary inflows, outflows, and waste inputs. For non-conservative (biodegradable) materials, such as BOD, nutrients, bacteria, volatile chemicals etc., the mass balance equation, assuming a first-order decay is

$$\frac{1}{A}\frac{d}{dx}(Qc) = -kc, \tag{4.30}$$

where Q is the flow rate, A is the average cross-sectional area of flow and k is a decay rate [T^{-1}]. If Q is constant, then,

$$u\frac{dc}{dx} + kc = 0, \tag{4.31}$$

where $u = Q/A$ is the average velocity of flow. Eq. (4.31), with the boundary condition $c = c_0$ at $x = 0$, is a linear first-order ordinary differential equation that has a solution of the form

$$c = c_0 e^{-\frac{kx}{u}}. \tag{4.32}$$

This may be written in logarithmic form,

$$\ln(c) = -kt^* + \ln(c_0), \tag{4.33}$$

where $t^* = \frac{x}{u}$ = time to travel a distance x at velocity u. This plots a straight line from which the system parameters can be estimated for a known concentration-time profile. For multiple-point sources, the principle of superposition can be used. The total effect is the sum of individual effects plus the effects due to boundary conditions.

4.3.2. *Unsteady state non-dispersive systems*

The unsteady state can be considered either as a non-dispersive system or as a dispersive system. In a non-dispersive system, there is no mixing in the longitudinal direction. This means that each parcel of water does not interfere with other parcels in front or behind. The condition is also called "plug flow." The governing equation is

$$\frac{\partial c}{\partial t} + u\frac{\partial c}{\partial x} + kc = 0. \tag{4.34}$$

The boundary condition is $c = c_0(t)$ at $x = 0$ which may be written as

$$c_0(t) = \frac{W(t)}{Q}. \tag{4.35}$$

The solution of Eq. (4.34), assuming that the waste load is a function of time, is (Li, 1962)

$$c(x,t) = \frac{W(t-t^*)}{Q} e^{-\frac{kx}{u}}, \tag{4.36}$$

where t^* is the travel time. The concentration change is due to dilution $\frac{W(t)}{Q}$ and decay $e^{-kx/u}$.

4.3.3. *Unsteady state dispersive systems*

In the real world, plug flow rarely exists. Instead, mixing of the waste load takes place along the longitudinal direction as well as in the vertical and lateral directions, primarily due to the respective velocity gradients. In addition, variations of geometrical parameters of the river channel also contribute to mixing. Longitudinal dispersion refers to the process of mixing in the longitudinal direction due mainly to velocity gradients. The governing equation has the form

$$\frac{\partial t}{\partial t} + u\frac{\partial c}{\partial x} + kc = D\frac{\partial^2 c}{\partial x^2}, \tag{4.37}$$

where the dispersion coefficient D is of dimension L^2T^{-1}. The solution to this equation depends upon the input type. For an impulse input of mass M, which is equivalent to a slug-type release, or sudden spill of a toxic material where the time of application is very short, the solution of Eq. (4.37) takes the form

$$c(x,t) = \frac{M}{A\sqrt{4\pi Dt}} e^{-\frac{(x-ut)^2}{4Dt} - kt}. \tag{4.38}$$

This is mathematically equivalent to the Gaussian probability density function with mean ut and variance $4Dt$.

Thus, if t is fixed, the concentration-distance profile is symmetric around its peak value. On the other hand, if the concentration-time profile is considered at different locations, they will not be symmetric. A measure of spread around its peak value is the variance σ^2 (or standard deviation, σ) which in this case is $4Dt$. As t increases, σ also increases. The concentration profiles flatten out as the waste material is carried downstream resulting in reduction of the peak concentration. If $D = 0$, then, $\sigma = 0$ and the result will be plug flow. The impulse response function can be used to determine the response to other inputs using the principles of superposition and proportionality. If the waste material is conservative ($k = 0$) or if there is no advection ($u = 0$), Eq. (4.38) can be simplified.

Another type of input is the step function which refers to an input over a fixed interval of time starting from zero, suddenly increasing to a fixed value, remaining at that value for a fixed interval of time and suddenly dropping back to zero. It has the shape of a rectangle. The general solution of the governing equation for a steady input with constant coefficients given by Thomann (1973) is

$$c(x,t) = \frac{c_0}{2} e^{-\frac{kx}{u}} \left(\text{erf}\left(\frac{x - u(t-\tau)(1+\eta)}{\sqrt{4D(t-\tau)}} \right) - \text{erf}\left(\frac{x - ut(1+\eta)}{\sqrt{4Dt}} \right) \right), \tag{4.39}$$

where c_0 is the concentration of the input after mixing over the cross section, τ is the time interval of the input, $\eta = \frac{kD}{u^2}$ (dimensionless),[a] and

$$\text{erf}(t) = \frac{2}{\sqrt{\pi}} \int_0^t e^{-z^2} dz, \quad \text{Note} : \text{erf}(-\text{t}) = -\text{erf}(\text{t}). \tag{4.40}$$

If $\tau \lll t^*$, (i.e., the time of application of the input is short in comparison to travel time), then, according to O'Loughlin and Bowmer (1975) and Rose (1977), the time of travel of the peak concentration a distance x is given by

$$t_p = \frac{x + u\tau(1+\eta)}{u(1+\eta)}, \tag{4.41}$$

and the peak concentration c_p at distance x and time t_p is

$$c_p(x,t_p) = \frac{c_0}{2} e^{-\frac{kx}{u}} \left(\text{erf}\left(\frac{x - u(t_p-\tau)(1+\eta)}{\sqrt{4D(t_p-\tau)}} \right) - \text{erf}\left(\frac{x - ut_p(1+\eta)}{\sqrt{4Dt_p}} \right) \right). \tag{4.42}$$

4.4. *General Purpose Water Quality Models*

Several water quality models that can simulate many different constituents are now available for general use. Among them, are the Enhanced Stream

[a]For upland streams, $\eta < 0.01$ (Thomann, 1973); for main drainage rivers, $\eta = 0.01 - 0.5$; for large rivers, $\eta = 0.5 - 1.0$, i.e., that longitudinal dispersion is not significant in upland streams.

Water Quality Model (QUAL2E), and its more recent version QUAL2K, both developed by the United States Environmental Protection Agency (USEPA), Water Quality Analysis Simulation Program (WASP), and the One-Dimensional Riverine Hydrodynamic and Water Quality Model (EPD-RIV1).

4.4.1. *Enhanced stream water quality model (QUAL2E)*

The Enhanced Stream Water Quality Model (QUAL2E) is applicable to well-mixed, dendritic streams. It simulates the major reactions of nutrient cycles, algal production, benthic and carbonaceous demand, atmospheric re-aeration and their effects on the dissolved oxygen balance. The model assumes that the major transport mechanisms, advection and dispersion, are significant only along the longitudinal direction of flow. It can predict the following 15 water quality constituent concentrations: Dissolved oxygen; biochemical oxygen demand; temperature; algae (as chlorophyll "a"); crganic nitrogen (as "N"); ammonia (as "N"); nitrite (as "N"); nitrate (as "N"); organic phosphorus (as "P"); dissolved phosphorus (as "P"); coliforms; one arbitrary non-conservative constituent; and three conservative constituents.

It is intended as a water quality planning tool for developing total maximum daily loads (TMDLs) and can also be used in conjunction with field sampling for identifying the magnitude and quality characteristics of non-point sources. By operating the model dynamically, the user can study diurnal dissolved oxygen variations and algal growth. However, the effects of dynamic forcing functions, such as headwater flows or point source loads, cannot be modeled with QUAL2E. The model assumes that the stream flow and waste inputs are constant during the simulation time periods. QUAL2EU is an enhancement that allows users to perform three types of uncertainty analyses: Sensitivity analysis, first-order error analysis, and Monte Carlo simulation. QUAL2K is an enhanced version of QUAL2E that takes into account the following: (i) unequally spaced river reaches and multiple loadings and abstractions in any reach; (ii) two forms of carbonaceous BOD (slowly oxidizing and rapidly oxidizing) to represent organic carbon as well as non-living particulate organic matter; (iii) anoxia by reducing oxidation reactions to zero at low oxygen levels; (iv) sediment-water interactions; (v) bottom algae; (vi) light extinction; (vii) pH; and (viii) pathogens.

The Windows interface provides input screens to facilitate preparing model inputs and executing the model. It also has help screens and provides graphical viewing of input data and model results. More details of the software can be found in the website http://www.epa.gov/OST/QUAL2E_WINDOWS, and in: "The Enhanced Stream Water Quality Models QUAL2E and QUAL2E-UNCAS: Documentation and User's Manual." (EPA 600/3-87-007). NTIS Accession Number: PB87 202 156.

4.4.2. *Water quality analysis simulation program (WASP)*

This program which is based on the work of several researchers can carry out dynamic compartment modeling of aquatic systems including the water column as well as the benthos. It can analyze a number of pollutant types in one, two or three dimensions. The program can also be linked to hydrodynamic and sediment transport models. The pollutants it can handle include nitrogen; phosphorus; dissolved oxygen; biochemical oxygen demand; sediment oxygen demand; algae; periphyton; organic chemicals; metals; mercury; pathogens; and temperature. More information about WASP can be found in the website: http//www.epa.gov/athens/wwqtsc/html/wasp.html.

4.4.3. *One dimensional riverine hydrodynamic and water quality model (EPD-RIV1)*

This is a system of programs that performs one-dimensional (cross-sectionally averaged) hydraulic and water quality simulations. The hydrodynamic model is first applied and the results are then used as inputs to the water quality model. The model can simulate the following state variables: Dissolved oxygen; temperature; nitrogenous biochemical oxygen demand (NBOD); carbonaceous oxygen demand (CBOD); phosphorus; algae; iron; manganese; coliform bacteria; and two arbitrary constituents. More information about EPD-RIV1 can be found in the website: http//www.epa.gov/athens/wwqtsc/html/epd-riv1.html.

5. Concluding Remarks

In this chapter, an attempt has been made to highlight the dynamics of the processes that lead to hydro-meteorological and environmental hazards and some of the approaches that are available for predicting their consequences. To supplement the material presented here, which by no means is exhaustive, a list of references as well as a bibliography is given for the interested readers to follow up.

References

Abbott, M., Bathurst, J., Cunge, J., *et al.* (1986). An introduction to the European Hydrological System — Systeme hydrologique Europeen "SHE" 1: History and philosophy of physically based, distributed modelling system. *J. Hydrol.* **87** pp. 45–59.

Adikari, Y., Yoshitani, J., Takemoto, N., *et al.* (2008). Technical report on the trends of global water-related disasters — a revised and updated version of 2005 report, Tech. rep., Public Works Research Institute Technical Report No. 4088., Tsukuba, Japan.

Anthes, R., Panofsky, H., Cahir, J., *et al.* (1978). *The Atmosphere* (Charten E Merrill Publishing Co. Columbus, Ohio).

Battan, L. J. (1984). *Fundamentals of Meteorology* (Prentice Hall Inc., Englewood Cliffs, NJ).

Box, G. and Jenkins, G. (1976). *Time Series Analysis: Forecasting and Control* (Holden-Day, Oakland, CA).

Committee on Sanitary Engineering Research (1960). Solubility of atmospheric oxygen in water, *Journal of the Sanitary Engineering Division, ASCE* **86**, 7, pp. 41–53.

Cotton, W. (1990). *Storms* (ASTeR press, Fort Collins, Co).

Das, P. (1972). *The Monsoons* (Edward Arnold, London).

Govindaraju, R. (2000). Artificial neural networks in hydrology ii: hydrologic applications, *J. Hydrol. Eng.* **5**, 2, pp. 24–137.

Hanwell, J. (1980). *Atmospheric Processes* (George Allen & Unwin, London).

Haykin, S. (1999). *Neural Networks: A Comprehensive Foundation* (Prentice Hall, New York).

Jayawardena, A. W. and Tian, Y. (2005). Flow modelling of Mekong River with variance in spatial scale, in S. Herath *et al.* (ed.), *Proceedings of the International Symposium on "Role of Water Sciences in Transboundary River Basin Management"* (held in Ubon Ratchathani, Thailand during March 10–12, 2005), pp. 147–154.

Jayawardena, A. W. and Zhou, M. (2000). A modified spatial soil moisture storage capacity distribution curve for the Xinanjiang model, *J. Hydrol.* **227**, 104, pp. 93–113.

Jayawardena, A. W. (2009). River flow prediction with artificial neural networks, in D. Palmer-Brown *et al.* (ed.), *Engineering Applications of Neural Networks*, EANN 2009, CCIS (Springer-Verlag, Berlin Heidelberg), pp. 463–471.

Jayawardena, A. and Fernando, D. A. K. (1998). Use of radial basis function type artificial neural networks for runoff simulation, *Computer-Aided Civil and Infrastructure Engineering* **13**, 2, pp. 91–99.

Jayawardena, A., Muttil, N. and Lee, J. (2006). Comparative analysis of a data-driven and a GIS based conceptual rainfall-runoff model, *J. Hydrol. Eng., ASCE* **11**, 1, pp. 1–11.

Li, W. H. (1962). Unsteady dissolved oxygen sag in a stream, *Journal of the Sanitary Engineering Division, ASCE* **88**, SA3, pp. 75–85.

Liang, X., Lettenmaier, D., Wood, E., *et al.* (1994). A simple hydrologically based model of land surface water and energy fluxes for general circulation models, *J. Geophys. Res.* **99**, D7, pp. 14415–14428.

Liang, X. and Xie, Z. (2003). Important factors in land-atmosphere interactions: surface runoff generations and interactions between surface and groundwater, *Global and Planetary Change* **38**, 1–2, pp. 101–114.

Lighthill, J. and Pearce, R. (eds) (1981). *Monsoon Dynamics* (Cambridge University Press, Cambridge).

Mcllveen, R. (1986). *Basic Meteorology — A Physical Outline* (Van Nostrand Reinhold (UK) Co. Ltd., London).

O'Loughlin, E. and Bowmer, K. (1975). Dilution and decay of aquatic herbicides in flowing channels, *J. Hydrol.* **26**, 3–4, pp. 217–235.

Raudkivi, R. J. (1979). *Hydrology — An advanced introduction to hydrological processing and modelling* (Pergamon Press, Oxford).

Refsgaard, J., Storm, B. and Singh, V. (1995). MIKE SHE, *Computer Models of Watershed Hydrology* **1**, pp. 809–846.

Rose, D. A. (1977). Dilution and decay of aquatic herbicides in flowing channels — comments, *J. Hydrol.* **32**, pp. 399–400.

Salas, J., Delleur, J., Yevjevich, V., *et al.* (1980). *Applied modelling of Hydrologic Time Series* (Water Resources Publications, Littleton, Colorado).

Streeter, H. and Phelps, E. (1925). A study of the pollution and natural purification of the Ohio River, *US Public Health Service Bulletin No. 146* .

Thomann, R. (1973). Effect of longitudinal dispersion on dynamic water quality of streams and rivers, *Water Resources Research* **9**, 2, pp. 355–366.

Thomann, R. V. and Mueller, J. (1987). *Principles of surface water quality modelling and control* (Harper & Row, New York).

Wood, E., Lettenmaier, D. and Zartarian, V. (1992). A land-surface hydrology parameterization with subgrid variability for general circulation models, *J. Geophys. Res.* **97**, D3, pp. 2717–2728.

Zhao, R., Zhang, Y., Fang, L., Liu, X. and Zhang, Q. (1980). The Xinanjiang model. Hydrological forecasting, in *Proceedings of Oxford Symposium*, Vol. 129 (IAHS Publication, Wallingford, UK), pp. 351–356.

Chapter 9

Tsunami Modeling and Forecasting Techniques

Pavel Tkalich* and Dao My Ha

Physical Oceanography Research Laboratory
National University of Singapore
**tmspt@nus.edu.sg*

Nonlinear waves are observed in all branches of science and engineering, and are present in different aspects of daily life. The great Sumatra-Andaman tsunami (December 2004) in the Indian Ocean provides an example of a series of dramatic events dominated by nonlinear wave dynamics. This chapter will review tsunami behavior at all stages, starting from a source in the open ocean, through trans-oceanic propagation, and up to breaking on shore. Major observed features of tsunami can be analyzed using mathematical wave models of various complexities, such as soliton theory and the Boussinesq approximation.

Several numerical models have been specifically developed for tsunami research and operational forecasts worldwide, and application of general hydrodynamic solvers is becoming common due to the availability of ever-increasing computational power and resources. Operational forecasters need instant solutions, which can as yet be achieved only using data-driven models, such as those based on Artificial Neural Networks or Empirical Orthogonal Functions. Operational procedures for tsunami forecasting and warning at major centers are reviewed in the light of lessons learned from the 2004 tsunami.

1. Introduction

Although tsunamis have been leaving tragic traces in human history from ancient times, scientific understanding of the phenomenon has been built up only during the past 150 years since the observations and experiments of the British hydraulic engineer Scott Russell (1885), and theoretical work by the French mathematician and physicist J. Boussinesq (1877). All earlier tsunami descriptions were based on anecdotal evidence of a few survivors, embedded in myth, folklore, and art.

Following the deadly 1946 Aleutian Island earthquake and tsunami, NOAA's Pacific Tsunami Warning Center (PTWC, Honolulu) was established in 1949; and the Japan Meteorological Agency (JMA) initiated tsunami warning services in 1952. Until about 1980, semi-empirical charts (connecting tsunami threats to seismic sources) were the only forecasting tools available. During the 80s and 90s, due to pioneering work of F. Imamura, N. Shuto, C. E. Synolakis and others, fast computers and efficient models have been employed for tsunami modeling. In the early stages of the computing era, it was not possible to solve the two-dimensional Boussinesq equations with nonlinear and dispersion terms; instead, simplified alternatives became popular. Due to the efforts of PTWC and JMA, most of the tsunami modeling and forecasting capabilities were focused on the Pacific Ocean; in other regions, tsunami science and awareness were not developed. Not surprisingly, the 2004 Indian Ocean tsunami caught off guard the coastal communities along the Indian Ocean shores, killing almost 230,000 people.

The modern development of worldwide tsunami research started with this devastating earthquake and tsunami which struck at 00.59 GMT 26 December 2004. Fifteen minutes after the earthquake, PTWC issued the Tsunami Bulletin #1. This document estimated the seismic source to be eight on the Richter scale, but revised it twice by the next day up to nine (i.e., ten times stronger in terms of released energy!). Implausibly, Bulletin #1 stated that the "*earthquake is located outside the Pacific. No destructive tsunami threat exists based on historical earthquake and tsunami data.*" An hour later, Tsunami Bulletin #2 made the revised forecast that "*no destructive tsunami threat exists for the Pacific basin based on historical earthquake and tsunami data.*" This was more accurate but still obviously problematic for countries along the Indian Ocean Rim. Forty hours after the deadly earthquake, the first quantitative description of the tsunami was released in Tsunami Bulletin #3, summarizing measurements rather than forecasts.

This tragic event drew attention to the lack of tsunami-warning infrastructure for the Indian Ocean, and triggered a worldwide movement to develop tsunami modeling and forecasting capabilities in countries adjacent to the Indian Ocean, as well as other regions. A United Nations conference in January 2005 in Kobe (Japan) initiated the creation of the Indian Ocean Tsunami Warning System, supported by the actions of involved nations in developing their own regional tsunami warning centers.

The number of scientists and students migrating from different areas into the tsunami field has increased significantly, resulting in a re-examination of established approaches and perceptions, and in the development of novel ideas and methods. In Singapore, a similar movement has led to the development of national earthquake and tsunami predictive capabilities, and of a tsunami warning system. This chapter highlights some of the most important steps and conclusions during the development stage, as well as providing a review of the historical milestones that have led to our modern understanding of tsunami behavior.

2. Tsunami Modeling

2.1. *The First Scientific Encounter of Solitons*

We may start the description of tsunami behavior using soliton theory, which is a simplified substitute for a full-scale tsunami model. In mathematics and physics, a soliton is a self-reinforcing solitary wave (a wave packet or pulse) that maintains its shape while it travels at constant speed. The soliton phenomenon was first described by John Scott Russell who observed a solitary wave in the Union Canal, Edinburgh (UK).

He reproduced the phenomenon in a wave tank and named it the "Wave of Translation" (Russell, 1885). The discovery is described here in his own words:

> *"I was observing the motion of a boat which was rapidly drawn along a narrow channel by a pair of horses, when the boat*

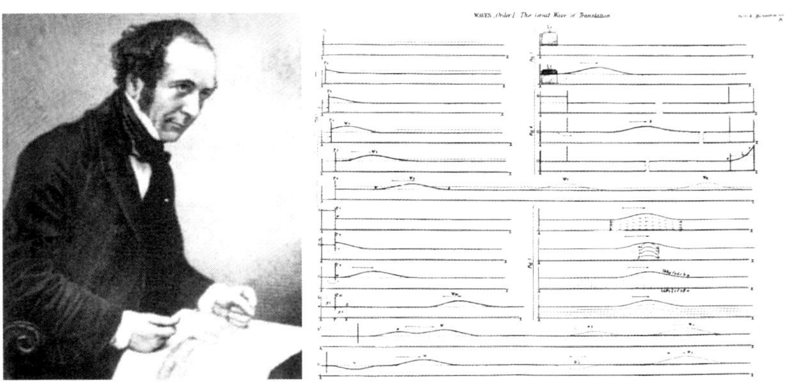

Fig. 1. John Scott Russell and his study on solitons.

> *suddenly stopped — not so the mass of water in the channel which it had put in motion; it accumulated round the prow of the vessel in a state of violent agitation, then suddenly leaving it behind, rolled forward with great velocity, assuming the form of a large solitary elevation, a rounded, smooth and well-defined heap of water, which continued its course along the channel apparently without change of form or diminution of speed. I followed it on horseback, and overtook it still rolling on at a rate of some eight or nine miles an hour [14 km/h], preserving its original figure some thirty feet [9 m] long and a foot to a foot and a half [300–450 mm] in height. Its height gradually diminished, and after a chase of one or two miles [2–3 km] I lost it in the windings of the channel. Such, in the month of August 1834, was my first chance interview with that singular and beautiful phenomenon which I have called the Wave of Translation."*

Following this discovery, Scott Russell built a 9 m wave tank in his garden and made observations of the properties of solitary waves, with the following conclusions (Russell, 1844):

- Solitary waves have the shape $a\ \mathrm{sech}^2(k(x - ct))$, where a is the wave height, k is the wavenumber, and c is the wave speed;
- A sufficiently large initial mass of water produces two or more independent solitary waves;
- Solitary waves can pass through each other without change of any kind;
- A wave of height a and traveling in a channel of depth h has a velocity given by the expression $c = \sqrt{g(a + h)}$, where g is the acceleration of gravity, implying that a large amplitude solitary wave travels faster than one of low amplitude.

Throughout his life Russell remained convinced that his "Wave of Translation" was of fundamental importance, but nineteenth- and early twentieth-century scientists thought otherwise, partly because his observations could not be explained by the then-existing theories of water waves.

The modern theory of soliton dates from the pioneering computer simulation of Zabusky and Kruskal (1965) of a nonlinear dispersion equation known as the Korteweg–de Vries equation (KdV). Zabusky and Kruskal carried out numerical experiments similar to those made by Russell in his wave tank and confirmed the above empirical findings. Solitonic behavior suggested that the KdV equation must have conservation laws beyond the obvious conservation laws of mass, energy, and momentum, and these were indeed discovered by Zabusky and Kruskal as well as by other authors at a later stage. Because of the particle-like properties of such a wave, they named it a "soliton," a term that has been used to describe the phenomenon ever since.

2.2. *Behavior of Solitons*

We may start to understand soliton behavior by means of a simple convective wave equation

$$\eta_t + c\eta_x = 0, \tag{2.1}$$

where the wave speed $c = c(\eta, x, t)$ could be generally a function of the surface elevation η, space x, and time t, as in Fig. 2.

If $c = \text{const}$, this equation has traveling wave solutions, and all waves propagate at the same speed c. Particular interest for the subsequent examples attaches to the initial condition illustrated in Fig. 3:

$$\eta(x, 0) = \text{sech}^2(x), \quad \text{where } \text{sech}(x) = 1/\cosh(x) = 2/(\mathrm{e}^x + \mathrm{e}^{-x}) \tag{2.2}$$

for which the exact solution of Eq. (2.1) at time t for $c = \text{const}$ is

$$\eta(x, t) = \text{sech}^2(x - ct) \tag{2.3}$$

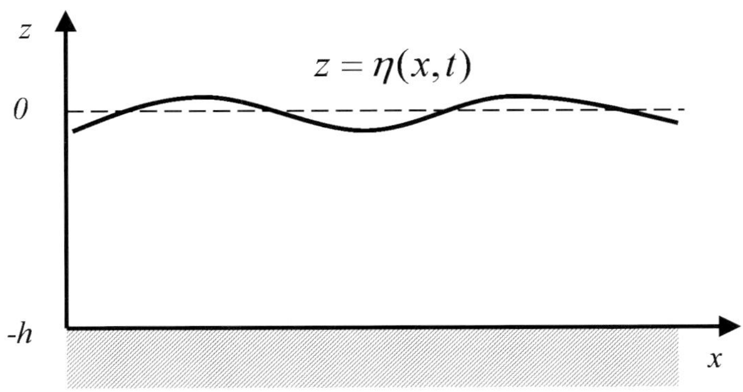

Fig. 2. Geometrical configuration for water waves.

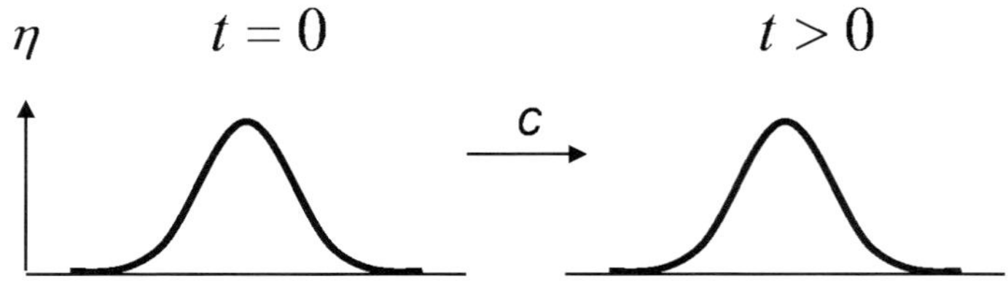

Fig. 3. Traveling wave solutions for linear wave equation.

If the wave speed is dependent on the wave elevation, initial wave profiles are generally not self-preserving. The simplest example is given by $c = \eta^p$ (where p is an integer greater than or equal to 1), which being substituted into the linear, non-dispersive wave Eq. (2.1) yields

$$\eta_t + \eta^p \eta_x = 0. \tag{2.4}$$

This equation governs a nonlinear wave propagation. Using the initial wave profile Eq. (2.2), solutions for $\eta(x, t)$ describe waves such that the profile eventually becomes multi-valued and gradient blowup occurs (Fig. 4(a)).

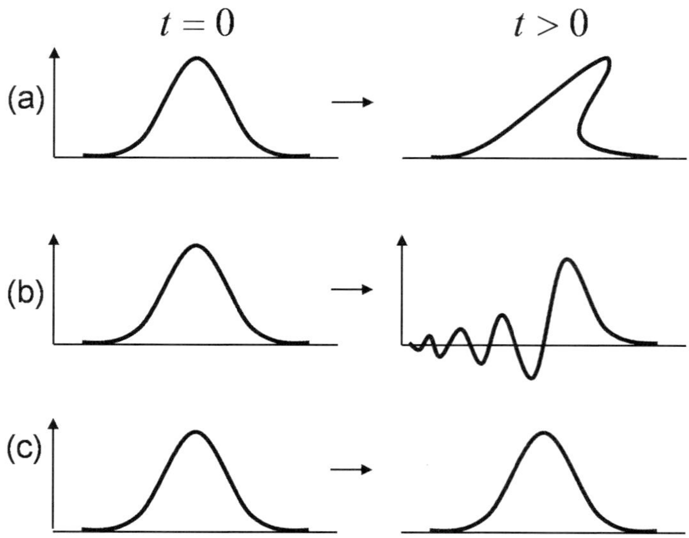

Fig. 4. Nonlinear and dispersive soliton behavior: (a) nonlinear term only; (b) dispersion term only; (c) nonlinear and dispersion terms balanced together.

At the next step, we will look into the dispersion behavior of the waves, described with a dispersive wave equation

$$\eta_t + \eta_{xxx} = 0. \tag{2.5}$$

This equation has traveling wave solutions

$$\eta(x,t) = \int_{-\infty}^{\infty} a(k)\exp(ik^3t + ikx)dk$$

where $a(k)$ is the component amplitude of the Fourier transform of the initial profile. If the initial wave profile is again in the form of Eq. (2.2), one can observe that a single propagating wave splits (disperses) from the tail and results in oscillatory waves of different frequency that continue to propagate at different speed as in Fig. 4(b). This behavior is explicitly embedded in the dispersive wave solution depicting shorter harmonics (with larger k) propagating left relative the peak of the wave. Hence, the solutions do not describe localized traveling waves of constant shape and speed.

Wave propagation exhibits both nonlinear and dispersive behavior if described with the Generalized Korteweg–de Vries (GKdV) equation:

$$\eta_t + \eta^p\eta_x + \eta_{xxx} = 0. \tag{2.6}$$

This equation has localized traveling wave solutions (solitary waves) in the form of

$$\eta(x,t) = \left[\frac{1}{2}(p+1)(p+2)c\,\mathrm{sech}^2(p\sqrt{c(x-ct)}/2)\right]^{1/p} \tag{2.7}$$

The GKdV equation (for $p = 1$) reduces to the Korteweg–de Vries equation, named after Korteweg and de Vries (Korteweg and de Vries, 1895), though the equation was in fact first derived by Boussinesq (1877). It was then understood that balancing dispersion against nonlinearity leads to traveling wave solutions (Fig. 4(c)) as earlier observed by Scott Russell, and this is precisely the physical feature of solitons.

For a tsunami propagating in the ocean, dispersion and nonlinearity are not necessarily in equilibrium. In somewhat simplistic terms, if nonlinearity dominates (usually nearshore) the incident soliton tends to break from the front side; whereas in deepwater conditions dispersion results in the soliton shedding waves from the tail. A tsunami can propagate across the ocean as a series of several solitons probably originating from a single wave at source.

2.3. *Derivation of Boussinesq-Type and KdV Equations*

To draw a more complete and accurate picture of tsunami behavior in the ocean, we need to start with the two-dimensional nonlinear water-wave model involving Laplace's equation combined with boundary conditions, nonlinear at the free-surface and linear at the sea bottom. This problem is complicated by the fact that the moving surface boundary is part of the solution. Direct numerical methods for solving the full equations exist, but they are extremely time-consuming and can only be applied to small-scale problems. As it is currently impracticable to compute a full solution valid over any significant domain such as the entire Indian or Pacific Ocean, approximations must be adopted, including the so-called Boussinesq-type formulations of the water-wave problem.

To understand the physics and assumptions embedded in the Boussinesq-type and KdV equations, it is advisable to follow the derivation in sufficient detail starting from the equations of motion, Eqs. (2.8)–(2.11), which are themselves obtained from the Euler equations of an ideal incompressible fluid (Dean and Dalrymple, 1984).

$$\nabla^2\phi = 0 \quad \text{in fluid} \tag{2.8}$$

$$\phi_t + \frac{1}{2}(\phi_x^2 + \phi_y^2 + \phi_z^2) + gz = 0 \quad \text{at } z = \eta \tag{2.9}$$

$$\eta_t + \phi_x\eta_x + \phi_y\eta_y - \phi_z = 0 \quad \text{at } z = \eta \tag{2.10}$$

$$\phi_x h_x + \phi_y h_y + \phi_z = 0 \quad \text{at } z = -h \tag{2.11}$$

Here ϕ is the velocity potential, giving fluid velocity components $u = \frac{\partial\phi}{\partial x}$, $v = \frac{\partial\phi}{\partial y}$, $w = \frac{\partial\phi}{\partial z}$. The Laplace equation (Eq. (2.8)) is deduced from the continuity equation, representing conservation of mass for irrotational incompressible fluids. One boundary condition at surface is given by the dynamic condition (Eq. (2.9)) derived from Bernoulli's equation. The kinematic condition at the surface (Eq. (2.10)) and bottom (Eq. (2.11)) are derived from the assumption that any fluid particle originating on the boundary will remain on it (this condition is violated when a wave breaks).

In the following text we follow a traditional derivation, as reviewed by Debnath (1994); however, some other methods obtaining of Boussinesq equations are available which might consider additional terms to account for a strong dispersion, nonlinearity, or varying anisotropic fluid properties. To differentiate from the original one-dimensional flat-bottom solution by Boussinesq (1877), researchers refer to the recent generalizations as Boussinesq-type equations.

For convenience of analysis non-dimensional variables are introduced as $(\tilde{x}, \tilde{y}) = \frac{1}{l}(x, y)$, $\tilde{z} = \frac{z}{h}$, $\tilde{t} = \frac{c}{l}t$, $\tilde{\eta} = \frac{\eta}{a}$, $\tilde{\phi} = \frac{h}{alc}\phi$. Here l is the typical horizontal scale, such as wave length; a is the wave amplitude; h is the typical water depth; $c = \sqrt{gh}$ is the dispersion relation for shallow water waves, connecting wave speed c with depth h (see also page 30–32 of this volume); and g is the gravitational acceleration.

If the horizontal length-scale of the sea-bed non-uniformities L is much larger than the wave length l (i.e., $\gamma \equiv l/L$, $\gamma \ll 1$), the sea bed is considered to have a "mild slope," and the gradient of the sea-bed shape can be neglected (i.e., h_x, $h_y \to 0$), as in Fig. 5.

Substituting the above non-dimensional variables into Eqs. (2.8)–(2.11) and dropping the tildes gives

$$\delta(\phi_{xx} + \phi_{yy}) + \phi_{zz} = 0 \quad \text{in fluid} \tag{2.12}$$

$$\phi_t + \frac{\epsilon}{2}(\phi_x^2 + \phi_y^2) + \frac{\epsilon}{2\delta}\phi_z^2 + \eta = 0 \quad \text{at } z = \epsilon\eta \tag{2.13}$$

$$\delta\left[\eta_t + \epsilon(\phi_x\eta_x + \phi_y\eta_y)\right] - \phi_z = 0 \quad \text{at } z = \epsilon\eta \tag{2.14}$$

$$\phi_z = 0 \quad \text{at } z = -1 \tag{2.15}$$

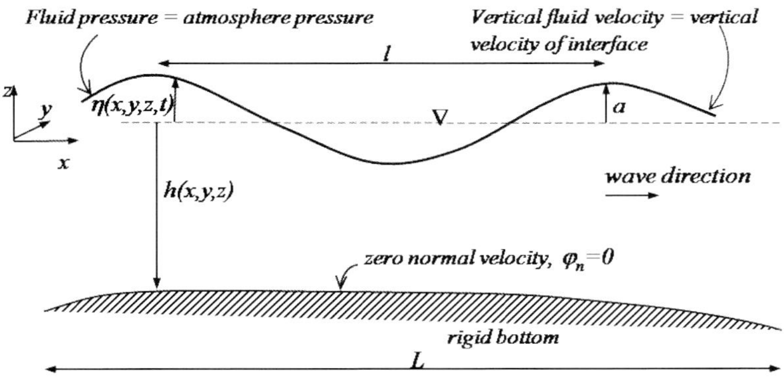

Fig. 5. Sketch of water column.

Here $\epsilon = a/h$ and $\delta = h^2/l^2$ are scale parameters introduced to represent nonlinearity and dispersion, respectively. The bottom slope terms in Eq. (2.15) are dropped out by assumption of small γ.

For the 2004 Indian Ocean tsunami, $a = 1$ m in the ocean, and up to 10 m nearshore; $h = 4000$ m and 10 m, respectively; $l = 400$ km and 50 km, respectively. Thus, the introduced scale parameters may have ranges: $\epsilon = 10^{-4}$ in the ocean and up to 1 nearshore; $\delta = 10^{-4}$ and 10^{-5}, respectively.

Following Boussinesq (1872), we expand the velocity potential in terms of δ without any assumption about ϵ:

$$\phi = \phi_0 + \delta\phi_1 + \delta^2\phi_2 + \cdots \tag{2.16}$$

and substitute into Eqs. (2.12)–(2.15) to derive the unknown terms

$$\begin{aligned} \phi_0 &= \phi_0(x, y, t), \quad u(x, y, t) \equiv (\phi_0)_x, \; v(x, y, t) \equiv (\phi_0)_y \\ \phi_1 &= -\frac{z^2}{2}(u_x + v_y) \\ \phi_2 &= \frac{1}{24}z^4((\nabla^2 u)_x + (\nabla^2 v)_y) \end{aligned} \tag{2.17}$$

The idea behind the Boussinesq approximation, Eqs. (2.16)–(2.17), was to incorporate the effects of non-hydrostatic pressure, while eliminating the vertical coordinate z, thus reducing the computational effort relative to the full three-dimensional problem. The assumption that the magnitude of the vertical velocity increases polynomially from the bottom to the free surface (Fig. 6), inevitably leads to some form of depth limitation in the accuracy of the embedded dispersive and nonlinear properties. Hence, Boussinesq-type equations are conventionally associated with relatively shallow water.

We next consider the free surface boundary conditions retaining all terms up to order δ, ϵ in Eq. (2.13) and δ^2, ϵ^2, $\delta\epsilon$ in Eq. (2.14) to obtain

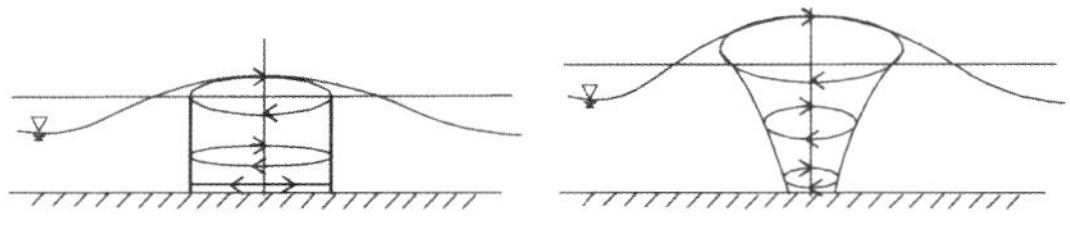

Fig. 6. Vertical structure of the water column beneath the waves.

two-dimensional Boussinesq-type equations

$$\eta_t + u(1+\epsilon)\eta_x + v(1+\epsilon)\eta_y - \frac{\delta}{6}\left(\left(\nabla^2 u\right)_x + \left(\nabla^2 v\right)_y\right) = 0 \tag{2.18}$$

$$u_t + \epsilon(uu_x + vu_y) + \eta_x - \frac{1}{2}\delta(u_{txx} + v_{txy}) = 0 \tag{2.19}$$

$$v_t + \epsilon(uv_x + vv_y) + \eta_y - \frac{1}{2}\delta(u_{txy} + v_{tyy}) = 0 \tag{2.20}$$

To simplify the set of Eqs. (2.18)–(2.20) to a single one, we assume a similar small scale for the introduced parameters, i.e., $\delta \sim \epsilon$, retain only one-dimension (x-dependence); eliminate u in linear terms of Eq. (2.18) using Eq. (2.19), and in nonlinear terms using linearized relationship $u = \eta + O(\gamma)$. The resulting expression (the Boussinesq equation), comprised of second- and higher-order derivatives, can be simplified further by letting $\delta \sim \epsilon \sim \gamma$. In physical terms the assumption $\gamma \ll 1$ impose wave parameters, such as height, length, and direction of propagation, to be slow varying at a distance of the wave length. In contrast with the Boussinesq equation, the condition allows us to consider the progressive wave solution traveling to one direction only, positive or negative with respect to x direction. For the positive direction we obtain a single equation, universally known as the Korteweg–de Vries (KdV) equation,

$$\eta_t + \left(1 + \frac{3}{2}\epsilon\right)\eta\eta_x + \frac{1}{6}\delta\eta_{xxx} = 0 \tag{2.21}$$

While deriving Eqs. (2.18)–(2.20) we have implicitly assumed that $\delta \ll 1$, $\epsilon \ll 1$, $\gamma \ll 1$ and $\delta \sim \epsilon$; therefore, the Boussinesq equations include only the lowest-order effects of frequency dispersion and nonlinearity. They can account for transfer of energy between different frequency components, changes in the shape of the individual waves, and the evolution of wave groups in the shoaling irregular wave train. However, the standard Boussinesq equations have two major limitations in their application to long waves on shallow water:

(1) The depth-averaged model describes poorly the frequency dispersion of wave propagation at intermediate depths and deep water (see also page 30–32 of this volume).
(2) The weakly nonlinear assumption is valid only for waves of small surface slope, and so there is a limit on the largest wave height that can be accurately modeled.

A number of attempts have been made to extend the range of applicability of the equations to deeper water by improving the dispersion characteristic of the equations, or to improve accuracy in the shallow water regime by improving nonlinear terms. Technically, this can be done by assuming different relationships between δ, ϵ, and γ; or equivalently, by including higher-order dispersion and/or nonlinear terms together with the conventional lower-order terms. Thus for example, the assumption $\delta > \epsilon$ relaxes limitation (1) allowing for deeper water to be considered; and condition $\epsilon^2 \sim \gamma$ permits simulation of higher (or shorter) waves as oppose to limitation (2). Taking into account the above scaling, Eqs. (2.18)–(2.20) eventually lead to the Stokes-type wave theory (Tkalich, 1986) describing both frequency and amplitude dispersion.

Witting (1984) used a different form of the exact, fully nonlinear, depth-integrated momentum equation for one horizontal dimension, expressed in terms of the velocity at the free surface. A Taylor-series-type expansion was used to relate the different velocity variables in the governing equations, the coefficients of the expansion being determined so as to yield the best linear dispersion characteristics. By retaining terms up to the fourth order in the dispersion parameter δ, Witting obtained relatively accurate results for both deep and shallow water waves. However, the expansions presented by Witting are only valid in water of constant depth. Murray (1989) and Madsen *et al.* (1991) examined the dispersion properties of various forms of the Boussinesq equations as well as Witting's (1984) Padé approximation of the linear dispersion relation for Airy waves. Based on the excellent properties of Padé approximants, the writers have introduced an additional third-order term in the momentum equation to improve the dispersion properties of the Boussinesq equations. The third-order is derived from the long-wave equations and reduces to zero in shallow water, resulting in the standard form of the equations in this case. The equations assume a constant water depth and are thus not applicable to shoaling waves. On the other hand, by defining the dependent variable as the velocity at an arbitrary depth, Nwogu (1993) achieved a rational polynomial approximation to the exact linear dispersion relationship without the need to add higher-order terms to the equations. Although the arbitrary location could be chosen to give a Padé approximation to the linear dispersion relationship, Nwogu chose an alternative value which minimized the error in the linear phase speed over a certain depth range. The equations obtained by both Madsen *et al.* (1991) and Nwogu (1993) give more accurate dispersion relation at intermediate water depths than do the standard Boussinesq equations. They have shown by examples that their extended equations are able to simulate wave propagation to shallow water from much deeper water.

Although higher-order Boussinesq equations for the improvement of the description of nonlinear and dispersive properties in water waves have been attempted and have been successful in certain respects, most of these attempts have involved numerous additional derivatives and hence made the accurate numerical solution increasingly difficult to obtain. In justification of such derivations of higher-order terms in the Boussinesq equations, preference has often been given to artificially constructed test cases having little (if any) correspondence with real tsunamis. The Northern Sumatra (December 2004) tsunami had provided a new test case for the various models. After several decades of intensive worldwide research, it is interesting to read the conclusion of Grilli *et al.* (2007) that "*in view of the apparently small dispersive effects, it could be argued that the use of a fully nonlinear Boussinesq equation model is overkill in the context of a general basin-scale tsunami model. However, it is our feeling that the generality of the modeling framework provided by the model is advantageous in that it automatically covers most of the range of effects of interest, from propagation out of the generation region, through propagation at ocean basin scale, to runup and inundation at affected shorelines.*"

Even the presence of the third-order derivative terms for dispersion in the standard Boussinesq equations (2.18)–(2.20) is considered challenging enough to be omitted in popular operational tsunami modeling codes, such as Tunami-N2 (Goto *et al.*, 1997). Boussinesq equations with omitted dispersion terms often are referred to as the Nonlinear Shallow Water Equations (NSWE). Alternative simplification suggested in MOST (Titov and Gonzalez, 1997) and COMCOT (Liu *et al.*, 1998) is to use NSWE, but implicitly include dispersion phenomenon by shaping a numerical approximation error in the form of the third-order derivatives (dispersion terms).

To avoid complex derivatives, Stelling and Zijlema (2003) proposed a semi-implicit finite difference model, which accounts for dispersion through a non-hydrostatic pressure term. In both the depth-integrated and multi-layer formulations, they decompose the pressure into hydrostatic and non-hydrostatic components. The solution to the hydrostatic problem remains explicit; the non-hydrostatic solution derives from an implicit scheme to the three-dimensional continuity equation. The depth-integrated governing equations are relatively simple and analogous to the nonlinear shallow water equations with the addition of a vertical momentum equation and non-hydrostatic terms in the horizontal momentum equations. Numerical results show that both depth-integrated models estimate the dispersive waves slightly better than the classical Boussinesq equations.

2.4. *Importance of Various Phenomena for Tsunami Propagation, A Sensitivity Analysis*

Modern tsunami research experiences two contradictory trends, one is to include more physical phenomena (previously neglected) into consideration, and another is to speed up the code to be used for the operational tsunami forecast. The optimal code for tsunami modeling must be sufficiently fast and accurate; however, the notions of speed and accuracy are quickly changing to reflect current understanding of tsunami physics as well as growing computational power. Hence, in order to assess parameters of the currently optimal code, established and new approaches need to be regularly re-evaluated to ensure that the most important (and yet computationally affordable) phenomena are taken into account. The importance of some phenomena, potentially capable of affecting tsunami propagation characteristics, has been recently quantitatively evaluated by Dao and Tkalich (2007).

To study the relative effects of various phenomena on tsunami wave propagation, we start with the Nonlinear Shallow Water Equations on a Cartesian grid, as commonly used in tsunami modeling, in particular the model "Tunami-N2." The NSWE could be obtained from the Boussinesq equations (2.18–2.20) by returning to dimensional units, omitting the dispersion terms, and adding bottom friction terms, as below

$$\begin{aligned} &\eta_t + M_x + N_y = 0 \\ &M_t + (M^2/D)_x + (MN/D)_y + gD\eta_x + \tau^x/\rho = 0 \\ &N_t + (MN/D)_x + (N^2/D)_y + gD\eta_y + \tau^y/\rho = 0 \end{aligned} \tag{2.22}$$

Here, D is the total water depth; ρ is the water density; τ^x and τ^y are the bottom friction in the x and y directions, respectively. The friction coefficient can be computed from Manning's roughness n as $\tau^x/\rho = (n^2/D^{7/3})M\sqrt{M^2+N^2}$ and $\tau^y/\rho = (n^2/D^{7/3})N\sqrt{M^2+N^2}$. Values of Manning's roughness for certain types of sea bottom are given in Table 1.

The water velocity fluxes in the x and y directions, M and N are defined as $M = \int_{-h}^{\eta} udz = u(h+\eta) = uD$ and $N = \int_{h}^{\eta} vdz = v(h+\eta) = vD$.

This model has been improved by Dao and Tkalich (2007) to utilize spherical coordinates, and to include Coriolis force and linear dispersion

Table 1. Values of Manning's roughness for certain types of seabottom.

Channel Material	n	Channel Material	n
Neat cement, smooth metal	0.010	Natural channels in good condition	0.025
Rubble masonry	0.017	Natural channels with stones and weeds	0.035
Smooth earth	0.018	Very poor natural channels	0.060

terms. The modified code (TUNAMI-N2-NUS) is applied to simulate the tsunami caused by the Northern Sumatra earthquake of 26 December 2004. The domain is discretized with a rectangular grid having 848×852 nodes and 2 minutes (~3.7 km) resolution. Bathymetry (i.e., sea floor topography data) is taken from the NGDC digital databases of sea-floor and land elevations on a 2 arc minute grid (etopo2, NGDC/NOAA). The earthquake fault parameters are adopted from Grilli *et al.* (2007). Five fault segments were identified in sequence to depict rupture propagation from south to north (Fig. 7, left pane). The initial surface elevation is assumed to be identical with the vertical instantaneous seismic displacement of the sea floor, given by Mansinha and Smylie (1971) for inclined strike-slip and dip-slip faults (see Sec. 3.1 below). The computed maximum tsunami height and arrival time of the first wave at 3.5 hours after the event are depicted in Fig. 7 (middle and right panes, respectively).

Computations show that the following phenomena have been important for the Northern Sumatra tsunami (in reducing order of importance).

Astronomical tide is one of the most important but often neglected phenomenon. A typical tsunami is much shorter (in duration) than

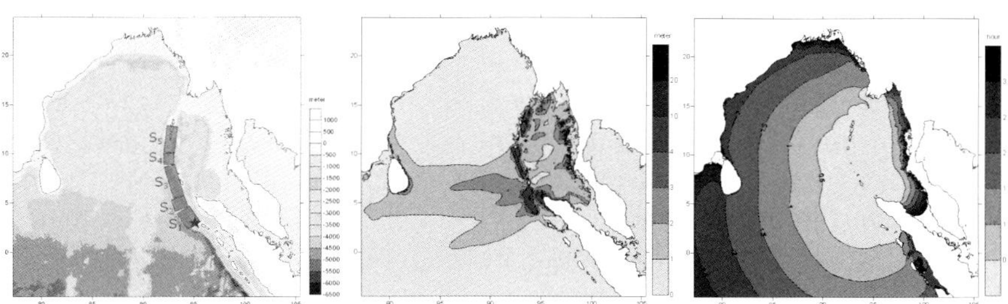

Fig. 7. Topography for computation domain and the fault segments S1–S5 (⋆ is the location of earthquake epicenter, left-hand pane), the maximum tsunami height (middle pane) and arrival time of the first wave (right-hand pane).

astronomically driven tides; the tidal range has therefore usually been neglected during tsunami modeling, the computed sea level dynamics being simply superimposed on the tidal dynamics after the computations. However, strong tidal activity in shallow areas may affect not only the magnitude of the inundation, but also the arrival time of a tsunami. It has been found that in coastal areas with a tidal range of about 3 m (as in most of the countries affected) the tsunami could be 0.5 m greater in amplitude during high tide (as compared with low tide) and could arrive 30 minutes earlier. In the past, discrepancies in wave height and arrival time in numerical simulations (as compared with the recorded observations) have been frequently attributed to local bathymetry features, but the error could be also due to neglect of this interaction with tides.

When tsunami waves enter shallow waters near the coast, friction effects in NSWE increase. To investigate the importance of this effect, different Manning's roughness of 0.025 and 0.011 were chosen for simulations. The results show some increases of 0.5–1.0 m in the maximum tsunami height nearshore, and the wave could approach six minutes earlier in the case with the lower bottom friction. In the deep ocean the effect of bottom friction is negligible.

Dispersion has a significant influence on tsunami simulations in deep water. Due to the frequency dispersion, longer waves travel faster and separate from the shorter waves, leading to a decrease of computed tsunami height; at the same time, since longer waves travel faster, they arrive earlier. The dispersion effect is stronger in the direction of tsunami propagation and toward deep waters where the wave speed is the largest. The analysis shows that dispersion can cause a drop of 0.4 m (40%) in the computed maximum tsunami height in these areas (see Fig. 8). A notable decrease of wave height also occurs near shorelines. No significant change in arrival time is observed.

Effects of the Earth's curvature and rotation (Coriolis force) could have influences on far-field tsunamis. The analysis shows that use of spherical coordinates may lead to a minor 0.1 m difference of computed maximum tsunami height, and just one minute difference in arrival time. Even though the effect of the Earth's curvature is small, this effect increases at higher latitudes or farther from the source in the main direction of the tsunami propagation. The Coriolis effect is expected to be larger at higher latitudes or for higher water velocity fluxes. The analysis particularly depicts a slight variation of maximum tsunami height nearshore or far from the equator.

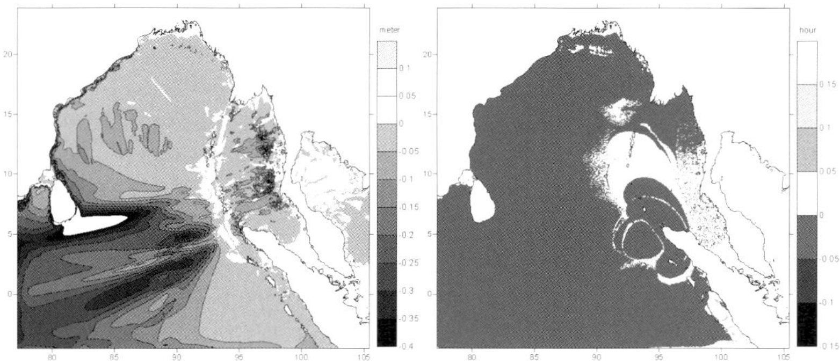

Fig. 8. Differences in maximum tsunami height (left) and arrival time (right) between simulations with and without dispersion terms.

The analyses imply that some of the above phenomena may cause significant changes of the tsunami propagation characteristics. Tide and bottom friction can alter significantly the waves near coastal areas and thus need to be included in research and operational codes when considering wave-shore interactions. Dispersion has a strong effect in deep water and little influence in shallow water, so should be included in trans-ocean tsunami simulations. The computation time required to solve the fully nonlinear dispersion model to gain a little accuracy locally may be impracticable for operational forecasts but still may be important for run-up simulation. The effects of curvature and Coriolis force are smaller than others, but can still be included for far-field tsunami modeling without sacrificing much computational resource. The final decision on what phenomena to include (and when) depends on available computational power and the purpose of a particular study or code. In view of the uncertainties involved, the simplest (and quickest) code may be appropriate for operational forecasts, whereas a research code can afford to include all the considered terms.

3. Tsunami Forecasting

Long before the modern instrumental era, people were trying to predict earthquakes and tsunamis using various nonscientific means (i.e., all that was then available). We will neither discuss nor dismiss the validity of unconventional methods; instead we focus on the scientifically based methods of Earth observation which are already sufficiently developed and utilized today (Bernard *et al.*, 2006), or could be developed in a short time frame (Rudloff *et al.*, 2009). Among the most recent ideas is the use of Global Positioning System (GPS) networks to monitor ionospheric disturbances in the atmosphere caused by tsunamis. Observational networks will never be dense because the ocean is vast, so establishing and maintaining monitoring stations is costly and difficult, especially in deep waters. However, sparse deep-ocean tsunami data combined with models (especially data driven) can provide timely and accurate forecasts for the entire ocean basin.

As most tsunamis are triggered by earthquakes, seismometers are the first obvious choice to trigger a tsunami warning system and to estimate the source parameters. Seismic signals from the near-real-time IRIS Global Seismographic Network (Fig. 9(a)) are commonly used to infer an earthquake's magnitude and epicenter location. If a tsunami has been generated, the waves propagate across the ocean eventually reaching one of the NOAA-developed DART buoys (Fig. 9(b)), which report sea-level measurements ("mareograms") back to the tsunami warning system (Fig. 9(c)). The information is processed to produce new, more refined, estimates of the tsunami source, which can then be used to compute a more accurate tsunami forecast. The speed and accuracy of the seismogram and mareogram inversion to the source are crucial for success of the tsunami forecast in the initial period.

Two auxiliary sources of tsunami information have to be mentioned, i.e., nearshore tide gauges and open-sea satellite altimetry. The tide gauge measurements are complicated by variations in local bathymetry and harbor shapes, which severely limit the effectiveness of the data for providing useful measurements for tsunami forecasting. Tide gauges can provide verification of tsunami forecasts, but they cannot provide the data necessary for efficient forecast itself, and definitely not for the coast where they are installed. Tsunami detection by satellite altimetry is similarly restricted by the high cost of imaging and low frequency of sampling.

3.1. *Tsunami Source Estimation*

The initial condition of a tsunami in a numerical model is often prescribed as a static elevation of sea level due to the displacement (rupture) of the sea bottom during an earthquake. For submarine earthquakes, a typical rupture may last for minutes, which can be considered as instantaneous comparing to the timescale of tsunami propagation (tens of minutes to a

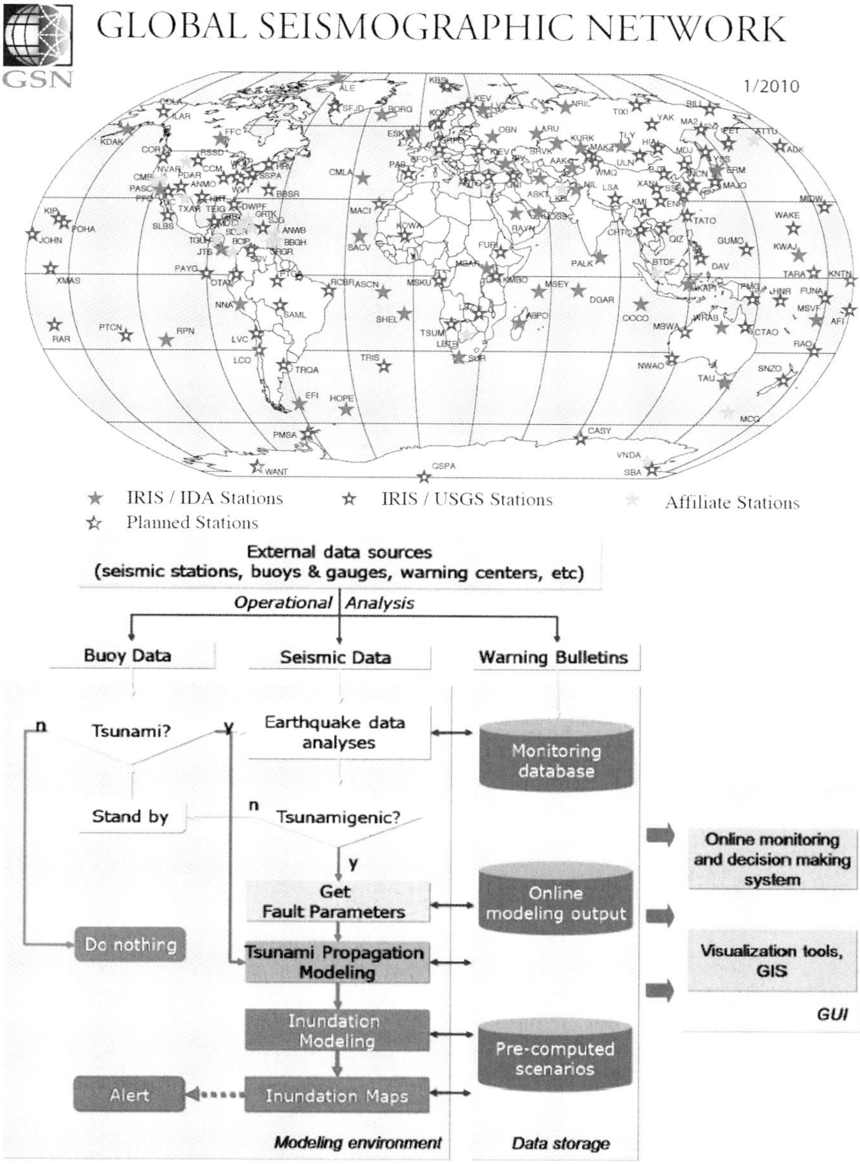

Fig. 9. IRIS seismographic observation network (upper pane), and structure of a tsunami forecasting system (lower pane).

few hours). The hydrodynamic effect is often neglected since the horizontal size of the wave profile (hundreds of kilometers) is sufficiently greater than the water depth (a few kilometers) at the source. Thus, a fault model is developed in which the initial surface wave is assumed to be identical to the vertical static co-seismic displacement of the sea floor, as given by Mansinha and Smylie (1971) for inclined strike-slip and dip-slip fault planes. In this fault model an earthquake is approximated by a displacement of an inclined plane. The magnitude of the earthquake is proportional to the size of the plane, and the displacement and rigidity of the earth at the earthquake center. A similar algorithm can be obtained from Okada (1985). Fault models of this type are simple, but very fast and accurate in most cases and have been used in various simulations. Parameters used to describe an earthquake in the fault model are: Position of the earthquake epicenter (x, y, z), size of the fault plane (length and width), direction of the fault plane (strike, dip, slip angle) and displacement of the fault plane (Fig. 10). The horizontal coordinates x, y are the longitude and latitude of the earthquake epicenter. The vertical position z is the depth of the epicenter measured downward from the sea bottom. The strike angle is the clockwise angle of the rupture direction from the geo-north direction. The dip angle is the angle between the fault plane and the horizontal plane in the direction perpendicular to the strike angle. The slip angle is the angle measured between the vector of displacement and the horizontal plane in the strike direction. The length of the fault plane is the length of the edge in the strike direction. The displacement is the magnitude of the vector of displacement. With very long ruptures, where the rupture occurs over an area more than 1000 km long and a few hundred km wide, such as for the 2004 Northern Sumatra earthquake, the instantaneous rupture

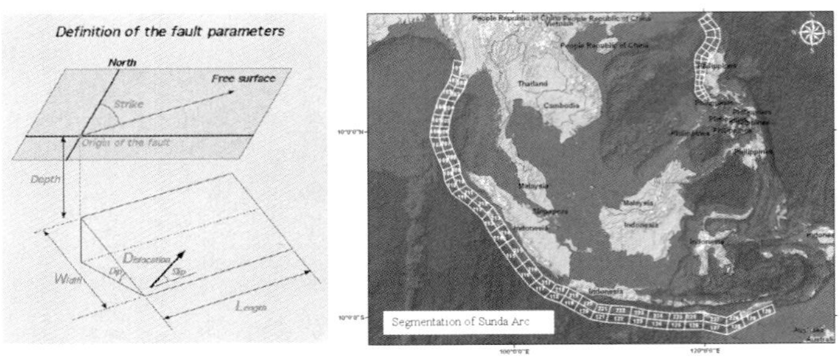

Fig. 10. The fault plane and associated parameters (left) and segmentation of Sunda Arc (right).

Fig. 11. A typical initial condition of a tsunami calculated by the fault model (the scale of the wave height is different from the horizontal scale, for clarity).

assumption in this fault model could lead to significant error. In order to model long earthquakes, the rupture area can be segmented into several rectangles. Each rectangle is treated as a fault plane, called a sub-fault. A time-lag can be imposed for each sub-fault. By using this multi-fault method, an earthquake that has a varying rupture direction can be easily modeled.

A typical initial condition for the 2004 Northern Sumatra earthquake calculated by this multi-fault model is shown in Fig. 7. Using this initial condition for tsunami computations, the highest amplitude and the tsunami arrival time are recorded at every computational node (as shown in Fig. 7), and then delivered to national forecasters worldwide as a part of standard output package.

3.2. *Quick Tsunami Forecasting Techniques*

Over the past few decades, accurate process-driven tsunami propagation models (based on Navier–Stokes and Boussinesq equations) have been developed and thoroughly tested. Most advanced models require significant computational resources at fine grid resolutions; hence, they cannot be used for operational tsunami forecasts. Accurate and computationally fast data-driven methods are found to be able to mimic the pattern of training data sets, which make them ideal for real-time operations. The use of data-driven methods can be extended to replace accurate but computationally demanding process-driven tsunami propagation models by means of training data-driven models with a large number of pre-computed tsunami scenarios. A simple data-driven tsunami forecast system consists of a database of pre-computed scenarios and a case selection routine with a conventional interpolation algorithm (such as those proposed by Whitmore and Sokolowski, 1996). In this method, the closest matching event from the database is identified by comparing the pre-computed scenarios with measured wave characteristics near the earthquake epicenter. Other researchers have proposed different approaches, e.g., the inversion methods of Wei *et al.* (2003) and Lee *et al.* (2005). These methods were constrained by the assumption that the tsunami wave propagation is linear to allow linear superposition of pre-computed data. Barman *et al.* (2006) used the Artificial Neural Networks (ANN) method in the prediction of the tsunami arrival time in the Indian Ocean. Srivichai *et al.* (2006) used the General Regression Neural Network (GRNN) method to forecast tsunami heights. This method allows the application of nonlinear process-driven tsunami models to build a database of scenarios, but the application was limited to only a few predefined discrete observation points.

In Romano *et al.* (2009), the ANN technique has been developed to provide a rapid and accurate prediction of maximum wave heights and arrival times for any location in the Singapore Region, and the data-driven model has become part of the Singapore tsunami warning system. The well-trained ANN models could mimic closely the performance of the TUNAMI-N2-NUS model within seconds. In the paper, plausible models for the rupture geometry and slip of the most important regional subduction zones have been used by the process-driven and ANN models in order to simulate tsunamis of varying slip magnitudes.

In recent years ANN methods have become a standard for tsunami forecasting whereas alternative methods based on empirical orthogonal functions (popular in other branches of fluid dynamics) were not utilized. One such method has been recently applied for tsunami prediction by Dao *et al.* (2008); a brief introduction and results are presented below. This method allows low computational cost by using a reduced-order representation of the output, i.e., by decreasing the number of unknowns to a computationally tractable number, say tens to hundreds. To derive the reduced-order representation, the output is expressed as a linear combination of q basis vectors

$$y(x,t,\mathbf{w}) = \sum_{j=1}^{q} a_j(t,\mathbf{w})\phi_j(x) \tag{3.1}$$

where the time-dependent coefficients a_j depend on the vector of parameters $\mathbf{w}$; and the basis vectors ϕ_j characterize the spatial variation in output.

Using Eq. (3.1), the size of the problem can be reduced from millions (size of output y) to tens or hundreds (size q of the vector of coefficients a_j). The decomposition Eq. (3.1) applies generally to other relevant choices of the pre-determined basis vectors ϕ_j, $j = 1, \ldots, q$. There are several different methods that could be used to define the basis vectors, such as Empirical Orthogonal Function (EOF), Principle Component Analysis (PCA) or Proper Orthogonal Decomposition (POD). The POD basis is a preferred choice, because for a given basis size it provides the optimal least-squares representation of a given data set. Additionally, through a wide range of applications, POD has been shown to provide an excellent low-order characterization of important features of flow dynamics.

Let $y^i = y(x, T, \mathbf{w}^i)$ denotes a solution corresponding to final time T and parameters $\mathbf{w}^i$, $i = 1, \ldots, m$. In tsunami forecasting, the final solution could be the map of maximum wave height and the parameters a set of fault parameters. A tsunami model is used to routinely simulate a series of tsunami scenarios according to the list of $\{\mathbf{w}^i\}_{i=1}^m$ to obtain $\{y^i\}_{i=1}^m$. The POD method is applied to the set of pre-computed solutions $\{y^i\}_{i=1}^m$ to obtain the orthonormal basis vectors $\{\phi_i\}_{i=1}^m$. Once the coefficients a_j are determined, the prediction of a tsunami solution is given by Eq. (3.1). There are different ways to calculate the coefficients a_j depending on the available information, such as fault (earthquake) parameters or tsunami wave record at observation stations. In the following, we present the method as used when the fault parameters are known.

As highlighted in the above section, rupture models have been developed from geodetic and seismic data and based on fault geometry, aerial maps and historical earthquake data (Fig. 10). For each set of fault parameters $\mathbf{w}^i$, the coefficient a^i_j describes the magnitude of the POD basis vector j needed to represent the ith solution, i.e., $y^i = \sum_{j=1}^m a^i_j \phi_j$, and is given by the projection $a^i_j = (\phi_j, y^i)$. When a set of fault parameter $\mathbf{w}$ is available for a tsunami that is actually occurring, the POD coefficients a_j associated with $\mathbf{w}$ ($\mathbf{w}$ being in the range, but not included in the list, of $\{\mathbf{w}^i\}_{i=1}^m$) can be found by interpolation among the a^i_j. This interpolation can be performed within seconds at the forecasting stage.

The performance of the method is shown in Fig. 12. In this example, a database of 40 tsunami solutions of maximum wave heights is built by varying the slip magnitudes of a multi-fault rupture. Slips are chosen from 10–22 m with an interval of 4 m. A sub-fault with no rupture has zero slip. Only ten basis vectors are selected. A predicted solution is compared with the exact one (computed by a full PDE model). The differences are less than

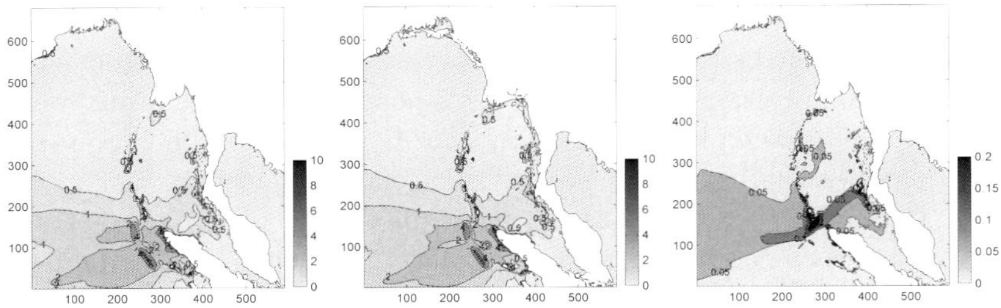

Fig. 12. Comparison of maximum wave amplitude. Left: Exact solution; Middle: Reduced-order model; Right: error. Color scale unit is one meter. Axes represent grid numbers, grid-cell size is two minutes.

0.2 m in some small areas near the earthquake source; and are very small elsewhere, including nearshore zones. Compared to the absolute values of tsunami amplitude in these areas, the error is acceptable (less than 10%), which is comparable with the discrepancy between a full PDE solution and field measurements. Moreover, prediction in the domain of interest (near coastlines) has a small error. The study also shows that with quite a small number of solutions in the database, POD can produce satisfactory results. Increasing the number of carefully selected scenarios will result in further improvement.

4. Conclusions

In this chapter we have reviewed the key developments leading to modern methods of tsunami modeling and forecasting. We started with the first observation of a soliton and the later experiments of Scott Russell, and followed with a derivation of the Boussinesq equations currently used worldwide for tsunami modeling. Difficulties associated with use of the Boussinesq equations are highlighted and alternatives are reviewed. For quick tsunami forecasting we have outlined the application of the latest data-driven methods, such as ANN and POD, and discussed the main functions and general structure of a typical regional tsunami warning system.

References

Barman, R., Kumar, B., Pandey, P., *et al.* (2006). Tsunami travel time prediction using neural networks, *Geophys. Res. Lett.* **33**, 16, L16612, doi:10.1029/2006GL026688.

Bernard, E., Mofjeld, H., Titov, V., *et al.* (2006). Tsunami: scientific frontiers, mitigation, forecasting and policy implications, *Phil. Trans. R. Soc. A* **364**, pp. 1989–2007, doi:10.1098/rsta.2006.1809.

Boussinesq, J. (1872). Théorie des ondes et des remous qui se propagent le long d'un canal rectangulaire horizontal, en communiquant au liquide contenu dans ce canal des vitesses sensiblement pareilles de la surface au fond, *J. de Math. Pures et Appl. Deuxiéme Série* **17**, pp. 55–108.

Boussinesq, J. (1877). Essai sur la théorie des eaux courantes, *Mémoires présentés par divers savants, à l'Acad. des Sc.* **23**(1), pp. 1–680.

Dao, M. and Tkalich, P. (2007). Tsunami propagation modelling — a sensitivity study, *Nat. Hazards Earth Syst. Sci.* **7**, pp. 741–754.

Dao, M., Tkalich, P. and Chan, E. S. (2008). Tsunami forecasting using proper orthogonal decomposition method, *J. Geophys. Res.* **113**, C06019, doi:10.1029/2007JC004583.

Dean, R. and Dalrymple, R. (1984). *Water Wave Mechanics for Engineers and Scientists* (Prentice Hall, Inc., Englewood Cliffs, NJ).

Debnath, L. (1994). *Nonlinear Water Waves* (Academic Press, San Diego).

Goto, C., Ogawa, Y., Shuto, N., *et al.* (1997). *Numerical Method of Tsunami Simulation with the Leap-frog Scheme (IUGG/IOC Time Project)*, IOC Manual, UNESCO, Paris.

Grilli, S., Ioualalen, M., Asavanant, J., *et al.* (2007). Source constraints and model simulation of the December 26, 2004 Indian Ocean tsunami, *J. Waterw. Port Ocean Coast. Eng.* **133**, 6, pp. 414–428.

Korteweg, D. J. and de Vries, G. (1895). On the change of form of long waves advancing in a rectangular canal, and on a new type of long stationary waves, *Philos. Mag.* **39**, pp. 422–443.

Lee, H., Cho, Y. and Woo, S. (2005). Quick tsunami forecasting based on database, *Advances in Natural and Technological Hazards Research, Part II*, pp. 231–240.

Liu, P. L.-F., Woo, S.-B. and Cho, Y.-S. (1998). *Computer Programs for Tsunami Propagation and Inundation* (Cornell University Press, Ithaca, N.Y.).

Madsen, P. A., Murray, R. and Sorensen, O. R. (1991). A new form of the Boussinesq equations with improved linear dispersion characteristics, *Coast. Eng.* **15**, pp. 371–388.

Mansinha, L. and Smylie, D. (1971). The displacement fields of inclined faults, *B. Seis. Soc. Am.* **61**, 5, pp. 1433–1440.

Murray, R. J. (1989). Short wave modelling using new equations of Boussinesq type, in *Proc. 9th Austr. Conf. on Coast and Ocean Engrg.* (Inst. of Engineers, Adelaide, Australia), pp. 331–336.

Nwogu, O. (1993). Alternative form of Boussinesq equations for nearshore wave propagation, *J. Waterw. Port Coast. Ocean Eng.* **119**, p. 6.

Okada (1985). Surface deformation due to shear and tensile faults in a half-space, *B. Seismol. Soc. Am.* **75**, pp. 1135–1154.

Romano, M., Liong, S., Vu, M., *et al.* (2009). Artificial neural network for tsunami forecasting, *J. Asian Earth Sci.* **36**, pp. 29–37.

Rudloff, A., Lauterjung, J., Munch, U., *et al.* (2009). The GITEWS Project (German-Indonesian Tsunami Early Warning System), *Nat. Hazards Earth Syst. Sci.* **9**, pp. 1381–1382.

Russell, J. S. (1844). Report on waves, fourteenth meeting of the British association for the advancement of science (London, John Murray).

Russell, J. S. (1885). The wave of translation in the oceans of water, air and ether (London, John Murray).

Srivichai, M., Supharatid, S. and Imamura, F. (2006). Developing of forecasted tsunami database along Thailand Andaman coastline, in *Proc. Asia Oceania Geosci. Soc. 3rd Ann. Meeting* (AGOS), p. 138.

Stelling, G. and Zijlema, M. (2003). An accurate and efficient finite-difference algorithm for non-hydrostatic free-surface flow with application to wave propagation, *Int. J. Numer. Meth. Fl.* **43**, 1, pp. 1–23.

Titov, V. and Gonzalez, F. (1997). Implementation and testing of the method of splitting tsunami (MOST) model, Technical memorandum erl pmel-112, NOAA.

Tkalich, P. (1986). *Propagation of Surface Waves in Fluids with Respect to Unsteady and Nonlinear Phenomena*, Ph.D. thesis, Institute of Hydromechanics, Kiev, Ukraine (in Russian).

Wei, Y., Cheung, K., Curtis, G. and McCreery, C. (2003). Inverse algorithm for tsunami forecasts, *J. Waterw. Port Coast. Ocean Eng.* **129**, 2, pp. 60–69.

Whitmore, P. and Sokolowski, T. (1996). Predicting tsunami heights along the North American coast from tsunamis generated in the Northwest Pacific Ocean during tsunami warnings, *Sci. Tsunami Haz.* **14**, 3, pp. 147–166.

Witting, J. (1984). A unified model for the evolution of nonlinear water waves, *J. Comput. Phys.* **56**, pp. 203–236.

Zabusky, N. and Kruskal, M. (1965). Interaction of "solitons" in a collisionless plasma and the recurrence of initial states, *Phys. Rev. Lett.* **15**, pp. 240–243.

Chapter 10

Rogue Waves

F. Dias

School of Mathematical Sciences
University College Dublin, Dublin, Ireland
frederic.dias@ucd.ie

T. J. Bridges

Department of Mathematics
University of Surrey, Guildford, UK
t.bridges@surrey.ac.uk

J. M. Dudley

Institut FEMTO-ST
University of Franche-Comté, Besançon, France
john.dudley@univ-fcomte.fr

Rogue waves are fascinating: once part of the folklore, they now make the news each time an observation is made. At the time of printing, the last example is that of the *Louis Majesty* cruise ship that was hit by an abnormal wave in March 2010, off the coast of Catalonia in the Mediterranean. The wave impact was immortalized by several videos taken by tourists on board the ship. Fortunately for travelers but unfortunately for scientists, rogue waves do not occur very often and their origin remains a mystery, even if the state-of-the-art in the understanding of rogue waves has witnessed some unprecedented progress in the last five years. Recently similar phenomena were observed in different fields of physics, in particular in optics. Is there hope to learn more on rogue waves from other fields or are these extreme events disconnected phenomena? This chapter provides a review on rogue waves, with an emphasis on the modulational instability and the absolute or convective character of this instability.

1. Introduction

The study of rogue waves is still relatively recent, even if this mysterious phenomenon has been known in various environments such as ocean waves for centuries. Undoubtedly, rogue waves have practical consequences and are not simply a theoretical subject. Views on rogue (or "freak") waves are sometimes controversial and even contradictory. Even the definition of a rogue wave is not so easy to explain. The standard approach is to call a wave a rogue wave whenever the wave height H (distance from trough to crest) exceeds a certain threshold related to the sea state. More precisely the common criterion states that a wave is a rogue wave when

$$H/H_s > 2, \tag{1.1}$$

where H_s is the significant wave height, here defined as four times the standard deviation of the surface elevation. Was this criterion satisfied during the *Louis Majesty* incident? The answer is no. Indeed, the wave height (in fact, there were three large waves) has been estimated to be 8 m, while the significant wave height was 5 m. Nevertheless, the waves were powerful enough to kill two tourists and to do quite a bit of damage.

Rogue waves arise in arbitrary water depth (in deep as well as shallow water), with or without currents. The observed probability of occurrence of freak waves in deep and shallow waters is approximately the same. It is important to remember that one is dealing here with rare events and consequently scientists have only few data available. However, the understanding of rogue waves is witnessing regular progress. Freak waves may have the shape of a solitary wave or correspond to a group of several waves. Various mechanisms have been proposed for rogue wave formation, either linear or nonlinear. Assuming that wind waves, at least in the framework of linear theory, can be considered as the sum of a large number of independent monochromatic waves with different frequencies and directions, a freak wave may arise in the process of spatial wave focusing (geometrical focusing) and spatio-temporal focusing (dispersion enhancement). The interaction between a wave and a counter-propagating current can also be at the origin of large wave events. Because freak waves are large-amplitude steep waves, one would expect nonlinearity to play an important role as well in the formation and the evolution of rogue waves. Nonlinearity modifies the linear focusing mechanisms, but does not destroy them. In fact linear mechanisms are more and more regarded as pre-conditioning for nonlinear focusing. It is now recognized that most focusing mechanisms are also robust with respect to random wave components.

There is one mechanism of freak wave formation which is suggested in the framework of nonlinear theory only: the modulational instability (MI), also referred to as the Benjamin–Feir (BF) instability in the hydrodynamics community.[a] A uniform train of relatively steep waves is unstable to side-band disturbances, that is disturbances whose frequencies deviate slightly from the fundamental frequency of the carrier waves. The BF instability increases the frequency of occurrence of freak waves in comparison with the linear theory. At the same time the randomness of the wave field reduces the BF instability. All the processes mentioned above can be investigated in the framework of weakly nonlinear models like the nonlinear Schrödinger (NLS) equation, the Davey–Stewartson system, the Korteweg–de Vries equation, and the Kadomtsev–Petviashvili equation. An excellent review is given in the recent book by Kharif *et al.* (2009). An earlier version was given in the review article by Kharif and Pelinovsky (2003); see also Dysthe *et al.* (2008). The state-of-the-art review on rogue waves can be found in the special issue of the European Physical Journal (Dudley *et al.*, 2010).

Since the BF instability is the main focus of this chapter, let us provide a brief review. A full account of the history of the BF instability can be found in Hunt's review article (2006). The discovery of the BF instability of traveling waves was a milestone in the history of water waves. Before 1960, the idea that a Stokes wave could be unstable does not appear to have been given much thought. The possibility that the Stokes wave could be unstable was pointed out in the late 1950s, but it was the seminal work of Benjamin and Feir (1967) that combined experimental evidence with a weakly nonlinear theory that convinced the scientific community.

Indeed, Benjamin and Feir started their experiments in 1963 assuming that Stokes waves were stable. After several frustrating years watching their waves disintegrate — in spite of equipment and laboratory changes and improvements — they finally came to the conclusion that they were witnessing a new kind of instability. The appearance of "sidebands" in the experiments suggested the form that the perturbations should take. The water wave community will celebrate soon the 50-year anniversary of the discovery of deep water wave instability but it is much more recently that

[a]In order to distinguish between nonlinear optics and hydrodynamics, we will use the MI terminology for optics and the BF terminology for water waves.

the community was convinced that this effect is able to generate a rogue wave in the real sea. However, there is still some controversy. Indeed the BF instability may be suppressed by various unfavorable conditions (Segur *et al.*, 2005; Bridges and Dias, 2007). Moreover it is well known that in two dimensions the BF instability does not occur in shallow water. This means that the BF instability is not necessarily the dominant mechanism causing rogue waves at least in the coastal zone where wave focusing and blocking due to bathymetry and current effects are important.

The BF instability applies to a plane wave on which a small perturbation is superimposed. Since ocean waves are characterized by a finite width spectrum, the concept of BF instability must be generalized. The Benjamin–Feir Index (BFI) was introduced by Janssen (2003). It measures the ratio between the wave steepness and the spectral bandwidth. Rigorous results for a broad-band spectrum are not straightforward, unless some hypotheses on the statistics (usually a quasi-Gaussian approximation) are introduced. Numerical results and recent experiments (Onorato *et al.*, 2009) show that sea states characterized by steep, long-crested waves are more likely to give rise to rogue waves as opposed to those characterized by a large directional spreading.

Several ship accidents have occurred in crossing sea conditions. The BF instability of a crossing sea was investigated by Onorato *et al.* (2006), who computed the growth rates based on two coupled NLS equations, and by Laine-Pearson (2010), who extended the analysis to the full water-wave problem. Tamura *et al.* (2009) investigated the sinking of the *Suwa-Maru* fishing boat east of Japan on 23 June 2008. Their retrospective result for sea-state conditions at the time of the incident indicated that a crossing sea state developed four hours before the accident. However, the wave condition was unimodal at the time of the accident and was favorable for the occurrence of freak waves according to quasi-resonance theory. Thus, for the case of the *Suwa-Maru* incident, the crossing sea was a “precursor” to the development of the narrow spectrum. Interactions between wind waves and swell took place as the wind speed increased and the sea state rapidly developed into a unimodal freakish state.

In 2007 a paper by Solli and Jalali (2007) led to a fundamental change in scientific thinking about rogue waves: Rogue waves are not restricted to ocean waves, they also occur in optics. More recently they have also been observed in capillary waves (Shats *et al.*, 2010) and conjectured in the atmosphere (Stenflo and Marklund, 2010). The Solli *et al.* paper prompted two of us (FD and JMD) to develop a new multidisciplinary project, the

MANUREVA project.[b] Can the optics community help the hydrodynamics community? A realistic ocean wave theory should be based on reasonable physical principles that can be formulated in mathematical terms and should generate useful predictions about what can happen, based on physically meaningful observations and parameters. It is a big challenge to find a way to satisfy both requirements. This is the purpose of the MANUREVA project. It is obvious that a useful theory that describes the statistics of rogue waves, regardless of how they are defined, must go beyond the linear stochastic Gaussian wave theory based on frequency decompositions. One must look for models that can reproduce steep and asymmetric waves. The MANUREVA project is still under way, but some conclusions have already been reached. They are summarized in the MANUREVA paper of the special issue of the European Journal of Physics mentioned above (Dudley *et al.*, 2010). Collisions appear to play a central role in the generation of large amplitude waves (Genty *et al.*, 2010). Indeed it is possible that the only real waves with statistics that can be characterized as "rogue" with genuine long tails arise from collisions. Collisions within NLS systems were proposed as ocean rogue wave generators previously. NLS-related dynamics are obvious from a mathematical viewpoint, but actually linking these effects to experiments is not so clear. For example, the discovery that Akhmediev breather theory and MI were linked experimentally was made only recently, quite surprisingly (Dudley *et al.*, 2009; Kibler *et al.*, 2010). Although the MI dynamics might be well known they can still seed a wide range of different behaviors because one is always in a perturbed NLS system.

In the context of rogue waves in optical fiber systems, Taki *et al.* (2010) provided theoretical and numerical evidence that optical rogue waves originate from convective modulational instabilities. This is an important observation because the BF instability for water waves is not convective as shown by Brevdo and Bridges (1997). This is reviewed in the next section.

2. The NLS Equation

The celebrated NLS equation usually includes cubic nonlinearity and second-order dispersion, at least in hydrodynamics. Here we consider the cubic NLS equation with additional third-order dispersion,

$$\mathrm{i}a_1 A_t + a_2 A_{xx} + \mathrm{i}a_3 A_{xxx} + a_4 |A|^2 A = 0.$$

[b]MANUREVA stands for "Mathematical modeling and experiments studying nonlinear instabilities, rogue waves and extreme phenomena."

Third-order dispersion is often considered in optics (Akhmediev *et al.*, 1990), together with Raman scattering and self-steepening. By scaling the time t, the space x, and the amplitude A, all the coefficients can be set to plus or minus one except the coefficient of A_{xxx}. Assuming the situation for BF instability, the canonical form of the equation is

$$\mathrm{i}A_t + A_{xx} + \mathrm{i}bA_{xxx} + |A|^2 A = 0, \tag{2.1}$$

where b is a real parameter.

The basic state which represents the Stokes wave is

$$A(x,t) = \xi \mathrm{e}^{\mathrm{i}\Omega t}, \quad \Omega = \|\xi\|^2. \tag{2.2}$$

Consider the linear stability of the Stokes wave (2.2); let

$$A(x,t) = (\xi + B(x,t))\mathrm{e}^{\mathrm{i}\Omega t}.$$

Substituting into Eq. (2.1) and linearizing about (2.2) yields

$$\mathrm{i}B_t + B_{xx} + \mathrm{i}bB_{xxx} + \xi^2\overline{B} + \|\xi\|^2 B = 0. \tag{2.3}$$

The general solution of Eq. (2.3) is

$$B(x,t) = B_1\,\mathrm{e}^{(\lambda t + \mathrm{i}kx)} + B_2\,\mathrm{e}^{(\overline{\lambda}t - \mathrm{i}kx)}.$$

Substitution into Eq. (2.3) gives

$$\mathrm{i}\lambda B_1 - k^2 B_1 + bk^3 B_1 + \xi^2\overline{B_2} + \|\xi\|^2 B_1 = 0$$
$$-\mathrm{i}\lambda\overline{B_2} - k^2\overline{B_2} - bk^3\overline{B_2} + \overline{\xi}^2 B_1 + \|\xi\|^2\overline{B_2} = 0.$$

Solutions exist if and only if the following condition is satisfied:

$$\det\begin{bmatrix} \mathrm{i}\lambda - k^2 + bk^3 + \|\xi\|^2 & \xi^2 \\ \overline{\xi}^2 & -\mathrm{i}\lambda - k^2 - bk^3 + \|\xi\|^2 \end{bmatrix} = 0, \tag{2.4}$$

or

$$\lambda = \mathrm{i}bk^3 \pm \sqrt{2k^2\|\xi\|^2 - k^4}.$$

When $b = 0$ we recover the usual plane-wave instability of NLS; when the amplitude $\|\xi\| > k/\sqrt{2}$ there is a real positive eigenvalue giving instability. For small $\|\xi\|$ or large $\|k\|$ the plane wave is stable.

Now, when $b \neq 0$ the main change is that λ becomes complex. Adding in the complex conjugate, there are four roots

$$\lambda = \pm\mathrm{i}bk^3 \pm \sqrt{2k^2\|\xi\|^2 - k^4}.$$

When $b \neq 0$, $k \neq 0$ and $2\|\xi\|^2 = k^2$ there is a collision of eigenvalues of opposite signature on the imaginary axis at $\lambda = \pm \mathrm{i} bk^3$.

In summary, with third-order dispersion the nature of the instability is different. The difference is explained in the next section.

3. Absolute and Convective Instabilities

An instability is absolute if the dispersion relation has an unstable saddle point, and the saddle point satisfies the pinching condition (Brevdo, 1988). An instability is convective if it is not absolute!

Saddle points play a central role when looking for absolute instabilities. Let $\lambda = -\mathrm{i}\omega$. Then the dispersion relation (2.4) can be written in the form

$$D(\omega, k) = -\omega^2 - 2bk^3\omega - 2k^2\|\xi\|^2 + k^4 - b^2k^6.$$

Saddle points satisfy

$$D = D_k = 0,$$

where

$$D_k = -6bk^2\omega - 4k\|\xi\|^2 + 4k^3 - 6b^2k^5,$$

and so, when $b \neq 0$ and $k \neq 0$, $D_k = 0$ gives

$$\omega = -\frac{2}{3bk}\|\xi\|^2 + \frac{2}{3b}k - bk^3. \tag{3.1}$$

Back substitution into the dispersion relation gives a polynomial in k. Assuming $k \neq 0$ and $b \neq 0$ this polynomial is

$$D(\omega, k) = k^2(k^2 - 2\|\xi\|^2) - \frac{4}{9b^2k^2}(k^2 - \|\xi\|^2)^2.$$

Simplifying and re-arranging gives $\delta(k) = 0$ with

$$\delta(k) = 4(\|\xi\|^2 - k^2)^2 + 9b^2k^4(2\|\xi\|^2 - k^2).$$

This is a polynomial of degree six in k. However, it is a polynomial of degree two in $\|\xi\|^2$. So let us solve for $\|\xi\|$ as a function of k. Assume b is non-zero and denote saddle points by (ω_0, k_0) with k_0 a root of $\delta(k_0) = 0$. Then solving for $\delta(k_0) = 0$ gives

$$\|\xi\|^2 = k_0^2 \frac{\sqrt{1-\theta^2}}{1+\sqrt{1-\theta^2}}, \quad \theta = \frac{2}{3bk_0}. \tag{3.2}$$

The frequency ω_0 is obtained by substituting (3.2) into Eq. (3.1)

$$\omega_0 = \frac{2k_0}{3b}\left(\frac{1}{1+\sqrt{1-\theta^2}}\right) - bk_0^3.$$

The only saddle points (ω_0, k_0) of interest are with $\|\xi\|$ real, since $\|\xi\|$ is the modulus of the amplitude of the Stokes wave. All saddle points giving real $\|\xi\|$ satisfy (3.2). However, note that real roots exist only if $\theta < 1$ or

$$k_0 > \frac{2}{3b}.$$

4. The Case with Only Second-order Dispersion

This is the case $b = 0$ and it has already been considered in Brevdo and Bridges (1997). In this case the dispersion relation simplifies to

$$D(\omega, k) = -\omega^2 - 2k^2\|\xi\|^2 + k^4.$$

The necessary condition for absolute instability is the existence of a pair (k_0, ω_0), with $\mathrm{Im}(\omega_0) > 0$ satisfying $D = D_k = 0$, that is, the existence of an unstable saddle point of $\omega := \omega(k)$. Now,

$$D_k = -4k\|\xi\|^2 + 4k^3 = 4k(k^2 - \|\xi\|^2).$$

The point $k = 0$ corresponds to $\omega = 0$ and so is a neutral saddle point. When $k \neq 0$ there are two roots

$$k_\pm = \pm\|\xi\|.$$

The corresponding values of ω are obtained from the dispersion relation

$$-\omega^2 - \|\xi\|^4 = 0 \quad \text{or} \quad \omega = \pm \mathrm{i}\|\xi\|^2,$$

and so the unstable saddle point is

$$\omega = \mathrm{i}\|\xi\|^2.$$

It is shown in Brevdo and Bridges (1997) that the pinching condition is satisfied in this case. Here is another proof that the pinching condition is satisfied. The pinching condition is defined as follows. Let (ω_0, k_0) be a saddle point. That is

$$D(\omega_0, k_0) = D_k(\omega_0, k_0) = 0.$$

Let $\omega = \omega_0 + \mathrm{i}y$ with y real and positive. Then look at the roots of $D(\omega_0 + \mathrm{i}y, k(y))$, with $k(0) = k_0$ the double root. The instability is absolute if k_0 splits into two roots $k^-(y)$ and $k^+(y)$ with

$$\mathrm{Im}(k^-(y)) < 0 \quad \text{and} \quad \mathrm{Im}(k^+(y)) > 0 \quad \text{as } y \to \infty.$$

A proof that the pinching condition is satisfied in the case $b = 0$ is as follows. In this case

$$\omega_0 = \mathrm{i}\, k_0^2 \quad \text{with} \quad k_0 = \|\xi\|,$$

and so

$$D(\omega, k) = -\omega^2 - 2k^2\|\xi\|^2 + k^4 = -(\omega^2 - \omega_0^2) + (k^2 - k_0^2)^2.$$

Now set $\omega = \omega_0 + \mathrm{i}y$. Then

$$D(\omega, k) = y^2 + 2k_0^2 y + (k^2 - k_0^2)^2.$$

Setting $D = 0$ then gives

$$k^{\pm}(y) = k_0 \pm \mathrm{i}\frac{\sqrt{y}}{2k_0} + \cdots,$$

where the ... represent terms which go to zero as $y \to \infty$. Clearly in the limit as $y \to \infty$, k_0 splits into two roots with imaginary parts of opposite sign. Hence the pinching condition is satisfied and the instability is absolute in the case $b = 0$.

5. Classifying the Instabilities in the Presence of Third-order Dispersion

In the case $b \neq 0$ the instability is convective for some values of b. To see this note that when $\theta^2 > 1$ then the only saddle points are associated with complex values of $\|\xi\|$. Hence there are no physical saddle points. Hence the instability cannot be absolute and is therefore convective.

One can check whether there are any transition points, where the instability goes from convective to absolute (or vice versa). According to Triantafyllou (1994) (see also de Langre, 2002), a change from absolute to convective instability (or vice versa) occurs when

$$D = D_k = D_{kk} = 0\,;$$

that is,

$$\begin{aligned} D &= -\omega^2 - 2bk^3\omega - 2k^2\|\xi\|^2 + k^4 - b^2k^6 = 0 \\ D_k &= -6bk^2\omega - 4k\|\xi\|^2 + 4k^3 - 6b^2k^5 = 0 \\ D_{kk} &= -12bk\omega - 4\|\xi\|^2 + 12k^2 - 30b^2k^4 = 0. \end{aligned}$$

Solving the latter two equations gives

$$\|\xi\|^2 = -k^2 + \frac{9}{2}b^2k^4 \qquad \text{and} \qquad \omega = \frac{4k}{3b} - 4bk^3.$$

Substituting these two expressions into D gives

$$D = -k^2\left(18b^2k^4 - 11k^2 + \frac{16}{9b^2}\right).$$

This equation has four complex roots

$$k^2 = \left(\frac{11 \pm \mathrm{i}\sqrt{7}}{36}\right)\frac{1}{b^2}.$$

These roots are complex and so it suggests that there is no change from absolute to convective instability, since when one of these values of k is substituted into the expression for $\|\xi\|^2$, it gives non-physical complex values of $\|\xi\|^2$.

6. Summary and Conclusions

It appears that the instability is absolute when $b = 0$ (second-order dispersion only) and convective for all $b \neq 0$. This change is not continuous. The discontinuity appears to be due to the fact that $b \neq 0$ is a singular perturbation. The character of the dispersion relation is dramatically changed when $b \neq 0$:

$$D(\omega, k, b) = -\omega^2 - 2k^2\|\xi\|^2 + k^4 - bk^3(2\omega + bk^3).$$

While in nonlinear optics the effects of higher-order dispersion are well understood, their significance is not so clear in hydrodynamics. We have found one major difference between the description of ocean waves and the description of waves in optical fibers. Although in each case the NLS equation seems to be a valid model, what we observe in reality is another matter. In optics one is measuring an averaged intensity, that is the square of the modulus of the envelope. The carrier frequency is usually forgotten. In the ocean one is observing the waves at the carrier frequency. Then an important parameter is the phase difference between the carrier and the envelope. The latter has been extensively discussed in optics when dealing with ultrashort pulses that contain only a few cycles but has not been discussed in the case of ocean waves. Meanwhile, if this parameter is small, one can observe higher amplitudes as opposed to the case when this parameter is close to π. Moreover, this parameter may change during wave propagation. Then one may see the specific property of rogue waves that appear from nowhere and disappear without a trace (Akhmediev *et al.*, 2009). In the ideal case, the effect would be periodic but in a chaotic wave field this may happen only once.

To conclude, one can state that rogue wave studies are the most mature in environments governed by the NLS equation (or its analogues), where efforts of experts with various scientific cultures have shaped the existing mechanisms and created a coherent picture about potential phenomena. However, it is easy to get lost in the mathematical complexity of the problem. It is essential to remain focused on trying to provide some concrete insight into the formation of rogue waves and prediction must be a priority. Going back to water waves, there is still a lack of laboratory experiments where the two-dimensional surface is measured in time. Moreover, the importance of wave breaking in the study of extreme waves is being more and more emphasized (Papadimitrakis and Dias, 2010). But is there a limiting process equivalent to wave breaking in optics?

Acknowledgments

We acknowledge support from the French Agence Nationale de la Recherche project MANUREVA ANR-08-SYSC-019 and the 2008 Framework Program for Research, Technological Development and Innovation of the Cyprus Research Promotion Foundation under the Project ΑΣΤΙ/0308(BE)/05.

References

Akhmediev, N., Ankiewicz, A. and Taki, M. (2009). Waves that appear from nowhere and disappear without a trace, *Physics Letters A* **373**, pp. 675–678.

Akhmediev, N. N., Korneev, V. I. and Mitskevich, N. V. (1990). Modulation instability of a continuous signal in an optical fiber taking into account third-order dispersion, *Izv. Radio* **33**, pp. 95–100.

Benjamin, T. and Feir, J. (1967). The disintegration of wavetrains on deep water. Part 1: theory, *J. Fluid Mech.* **27**, pp. 417–430.

Brevdo, L. (1988). A study of absolute and convective instabilities with an application to the Eady model, *Geophys. Astro. Fluid Dyn.* **40**, pp. 1–92.

Brevdo, L. and Bridges, T. (1997). Absolute and convective instabilities of temporally oscillating flows, *Z. Angew. Math. Phys.* **48**, pp. 290–309.

Bridges, T. and Dias, F. (2007). Enhancement of the Benjamin–Feir instability with dissipation, *Phys. Fluids* **19**, p. 104104.

de Langre, E. (2002). Absolutely unstable waves in inviscid hydroelastic systems, *J. Sound Vib.* **256**, pp. 299–317.

Dudley, J. M., Finot, C., Millot, G., *et al.* (2010). Extreme events in optics: challenges of the MANUREVA project, *Eur. Phys. J. Special Topics* **185**, pp. 125–133.

Dudley, J. M., Genty, G., Dias, F., *et al.* (2009). Modulation instability, Akhmediev Breathers and continuous wave supercontinuum generation, *Optics Express* **17**, pp. 21497–21508.

Dysthe, K., Krogstad, H. and Müller, P. (2008). Oceanic rogue waves, *Annu. Rev. Fluid Mech.* **40**, pp. 287–310.

Genty, G., de Sterke, C. M., Bang, O., *et al.* (2010). Collisions and turbulence in optical rogue wave formation, *Physics Letters A* **374**, pp. 989–996.

Hunt, J. (2006). Nonlinear and wave theory contributions of T. Brooke Benjamin (1929–1995), *Annu. Rev. Fluid Mech.* **38**, pp. 1–25.

Janssen, P. (2003). Nonlinear four-wave interactions and freak waves, *J. Phys. Oceanography* **33**, pp. 863–884.

Kharif, C. and Pelinovsky, E. (2003). Physical mechanisms of the rogue wave phenomenon, *Eur. J. Mech. B/Fluids* **22**, pp. 603–634.

Kharif, C., Pelinovsky, E. and Slunyaev, A. (2009). *Rogue Waves in the Ocean* (Springer, Berlin).

Kibler, B., Fatome, J., Finot, C., *et al.* (2010). The Peregrine Soliton in nonlinear fibre optics, *Nature Physics* **6**, pp. 790–795.

Laine-Pearson, F. (2010). Instability growth rates of crossing sea states, *Phys. Rev. E* **81**, p. 036316.

Onorato, M., Osborne, A. and Serio, M. (2006). Modulational instability in crossing sea states: a possible mechanism for the formation of freak waves, *Physical Review Letters* **96**, p. 014503.

Onorato, M., Waseda, T., Toffoli, A., *et al.* (2009). Statistical properties of directional ocean waves: the role of the modulational instability in the formation of extreme events, *Phys. Rev. Lett.* **102**, p. 114502.

Papadimitrakis, I. and Dias, F. (2010). Occurrence and breaking of extreme waves in deep water. A stochastic approach revisit, *Open Ocean Eng. J.* (in press).

Segur, H., Henderson, D., Carter, J., *et al.* (2005). Stabilizing the Benjamin–Feir instability, *J. Fluid Mech.* **539**, p. 229.

Shats, M., Punzmann, H. and Xia, H. (2010). Capillary rogue waves, *Phys. Rev. Lett.* **104**, p. 104503.

Solli, D. R., P. K. Ropers, C. and Jalali, B. (2007). Optical rogue waves, *Nature* **450**, pp. 1054–1057.

Stenflo, L. and Marklund, M. (2010). Rogue waves in the atmosphere, *J. Plasma Phys.* **76**, pp. 293–295.

Taki, M., Mussot, A., Kudlinski, A., *et al.* (2010). Third-order dispersion for generating optical rogue solitons, *Phys. Lett. A* **374**, pp. 691–695.

Tamura, H., Waseda, T. and Miyazawa, Y. (2009). Freakish sea state and swell-windsea coupling: numerical study of the *Suwa-Maru* incident, *Geophys. Res. Lett.* **36**, p. L01607.

Triantafyllou, G. (1994). Note on the Kelvin–Helmholtz instability of stratified fluids, *Phys. Fluids* **6**, pp. 164–171.

Index